Policy Indicators

Urban and Regional Policy and Development Studies

Michael A. Stegman, General Editor

Policy Indicators
Links between Social Science and Public Debate

Duncan MacRae, Jr.

The University of North Carolina Press

Chapel Hill and London

Manufactured in the United States of America

Library of Congress Cataloging in Publication Data

MacRae, Duncan.
Policy indicators.

(Urban and regional policy and development studies)
"A continuation of . . . two earlier works, The social function of social science, and (with James Wilde) Policy analysis for public decisions"—P.
Bibliography: p.
Includes index.
1. Policy-sciences. 2. Social sciences. 3. Statistical services. 4. Social indicators. 5. Economic indicators. I. Title. II. Series.
H97.M33 1985 361.6'1 84-17294
ISBN 0-8078-6565-6

To Edie and Amy

Contents

Tables and Figures

Tables

Figures

Preface

In the fall of 1979 Richard Rockwell invited me, on behalf of the Social Science Research Council's Advisory and Planning Committee on Social Indicators, to write a monograph on the relation between values and social indicators. Having done some preparatory work on the related topic of "tradeoff indicators," I agreed. The work has taken me on a tortuous path through a literature that extends into a number of fields.

One way of describing the result is to say that it is an analysis of information policy, asking what sort of statistics should be provided to citizens for making policy judgments. This is a type of policy analysis that does not easily fit the benefit-cost approach, since we cannot easily evaluate this sort of information or its uses quantitatively. Nor is it typical of the analysis of communications policy, which usually deals with such matters as censorship or government regulation of channels of communication, but not with choosing communications that originate from government itself. My recommendations are also abstract and imprecise, a quality not usually esteemed in policy analysis.

The early social indicators movement, concerned with values, policy, and program evaluation, posed a large task for social scientists within it: to work with political communities to choose measures of "how well off we really are," so as to aid those communities to act wisely collectively. By the 1970s, this task had been largely rejected as impossible or at least ill conceived. The field of social indicators then came to be studied according to the canons of basic science, or in partial imitation of macroeconomics; but the problems of its initial policy-oriented tendency remained. They are complex because they include specifying the community's values; reconciling them to some extent; measuring valued conditions; learning how to act so as to promote them (by means of causal models); and reorganizing expert communities so as to aid these tasks. These problems require us to link social science with public debate.

My analysis is a continuation of my two earlier works: *The Social Function of Social Science*, and (with James Wilde) *Policy Analysis for Public Decisions*. The former proposed that the latent values of the social sciences be made explicit and systematized. The "ethical hypotheses" that it suggested, as ways of organizing one's ethical system, correspond roughly to the "general end-values" that I here suggest as indicators. But here they are placed in the arena of public discussion, rather than being used in ethical arguments among social scientists.

The Social Function of Social Science also suggests that an explicit valuative concern among social scientists might take the form of an applied discipline of policy analysis. Such a discipline has continued to develop and is a major example of the "technical communities" that I here advocate as a means of developing policy models.

Policy Analysis for Public Decisions was intended to teach undergraduates as citizens to be policy analysts, choosing policies intelligently on the basis of their own systematized notions of the general welfare. Its idealized notion of democratic citizenship—perhaps attainable only by a minority of citizens—constitutes a standard for public discussion toward which I hope we can strive. It differs from our behavioral-science notions of citizenship in its presumed level of information and in its foundation in public rather than private values.

From the viewpoint of values and citizenship, a new element in the present book is its concern with reconciliation of values in democracies (Chap. 2). Both the earlier books centered about the systematized values of social scientists or of individual citizens; though recognizing that those values had to be incorporated in a larger system of argument and politics, they did not carry that involvement through in detail. Here I hope to have gone a step further, though many other steps remain.

This book continues a major theme of the two previous works—that the prestigious domain of basic science as practiced in American universities is not optimally organized to deal with questions of policy and value; rather, it needs to be supplemented by different sorts of communities pursuing policy-related practical science. But having proposed this alternative path, I find that following it involves some complications. It has led me to coin or redefine a set of terms that are central to the argument: policy indicators, end-values, contributory variables, policy models, exact and less exact sciences, subjectively social sciences, subjective variables, motivated bias, practical—as distinguished from applied—science, and technical communities. I have chosen these redefinitions reluctantly, in the hope that they will clarify the argument rather than obscure it.

As in previous writings, I have taken the position that valuative discourse is meaningful, challenging a tenet of logical positivism that may seem to have been amply criticized. Yet aspects of that tenet remain and limit valuative discourse in social science. The value-free stance of much of social science is partly attributable to this view, though it also derives from political prudence about the conditions for maintenance of internal consensus and external support. One earlier line of argument in political science, claiming the "public interest" to be meaningless, is challenged in an unusual way here; I am proposing that a major emphasis in development of policy indicators should be to provide actual measures of notions of the public interest, in the form of end-value indicator statistics.

My approach may seem to some to be ahistorical. I write during the Reagan administration, under which there have been cutbacks in the national statistical system as well as in other governmental activities. Yet I am trying to take a somewhat longer view. It is important to recognize that some indicator series will and should be discontinued. Termination is not necessarily wrong, so long as we find ways to guide it by careful consideration of its consequences.

In another sense my analysis is quite specific as regards time and place. It takes for granted the structure of contemporary universities, centered about disciplines and departments, and seeks ways to modify this structure incrementally. It does not question the decentralized system of public and private statistics that has been customary in the United States. It looks to the situation of some technical communities that have been recently organized and to their growth.

Because this analysis attempts to cover so much ground—even while omitting important aspects of indicator design and public response—it will necessarily be open to criticism. Even if readers should disagree with some of my interpretations and conclusions, however, I hope they will consider the problems addressed here to be significant and worthy of further study.

Two books stand out in the United States literature in the field I analyze here: Edgar S. Dunn's *Social Information Processing and Statistical Systems—Change and Reform* (1974) and Judith I. de Neufville's *Social Indicators and Public Policy* (1975). They deal with the practical problems of adding to or reforming public governmental statistical systems. Somewhat different in emphasis, both deal with wide-ranging aspects of the problem as to how concepts and measures should be chosen, and how they can be incorporated in these systems. Together, they tell the reader forcefully that information achieves its value through use, and that the study of how to realize this value requires consideration of many specialties.

My intended audience includes scholars and students in any of the variety of fields on which policy indicators and models must draw. These readers will undoubtedly find most to criticize in the parts of the book that they know best. But I should like each of them also to be able to understand the aspects of policy indicators that he or she finds less familiar, and for this reason I have included relatively elementary reviews of existing principles and literature in a number of places.

I am indebted to a number of persons for helpful suggestions; none, however, is responsible for my errors and some will disagree with my arguments. They include Erik Allardt, Richard A. Berk, Angell G. Beza, Stephen L.

Darwall, Judith I. de Neufville, Norman L. Johnson, Denis F. Johnston, Robert C. Kelner, Edith K. MacRae, Peter V. Marsden, Eugene J. Meehan, E. Sam Overman, Robert Parke, Solomon W. Polachek, Richard C. Rockwell, Rachel A. Rosenfeld, Peter H. Rossi, C. Ford Runge, Stephen H. Schneider, Eleanor Bernert Sheldon, Henry West, James A. Wilde, and an anonymous reviewer. The students in my seminar on social indicators and public policy in Fall 1983 also raised stimulating questions.

For encouraging this study I am indebted to the Social Science Research Council's Advisory and Planning Committee on Social Indicators and to the staff of its Center for Coordination of Research on Social Indicators. Richard C. Rockwell and Robert Parke have provided contacts with persons working in this field, references to literature, and a bibliography on social indicators prepared by Howard D. White. The typing of many drafts and revisions of the manuscript has been done with care, first by Gert H. Rippy and Bruce Geyer, and later by Stacy C. Reynolds, Florine P. Purdie, and Charlene LeGrand. The Bush Institute for Child and Family Policy was helpful in making its word processor available. Nancy H. Margolis edited the manuscript skillfully.

A Kenan leave from the University of North Carolina at Chapel Hill in Fall 1982 was of great value in furthering this work, as was the use of studies in Wilson and in Davis libraries.

I am indebted to the JAI Press for permission to make use of material from my paper, "Policy Indicators and Public Statistics: Democratic Information Systems," to be published in *Policy Analysis, Concepts and Methods,* edited by William N. Dunn, vol. 1 of *Public Policy Studies: A Multi-Volume Treatise,* edited by Stuart S. Nagel; and to the American Political Science Association for permission to use a section of my paper, "Present and Future in the Valuation of Life," presented at their August 1980 annual meeting.

Part I.
Statistical Systems for Democratic Policy Choice

Introduction: The Need for Policy Indicators

The Preamble to the United States Constitution states as two of the purposes of the Constitution to "promote the general welfare" and to "establish justice." Amid the play of private interests in democratic politics, these general ethical values remain as bases to which we return in trying to persuade our fellow citizens of the rightness of our arguments in public debate. They must enter into citizens' and leaders' public justification of proposed policies in all democracies and representative political systems.[1]

To claim that a proposed policy promotes the general welfare or justice (for which we later use the term "equity"), we must give reasons why we believe that it will do so. In assessing possible policies, we also often wish to survey the present and past states of well-being of the population in order to see where we have made progress or lost ground or to see where we can be most effective. We thus need information on the current state of well-being of the population, its trends, its distribution, and their causes.[2]

The complex public statistical systems developed by modern governments contain much information about the well-being of their populations, and have been supplemented by information from nongovernmental sources. This combined body of public statistics[3] includes meteorological, environmental, eco-

1. The importance of public debate in reflecting and developing shared ethical notions concerning public policy has also been recognized as central to the use of social indicators by Johansson (1976: 75–78) who stresses the role of public debate in evolving collective notions of "how it ought to be" and "what should be done."

2. By "policy" we here mean any action taken by government (Dye 1981: 1); but ordinarily it is only those actions with widespread effects that citizens debate and analyze. The diversity of notions of the general welfare in democracies is discussed in Chap. 2. In considering distribution we extend our concern to justice or equity, a public value that is not included in a simple summation of welfare.

3. We refer to statistics available to the public, whether from governmental or nongovernmental sources, as "public statistics." We also require that statistics so designated meet standards of quality set by expert communities (Chap. 10). This and other definitions are summarized in Table 1–2. Governments also produce extensive bodies of statistics for internal use; public use

nomic, psychological, and social information. Statistics of the last three types have benefited from advances in the social sciences (Simon 1980: 72). Among them demographic and economic statistics were generally compiled first, including data on population, mortality, production, and employment.

We need to choose the topics of these statistics wisely in view of the benefits they may bring, the costs they impose, and the distribution of both. The aim of this book is to set forth general principles that will aid in this choice. These principles will concern:

1. The values related to the general welfare, and justice or equity, that are sought within a particular representative political community—nation, state, or locality. Statistics can also be used to further private and self-interested values, but we shall stress the role of general values used for public justification (Chaps. 1, 5–8).

2. The part played by specific and general values in public debate (Chap. 2).

3. Our knowledge as to what public policies might be means to these values (causal policy models) (Chaps. 3, 4).

4. The ways in which we expect public statistics to be used (Chap. 9).

5. The organization of expert communities for proposing and reviewing public statistics, developing policy models, and facilitating their use (Chap. 10).

Social Indicators and Policy Indicators

Since the mid-1960s a concern has arisen for improving statistics on the noneconomic aspects of public well-being and of social conditions. This concern has been expressed in advocacy of *social indicators*, usually understood as time series of national statistics on subjects such as health, education, housing, and crime, to be presented in periodic social reports. Social indicators have been defined in two principal ways: one more valuative and policy-related, the other stressing a scientific approach to the analysis and measurement of social change. The scientific emphasis on social change now seems to predominate in the United States, especially among sociologists (H. D. White 1983), and the valuative or policy-related emphasis in

statistics recently took less than a third of the U.S. federal statistical budget (President's Commission 1971: I, 41). Numerous nongovernmental statistics, including publications in the social science literature, can inform public judgments. But these, like governmental statistics, can vary in their real accessibility to the public at any given time even when they are in the public domain. We thus face the questions whether public statistics should be subdivided according to their intended audiences and whether separate organizations should collect and present statistics for different audiences (Biderman 1970: 219). We nevertheless refer to this body of information as a "system," as it is linked together by the interrelations among values, by causal models, and by the public's interpretations.

Europe. The coexistence of these two distinct but overlapping definitions has led some observers of the social indicators movement to regard definitions of social indicators as diverse (Carley 1981: Chap. 2) and ambiguous (Dunn 1974: 103).

This book is concerned with the first or valuative approach, and asks how public statistics should be planned so as to aid democratic policymaking. To clarify this approach I propose to call the measures with which it deals *policy indicators*,[4] supply a new vocabulary for discussing it, extend its scope, and suggest long-term priorities for the values it measures and for the steps to be taken in relating it to public policy.

The two distinct meanings of social indicators—concerned with values and public policy on the one hand and social change on the other—have arisen through a debate between two groups within a broader movement. To understand why a new definition is needed, it is useful to review this debate, so as to see how the initial valuative definition lost favor and how it can be reformulated so as to be more defensible. A valuative[5] or normative definition of social indicators was set forth by Bauer (1966: 1) in the pioneering volume *Social Indicators*: "This volume . . . is devoted to the topic of social indicators—statistics, statistical series, and all other forms of evidence—that enable us to assess where we stand and are going with respect to our values and goals, and to evaluate specific programs and determine their impact."

The basic assumption of the volume was that "For many of the important topics on which social critics blithely pass judgment, and on which policies are made, there are no yardsticks to know if things are getting better or worse" (20).

This theme was echoed in *Toward a Social Report* (U.S. HEW 1969: 97, xii-xiii), the author of which (Mancur Olson) defined social indicators as "statistic[s] of direct normative interest" that can tell us "how well off we really are." They would thus be measures of the general welfare. A social report "could give social problems more visibility," thus changing priorities on the public agenda, and "might ultimately make possible a better evaluation of what public programs are accomplishing." Recent usage in the United States has moved away from this practical and valuative definition, but some concern for values remains. As a distinguishing feature of social indicators, Johnston (1978: 292), for example, has still stressed "normatively significant statistics," pertaining to values, whose significance "lies in the sphere of human action."

These arguments for public statistics continued a theme that dates back to

4. We shall define policy indicators (Chap. 1) as measures of those variables that are to be included in a broadly policy-relevant system of public statistics. The term has been used similarly by Rose (1972: 123) and de Neufville (1978).

5. Although "normative" is used in the literature cited, "valuative" avoids an ambiguity in sociological terminology (MacRae 1976a: 5n).

the founding of the Republic. James Madison proposed in 1790 "to gather basic economic data so that 'the description of the several classes into which the community is divided should be accurately known.' This would permit the Congress 'to make proper provision for the agricultural, commercial, and manufacturing interests' in its legislation. . . . His argument that statistics were vital to legislation became the standard plea for the enlargement of the census, and his desire to monitor the growth of the new nation was repeated by scores of voices in the years to come." Jefferson, in turn, "revealed . . . theoretical as well as practical aims for the census," which he concluded "should provide a practical guide to public policy and, at the same time, might perhaps supply the basis for understanding economic laws" (Davis 1972: 154–56). This practical emphasis has continued within economics to the present day, but the theoretical goal has been more prominent than the practical among the proponents of social indicators.

This valuative definition of social indicators and public statistics has been largely replaced, however, by an alternative definition centered about the analysis and measurement of social change, guided more by the standards of basic science than by the requirements of practical application. Prior to Bauer's (1966) volume on social indicators, Moore and Sheldon (1965) advocated the use of statistics for monitoring of social change. In addition, Sheldon and several coauthors have advanced a number of specific criticisms of the more valuative and policy-related view to which we must respond. In each I shall designate the argument for a scientific approach to social change as (S), and a response by an advocate of policy indicators as (P).

1. (S) The initial claims of the proponents of using social indicators for policy purposes were excessive; "their promises far exceeded any possibility for realistic attainment" (Sheldon and Land 1972: 138–39).

(P) We need not promise that policy indicators will all affect policy in the short run. There have indeed been large and unrealized claims for various rationalized systems for information and decision (Wildavsky 1979: Chap. 1). We can nevertheless aim our efforts in this field in a more practical direction, hoping to accomplish more practically than if we had ignored the problems of practical application or postponed them in favor of basic science.

2. (S) Variables of "direct normative interest" are of ephemeral interest, as "what is salient today may not be so next year and vice versa." Moreover, the restriction of indicators to direct normative variables would exclude data on means to them, such as numbers of doctors or policemen (U.S. HEW 1969: 97; Sheldon and Freeman 1970: 98).

(P) If we focus attention on intrinsic values rather than the temporary means to them, and on the continuing social goals associated with major institutions, we shall find that they are of lasting interest. Intrinsic values, in fact, include measures of subjective well-being that Sheldon and Parke (1975: 697) have suggested for conceptual and empirical development. We must consider including indicators not only of intrinsic values, but also of variables that contribute to them, among policy indicators.

3. (S) Social indicator statistics cannot be used to set national priorities, since "priorities and goals are more dependent on national values than on assembled data" (Sheldon and Land 1972: 139). National values and goals are set by a political process. Indicator statistics may encourage a concern for certain problems, but we need to ask whether this concern is justified. The problems that are given more visibility by these statistics are in many instances already visible (Sheldon and Freeman 1970: 100). And finally, Sheldon and Freeman note (1970: 108), citing Merton, that "there are no agreed-upon bases for rigorously appraising the comparative magnitude of different social problems."

(P) Whether indicators should and can play a part in setting the priorities of nations (or states or localities) depends on the nature of the indicators and their relation to the values of citizens and decision makers in those political systems. Measures of net economic benefit do sometimes induce people to change the priorities that they place on specific programs. Such a change of priority can be persuasive only if the audience agrees sufficiently with the underlying values, such as those embodied in benefit-cost analysis. A community of experts, engaged in design of statistical systems and in relevant research, can center its work about value concepts that are consistent with, or reformulations of, broad values existing in their society. Moreover, when their work impinges on public debates about policy choice, they should expect their fellow citizens to examine the experts' valuative assumptions critically.

4. (S) Social indicators cannot be used for program evaluation. National time series do not allow us to distinguish between causation and mere association, as they fail to separate program effects from those of "uncontrolled extraneous variables" (Sheldon and Land 1972: 139; Sheldon and Freeman 1970: 100–01). A distinction should be made between "the descriptive and analytical time series needed to monitor and analyze social change" and "the experimental designs by which government programs may be evaluated" (Sheldon and Parke 1975: 694).

(P) The alleged separation between social indicators and program evaluation rests on a semantic problem. The term social indicators is in fact overdefined, i.e., defined by an unnecessarily large set of attributes. It conventionally combines a set of concepts and measures, reasons for choosing them, a population domain (national), and a temporal period (time series extending over a decade or more). The same concepts and measures can be used, however, in studies of both general national samples and local or client populations. They can also be used in experimental or quasi-experimental studies in which every effort is made to devise adequately specified models and to avoid spurious causal inferences. The time series in which some policy-related public statistics are collected need not extend over decades, but may well extend only over a much shorter period related to the expected effects of a particular program. In the earlier literature on social indicators it is true that "the linkage between actual policy options and the . . . measured state of society" was typically left implicit (Juster et al. 1981: 27). Our task, however, is to link problem-defining statistics for general populations with causal models by choosing concepts to be used in both.

5. (S) In order to make informed policy interventions, we require models of the social system, analogous to macroeconomic models of the entire economy. We therefore need "analytical studies of social change" (Sheldon and Freeman 1970: 105), and a capacity to predict. We need basic model construction[6] (Sheldon and Land 1972: 141). We need to concentrate our efforts on the measurement of social change (Sheldon and Parke 1975: 694). In this perspective, policy development is only a "long-run goal" (Land 1983:11).

(P) The argument that we must forego immediate application until basic research has provided the necessary models is also drawn too sharply by Sheldon and her coauthors. If all our social policy choices depended on full-scale models of social systems, we might have no choice but to retreat into basic research exclusively (Klages 1973: 260). But the models of causation needed for policy inferences actually range from the general to the highly specific. Indeed, in the realm of program evaluation, general social-system models are usually absent. Many of the models used in policy choice are quite specific, and concern the relative merits of particular programs. The dependent variables in these studies may nevertheless be useful for public indicator statistics, even while in research on causal models they are used with specific target populations rather than with national samples. We must thus distinguish in our terminology between aggregate indicator statistics and the variables and measures on which they are based.

6. (S) One view of the role of social indicators "is to contribute to problem solving through their application in goal-oriented analyses, of the kind developed in economics and operations research," but:

> The flaw in this definition is that it tends to limit their scope to matters covered by the authority, competence, and objectives of administrators. No one will deny the importance of such matters. But this is no justification for our letting the agenda of work on social indica-

6. "Basic research" refers here to research that is guided by the autonomous theoretical requirements of scientific communities rather than by consideration of practice (Lazarsfeld et al. 1967: xxiv-xxv). In contrast, we shall refer to research guided by the needs of practice as "practical research." We avoid the term "applied" because practical research can develop independently of basic research; as Meehan (1982: 69) points out, "there is no need to know the origin of every human situation in order to improve it." Practical research can, however, be general or fundamental (Levins 1973; Stokes 1982). In much of the basic research on social change, carried out by proponents of social indicators, the dependent variables used have valuative connotations related to well-being.

> tors be governed by the perceived information requirements imposed by a social engineering approach or letting it be limited to "policy-manipulable" variables—that is, those subject to the control of the agencies responsible. . . .
>
> In social engineering, one starts with an agency objective, and the information on social behavior and social conditions is limited to that which is deemed relevant and appropriate, given an agency's authority, traditions, and the tools available. By contrast, a social scientific approach to indicators, in our view, starts with social behavior, and seeks to comprehend and measure it and account for changes in it. Adequate description, measurement, and explanation serve the needs of science and also provide the "enlightenment" function to guide policy (Sheldon and Parke 1975: 695–96).

(P) Policy-relevant information need not be limited to the needs of public agencies. Policy initiatives in a democracy can come from citizens and their leaders and be judged by them. Although basic social science can contribute to the enlightenment of citizens, it is not always the only or the best means for doing so; education of citizens in public policy analysis (MacRae and Wilde 1979), related to criteria embodied in policy indicators, may be more useful. In seeking general enlightenment, therefore, we need to inquire in more specific terms as to how citizens can use information in assessing public policies.

Although the criticisms by "S" have some merit and the analysis of social change is an important topic in basic social science, "P" thus argues that we can also adopt a different approach so as to relate public statistics usefully to valuative variables, policy choice, and program evaluation. Nor need we limit policy indicators to the social domain. Economic indicators should be joined in a single inclusive domain with the social and related to them. Indicators based on the natural sciences (meteorological, environmental) should also be treated in this common framework. I shall emphasize the contributions of social science, but leave these additional possibilities open.

Outcome Values as the Test of Information Systems

The design of information systems must be guided by an explicit concern for their purposes (Bauer 1966: 39; Johnson and Lewin 1984); the benefit from any additional element in an information system is the additional value resulting from its use.[7] Conversely, the elimination of unneeded

7. Knowledge can have value in itself, in that understanding can be a component of well-being. Some "enlightenment" can also have long-run practical value that is difficult to predict; this has been an argument for pure theoretical science. Yet these arguments provide only limited

information should also be guided by its net contribution, including its benefit and burden (Morss and Rich 1980: Chap. 5). This principle embeds information in a context of valuation and reminds us that an information system is not simply good or bad in itself but must be judged in terms of its use. We shall illustrate the principle with two different types of examples, involving organisms and decision trees.

The information systems of individual organisms are their sense organs, and the value of the information provided by these organs derives from the organisms' resulting capacity for response and action. Every species has sensory mechanisms for sensing both its external environment and its internal state. Plants respond to information they receive about temperature, humidity, light and gravity; they also respond to the presence of nutrients and hormones within their cells. Animals respond to the environment, including other forms of life about them; they can move, attack, feed, mate, or nest in relation to signals they receive from their senses. They also respond to internal signals corresponding to temperature, pain, excitement, hunger, thirst, or fatigue.

The distinct sensory system that each species evolves relates to its particular environment and the corresponding responses or actions that it requires to survive. In humans, eye and hand work together, aided by the nervous system and by learned causal connections in the brain;[8] either eye or hand without the other loses much of its function. A mole and a bird require different types of sense organs because they face different requirements for survival in their environments. Birds require distance vision, while moles require touch and hearing. For similar reasons, the types of information needed by different sorts of political systems can differ (Chap. 9). Information and action are parts of a single system.

Human beings are organisms with special potential for generating and using information. They can supplement their bodies' capacities for using information by making use of material technology and social organization. Technologies of prediction and communication provide information about their external environment, including indications of the weather, economic conditions, or dangers such as floods or earthquakes. They may also seek information about themselves, by using mirrors, timepieces for pulse counts, or devices for measuring blood pressure—but also use eyeglasses or education as aids in reading meters or interpreting the readings. When people extend

bases for the choice of variables in public statistics or for the allocation of public resources among types of research.

8. The importance of our knowing these causal connections may have been underestimated in the early policy-oriented literature on social indicators. Proponents of this approach may have assumed that societies and polities would automatically "know" what to do when an indicator statistic changed. Knowledge of causation is an essential accompaniment of indicator-guided collective action, however, and must be considered explicitly.

Fig. In-1. Decision Tree for Estimating Value of Information*

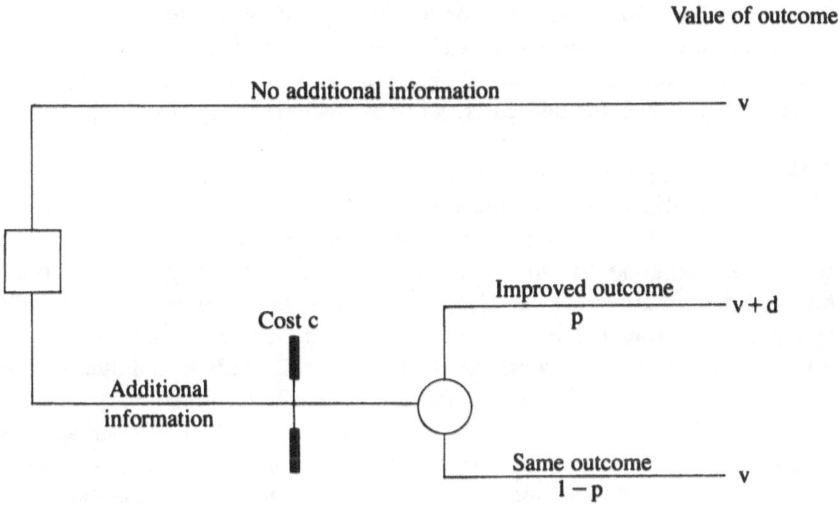

*Adapted from Behn and Vaupel (1982: 363).

their capacities in this way, they benefit from the work of others including designers, producers, and sellers of goods; or organizations and professions concerned with knowledge and learning as well as action. The design of information systems thus returns to our attention when the gathering and use of information move from the realm of evolutionary development to that of conscious planning and choice.

An alternative perspective on the use-value of information is given by a process we can use to decide whether a given piece of information is worth its cost, summarized in the decision tree of Figure In-1. We begin at the square box (decision fork) on the left, where we face the choice between foregoing the additional piece of information (top branch) and buying it at the cost c (indicated by the "toll gate" in the lower branch). If we forego it, the value of our action is then v (right-hand end of top branch). If we use the information, the outcome is uncertain; we find ourselves at a "chance fork" (circle) with two possible outcomes. With probability p we obtain an improved outcome with value $v+d$ (right-hand end of upper branch from chance fork); thus d is the difference that the information makes if it is helpful, and knowledge about this difference (a causal model) is essential to proper choice. We assume that the only other possibility is that the information makes no difference and that in that event we again obtain the value v (right-hand end of lower branch from chance fork).

It can be shown easily[9] that if our criterion of choice is expected value, the condition for our buying the additional information is $pd > c$. This means that the cost of the information must be less than pd if we are to buy it; the expected value of the information is then pd, the probability that it will make a difference times the amount of that difference. If we are calculating in terms of social benefits and costs, we must include those costs imposed on respondents.

We have again illustrated that information gains its value from use.[10] But the difference that a piece of information makes may well depend on the user. Information on the price of a good, for example, is of more value to a person who needs that good than to one who does not. The probability that information will make a difference can also vary among potential users. The mole will rarely benefit from the information provided by distance vision (if above ground and threatened by a predator) while the bird needs it continually. The different expected values that potential users place on information, because of their respective p's and d's, parallel the survival values that different species obtain from a given type of sensory capacity. We cannot so easily estimate the evolutionary cost of developing one type of sense organ rather than another, but the principles in both examples are similar. Both illustrate by analogy that the value of a particular type of information can vary among political systems.

The examples of the organisms and the decision tree are illuminating, but they do not directly tell us how to design information systems. They tell us to ask what additional benefits may be expected and with what probability, and how much the information will cost. They alert us to the different information requirements of different political systems (Chap. 9), and of different practical problems. Thus we may benefit more from frequent observation of health conditions in tracing an epidemic than in assessing the effects of longer-term public health measures; or we may need to sample a different population in order to trace causation from that required to survey well-being. The information needed depends on its uses and the expected benefits from them. Even if we cannot easily quantify the benefits and costs or their distribution (Downs 1967), this type of reasoning can be useful.

9. We buy the additional information if the expected value of the lower branch at the decision fork is greater than that of the upper branch, i.e., if:

$$-c + p(v + d) + (1 - p)v > v.$$

The terms in v on both sides of the inequality are equal, leaving

$$-c + pd > 0, \quad \text{or} \quad pd > c.$$

10. An instructive and more detailed example is given by Raiffa (1968: Chaps. 1–2), showing how the possession of information can alter the expected values of our choices by modifying the posterior probabilities of events following chance forks. We have stressed tangible outcomes as tests of the value of information, but sometimes we seek less tangible benefits such as those from general enlightenment, the better definition of problems, or the stimulation of research.

The outcome values of our actions are measured in terms of our basic valuative criteria. For individual action these may simply be personal survival or well-being. In the domain of policy indicators, centering about governmental choice in view of the general welfare and justice, the particular meanings we choose for these values will then become the criteria for goodness of our information system. We shall also give special consideration to including measures of welfare and justice in that system. A system that aimed to maximize the national income would differ from one that aimed to maximize aggregate subjective well-being, and these in turn from one that aimed at human perfection or justice. Our task as a political community, with members and groups valuing all these things, may be to provide "sensory modalities" for them all, and then to decide among ourselves how to act together on the information they provide. But "internal sensing" by a society about itself no longer involves a single organism seeking a single set of goals; internal conflict as to which part of society realizes the benefits is also part of the problem (Chap. 2). A society, too, can redesign its information systems, in ways that an animal cannot.

The domain of policy indicators, as we have defined it, is guided by distinct criteria from those of basic science and from the study of social change within basic science. Yet our decision tree example also illustrates a benefit often claimed for basic science: findings that contribute to verified general theory can have wide-ranging effects when applied, thus greatly increasing d and thus the value of the information they provide. An indicator with wide-ranging potential uses is desirable for the same reason. We shall return to this feature of basic science, eventually proposing a general approach to practical problems through research on models of causation of value concepts (Chap. 4).

We shall proceed to more specific ways of assessing policy indicators in terms of their uses and general social values.

1 Values in Public Statistical Systems

The designer of policy indicator systems must assess numerous variables to judge whether they should be included in public statistics that aid policy choice. These variables include those used in economic and census statistics; various social goal variables that have been proposed as social indicators, such as those characterizing health, education, crime, and the physical environment; and variables describing means to social and economic goals, including expenditures and aspects of personnel and organizations. Statistics on these variables may be presented for national, state, or local populations, and their relative usefulness may vary among levels of government. Statistics may also vary in their intelligibility, being accessible to wider or narrower segments of the public, and the designer must judge their value in these terms as well. These judgments are complex and often rest on experience rather than on specific researches. Our task here is to present an organized framework for making such judgments.

The variables chosen for public statistics can be used not only in debates about public policies but also for private adaptation by individuals and groups to specific conditions. Information on business fluctuations or population trends can be used by businesses planning their production and marketing as well as by citizens and officials considering public policies. If these private uses are of social value, they should count in favor of the inclusion of a given variable in public statistics. However, because this book is concerned only with uses in public debates on policy, we shall set aside private, nonpolicy uses, which the designer must assess on grounds other than those discussed here.

To assess the value of a variable as a policy indicator, we must consider the different contributions that it may make at various stages in a cyclical process of policy choice: the initial definition of a problem; the choice of policies; the monitoring of results of those policies; and the redefinition of the problem, returning to the initial stage. In the simplest form that this cycle may take, the relevant political community is concerned with a single criterion variable or value, such as an aspect of education or crime, throughout. This same variable will be used in all the stages, but in different ways. In problem definition or redefinition, it will be used in statistics on the general population of concern to the political community. In the causal models used for policy choice, it will enter into research on particular populations that have been exposed to relevant treatments or conditions. In monitoring, it will be used for populations affected by policies, but often in conjunction with more specific "input" variables believed to affect it, such as expenditures and personnel characteristics. The value of a variable for policy indicator statistics depends on the combined values from its use in all these stages.

The assessment of a possible policy indicator variable thus depends on the overall value that we expect from its use. But if we wish to set priorities among policy indicators, we must compare values that result from the uses of statistics in various goal areas such as health, education, and crime. The same type of comparisons are necessary when we wish to compare specific policies in these areas and decide how much of our collective resources to devote to each. These comparisons may be made intuitively, or through the interaction of political pressures, and sometimes we can do no better than this. But if we wish to assess the adequacy of our political processes, or of leaders' intuition, we may wish to make the comparisons more systematically. We may then seek one or more common denominators—more general or more intrinsic values—in terms of which to assess specific goals. A central concern of this book will be a set of "general end-values" that can guide our choices among more specific goals: net economic benefit, subjective well-being (together with its time dimension), and equity. We shall propose that end-values enter into public debate as guides to reconciliation or tradeoffs among more specific values; that they be considered as possible policy indicators; and that they link together the political and expert communities in a common discourse, through adaptation by each of these communities to the requirements of the other. Such a proposal may well be controversial in both communities—in the political, because it favors greater guidance of debate by the logic of abstract values; and in the academic, because it favors greater concern with practical questions and communication with citizens and public leaders.

Four Types of Action Aided by Public Statistics

The information provided by public statistics can be used to aid several different types of action, all of which are relevant to the planning of an entire statistical system but only some of which relate to policy choice (Biderman 1966b). These types of action differ from one another as regards who makes the ultimate decision to act (Is the actor acting alone or participating in a collective decision?)[1] and as regards the motive for action (Is it done

1. The actor may be an individual, group, or organization. A collective decision is a decision taken by a collection of actors and binding on the particular actor; for our purposes the most

Table 1–1. Types of Decisions and Motives

		Locus of Decision: Actor Is—	
		Acting Alone	Participating in a Collective Decision
Motive	Self-interest (private interest)	a. "Economic" actions	b. Pursuit of particular interests through collective decisions
	Ethical (general welfare or justice)	c. Private altruism	d. Pursuit of ethical values through collective decisions

in private self-interest, or for ethical reasons connected with the general welfare or justice?).[2] On these bases we distinguish four types of action as shown in Table 1–1. We shall discuss cells *a, b, c,* and *d* in turn; the last of these will be the domain of *policy indicators*.

a. Information may be used in private market or nonmarket decisions (*"economic" actions*), in which the actor acts alone, seeks self-interest, and thus makes no effort to justify the action ethically. An increase in crime may lead individuals to buy burglar alarms or exercise caution in walking at night; an impending energy shortage may lead individuals to buy energy-efficient automobiles or to lower their home temperatures in winter; an expected economic downturn may lead firms to reduce their inventories, or couples to change their plans about marriage or the timing of births. Jewelers' need to anticipate demand for rings was an early source of demand for marriage statistics (Hauser 1975: 62). These private decisions usually involve adaptation

important example is a governmental decision. A group as an actor may act alone, as in deciding to produce some good; or as a participant in a collective decision, as in lobbying. Our classification of motives, like that of economics, tends to omit malevolence.

2. By "ethical reasons" we mean reasons that can be used to justify a proposed act in terms of what ought to be done. These include teleological justifications in terms of the consequences of acts, such as effects on the general welfare, and nonteleological justifications such as those based on rights or moral prohibitions related to the act itself. If an actor gives ethical reasons, his need for statistics in justification will be much the same regardless of whether the reason is sincere or hypocritical.

to the conditions revealed by the statistics, rather than efforts to change these conditions through collective action.[3] In aiding private actions, a given statistic may describe a condition that different people value differently; a prediction of rain may be useful to farmers for planting and to vacationers for rescheduling their trips. Manufacturers of various products may use disaggregated statistics in different ways to plan their marketing. Parents choosing schools in a voucher system may use school data to choose in favor of art or science. The private values or tastes that enter into decisions of this sort are usually taken as given and do not enter into public debate. Those who use this information need also to know the causal relations between their actions and the consequences, but the models of causal relations that they need for adapting to a given state of affairs are generally simpler than those involved in public policies for changing it.

These adaptations can create values (and disvalues) that contribute to the overall net benefit of providing the information. The private values served may be disparate and not subject to valuative discourse, but we (and our fellow citizens) can aggregate them, trade them off, examine their distribution, and discuss their contributions to the general welfare or justice. An economic approach to knowledge and information is thus possible (Machlup 1962, 1980). In this respect, information policy is public policy affecting private actions.

b. Actors can use information to pursue their *particular interests* (private, nonethical values) through *collective decisions* such as the formation of public policy. A decline in the real wages of an employee group may lead to their advocacy of public support for wage increases. A decline in the economic position of a domestic industry may lead to an appeal by the industry for price supports or tariffs.[4] A voter facing a choice about a town budget may "want only to understand 'What's in it for me?'" (Lindblom and Cohen 1979: 59). The success of the Census Bureau, and later of the Bureau of Labor Statistics, in measuring the unemployment rate has depended in part on relations with opposed constituencies—business and labor—who use this information to serve their disparate interests (de Neufville 1975: 70–86). The Census Bureau performs the same sort of service to opposed groups in furnishing population statistics that are used in allocating public resources, though it experiences increasing pressure as a result. Public officials may also respond to indicator statistics in seeking private goals such as material gain or reelection.

3. Some governmental decisions can be adaptive, such as the response of public school planning to demographic trends (Parke and Sheldon 1973: 107–08). When private adaptation takes the form of a market decision, the cumulative effect of such decisions can alter what is produced; producers may then be motivated to bias the information (Chap. 9).

4. The individual member of such a group is then acting in the group's interest rather than simply in self-interest. We nevertheless refer to the group's motive in such cases as "private," in contrast to the wider public interest.

Johnson and Lewin (1984) have proposed that municipalities publish periodic reports with specific attention to the interests of diverse user groups.

These private actions affecting public policy may or may not further the general welfare. Exploitation of the public purse by public officials or private groups will presumably not do so. But when a group argues persuasively that it is needy or that its claims are just, and when this argument is supported by others, we may have difficulty in classifying its action. Marketlike competition and exchange by diverse interests in politics have also been seen by some as means for furthering the general welfare (Coleman 1970).

Members of a group seeking its particular interests through politics may support the group's actions by arguments within the group based on values that they share with their fellow members ("It will benefit us"); but when they seek to persuade outsiders in public terms rather than through private bargains, they must invoke more general ethical values, such as the general welfare or justice, in their arguments (Dibble 1962). The use of these arguments requires statistics similar to those needed in cell *d*.

c. Actors acting alone and seeking to benefit others without increasing their own well-being *(private altruism)* may also make use of information. Information about poverty may increase charitable donations. Knowledge of an impending disaster may lead to community aid. Information on discrimination in employment can aid private as well as public efforts to remedy it. Those who act in this way may try to persuade others by invoking general ethical values; but this persuasion is directed toward private action rather than toward voting or other participation in collective decisions. In this respect expert communities, universities, and foundations can use information in choices affecting the general welfare or justice without entering into the collective decisions of the larger political community. The argument of this book, recommending indicator variables and directions for research, is aimed in considerable measure at these actors and not simply at governmental decisions.

d. Finally, information can be used by actors in their pursuit of *ethical values* such as the general welfare or justice, through *collective decisions,* e.g., through public policy.[5] This use of information concerns what should be done for the good of society or as right action, in more or less general terms. In specific terms it can relate to particular values that are widely accepted in the community (e.g., traffic safety). In general terms it can call for comparisons or tradeoffs in terms of the general good among various particular values (How much health is worth how much education?) and groups (What benefit to the nonpoor is equivalent to a given benefit to the poor? What benefit to

5. The planning of statistical systems to aid these decisions depends not only on the nature of the collective decisions made in a political system, but also on the desired distribution of decisions between the public and private sectors. Steinhausen (1975: 31ff.) points out that this distribution can be justified on economic and other grounds, and can itself be influenced by the availability of statistics.

those now living is equivalent to a given benefit to the next generation?). These tradeoffs, in addition to the extensive causal information required, complicate the calculations required relative to those needed in cell *a*.

When public policy choice aimed at ethical values is involved, we shall refer to the public statistics that are useful for this choice as *policy indicator statistics*.[6] In the following sections of this chapter we shall distinguish between types of variables, measures of them, populations, and time sequences of measurement that can characterize these statistics. The choice among public policies, like all the other types of action discussed here, requires not only information on conditions of well-being or justice but also causal knowledge as to how to promote them. Causal models, whether derived from popular belief or scientific research, are therefore an indispensable accompaniment of policy indicator statistics, even though not part of the body of statistics themselves; and we shall discuss such models throughout the book.

All four of these types of action may be aided not only by information directly relevant to them, but also by the general enlightenment that results from increased understanding of causal relations (Lindblom and Cohen 1979: 73–81; Janowitz 1970: 250–54). In planning information systems, however, we must try to judge whether one type of information is more useful than another. Even if both provide general enlightenment to some degree, we must try to be more specific in judging how effectively they do so.

In any practical decision whether to use a variable in a public statistical system, we would try to weigh the benefits and costs of including that variable as they related to all four of these types of uses. Surveys of types of users of government statistics (Hauser 1975: 55, 64) might be used for approximate judgments. Practical considerations such as the cost of changing existing statistics would also enter our calculation. In this book, however, we are narrowing our concern to that part of the choice that reflects the initial concerns of Bauer (1966: 1) with "where we stand . . . with respect to our values and goals," and of *Toward a Social Report* (U.S. HEW 1969: xii) with "how well off we really are." These are concerns with ethical values and with possible collective actions affecting them. They lead us to the fourth of our categories (cell *d*), and to the general notion of policy indicators, which will include these earlier authors' concerns and extend beyond them.

Choice or advocacy of public policies in view of ethical values is of special interest because it involves cooperative discourse between the political community (discussing these values and means to them) and expert communities. Self-interested decisions, in contrast, do not require public debate. For self-

6. Diverse notions of the general welfare or of justice, for example, would give rise to diverse sets of policy indicators, the reconciliation of which might have to be done politically (Chap. 2). Choices among candidates for office, if they correspond to sets of anticipated policies, may also be informed by these statistics. Government statistics for internal use, not available to the public, might be used for assessing policies but would not be included in our definition.

interested decisions, some of the information required is produced in response to market demand, as when newspapers publish price information or when candidates purchase opinion-survey data. Other information used in self-interested decisions is supplied partly by the market, as when private data processing organizations (e.g., Data Resources, Inc.) reinterpret and re-analyze government statistics for profit, or when the media publish them. But much of the collection and preparation of these data must be undertaken or supported by government; as a collective good they are inadequately supplied by the private market. Systematic monitoring of conditions such as those of the environment is also unlikely to be done by basic scientific communities (Brooks 1982: 21).

We are thus engaged with ethical values at two levels. At the most abstract level we are comparing information policies in view of these values; but we are also concentrating on those aspects of information policy that aid other citizens in comparing possible policies in terms of their ethical values.

Indicator Statistics, Models, and the Cycle of Policy Choice

We have given a preliminary definition of policy indicator statistics as public statistics useful for public policy choice in view of ethical values; but this does not yet say what kinds of statistics, or other related information, will be most useful for this purpose. Descriptive statistics on valued characteristics of general populations of political units (national, state, or local), repeated in time series, have been a feature of social reports. These time series have some value for defining problems, but cannot by themselves guide us in choice among policies to deal with those problems. For this guidance we require causal models that will enable us to compare alternative policies by predicting their differential consequences in terms of the values we seek. Any public argument for a policy in terms of its expected results implies at least an intuitive causal model; but such models may be improved by research. These improved models may be sought by many types of research, among which studies of such general populations play only a small part.

The common element connecting general population time series and models of causation is the valued or disvalued variable entering into both and allowing them to supplement one another. This connection works through a repeated cyclical sequence in which public problems and policies are considered:[7]

7. We here follow an abbreviated sequence of the elements of policy analysis (MacRae and Wilde 1979), which correspond to similar stages within evaluation research, broadly defined (Rossi and Freeman 1982: 20). A similar but more detailed sequence involving social indicators is proposed by Horn (1980: 432). Cycles including action or experiment, review, and further action have also been discussed by Etzioni (1968) and D. Campbell (1969); Etzioni's notion of mixed-

a. Definition of the problem (general population time series useful);
b. Choice of policies (models of causation needed);
c. Program monitoring (special population time series useful);
d. Return to a redefinition of the problem with information on the general population (general population time series).

If a problem is defined by a value measure for a general population time series (e.g., an increase in unemployment), then in principle our efforts to deal with that problem should focus on measures of that same value. Evaluation research—one means of seeking causal models—is sometimes believed to be separate from public statistical systems because of its specificity, focusing on particular program goals rather than on the general values that brought the program into being. But in moving from program evaluation to the analysis of policy choices for the future—as we must—we should return to the more general variables. If, for example, national series of measures of health led us to define a problem and to institute a program requiring health personnel, we should not end our evaluation with evidence that the personnel have been supplied, but should continue it to study the program's effects on health.

If we concentrate attention on the same values (social, economic, or environmental) throughout this sequence of steps, we can connect several concerns of public debate: the identification of problems, the choice of policies, and the judgment as to whether our policies have dealt with the problems. To show how this can occur, as well as to note some of the limitations of this approach, let us consider each of the above steps in turn. We must recognize that political considerations are of major importance throughout the sequence; but here we stress the incremental contribution of policy indicators and of arguments related to them.

a. Definition of the problem. We consider a problem, as defined within this sequence, to be a shared judgment of the existence of a "social problem," i.e., a contrast between an observed state of affairs and a valued expectation shared by a group (MacRae and Wilde 1979: 23). An example is the recently increased concern in the United States with educational outcomes in public schools. Concerns may also be raised by anticipation of undesirable states of affairs (Brewer and deLeon 1983: Chap. 2), based on monitoring of social or environmental conditions together with predictive models. The definition of a social problem can take place in any social system; but insofar as the definition of a problem leads to collective public action, this action must ordinarily be chosen in a political jurisdiction such as a nation, state, or city, in which there are governmental authorities entitled to make policies. A problem may

---

scanning (1968: Chap. 12) is consistent with an alternation between the evaluation of particular programs and the analysis of more general policy alternatives. In actuality, the steps in such a cycle can be combined and occur together. In addition, if a policy has side effects, the cycle may bring in values not considered at the start.

be defined, for example, in a metropolitan area; but only political subunits that are parts of that area may be able to act on it directly. For purposes of problem definition in a particular political jurisdiction, the relevant policy indicator statistics will presumably be presented as aggregates or distributions in that jurisdiction; we may need to know the income distribution in a city, the infant mortality rate of a state, or the conditions of minorities or the elderly in those areas. The principal justification for basing public statistics on general populations lies in their use for collective definition of problems. At the same time, however, problems of special populations or subareas may also be identified.

The groups that share these concerns may be diverse. At one extreme, an entire community or nation may be concerned about problems such as those of the environment or the economy. At the other extreme, only a small group may be concerned, either because they are affected or because they have special knowledge or beliefs about the situation. Such a group may then try to persuade others to share their concern by publicizing the condition and relating it to generally shared values. As they do so, they may also propose remedial action, the next step that we shall consider. They may join with other groups and redefine the problem more inclusively (e.g., problems of minorities in general, rather than those of particular minorities). They may also encounter opposition and provoke other groups, who see themselves as threatened, to propose counterdefinitions of *their* problems (Rule 1978: 174). Small elite groups can also sometimes influence public action without involving the public or the mass media; initiatives in specialized areas such as foreign policy or technology development may be proposed in this way.

Descriptive statistics about the level or distribution of a valued or disvalued condition (e.g., a crime rate) can thus contribute to the definition of a problem in relation to a valued expectation of that level.[8] Such expectations may be set either by earlier levels (one use of time series) or by parallel levels in comparable populations (MacRae and Wilde 1979: 26–30). Comparisons may be "intergroup, intertemporal, and international" (OECD 1976: 159). In addition, we may compare the present situation with goals that have been set in

8. Such statistics may lead individuals to redefine their personal hardships as general and shared by others, thus encouraging collective rather than individual action. If the statistics define a condition to which we must adapt, such as a declining number of school-age children, they can be translated into valued or disvalued conditions that are manipulable, such as excessive expenditures for buildings and teachers. Problems of adaptation can arise from fundamental social change over which we have little control. There may also be disagreement within a political community as to what values are mainly threatened, especially when questions of equity arise (Chap. 7); diverse indicators and interpretations of them can then be involved.

The types of statistics needed can vary depending on the specificity or generality of the problem and the position of the initial problem definer in the political system—line operational units, administration and management, or the general public (Biderman 1970: 219–20); but even specific administrative statistics can be of interest to the public.

advance; examples are minimal desired levels with respect to which "needs" are assessed (Rossi and Freeman 1982: Chap. 3), and planning goals. Thus, when levels of aggregate value measures are made public, they typically arouse or allay concern depending on whether positively valued conditions are smaller or larger relative to these expectations (and conversely for negatively valued conditions). By extension, we may use projections or predictions to anticipate future problems. The statistics used to define problems in this way may reflect general ethical values, contributory variables that are surrogates for them, or others that serve as early-warning predictors of them (Meehan 1975: 44). They need not always be presented in time series.

The definition of problems in this way, by contrasting observed conditions with generally valued expectations, is a common way of choosing priorities for attention and proposing policy alternatives. It is not always the best way, however. To choose actions, we need to know whether those actions will be effective, and this is not the same as knowing that some valued condition has fallen short of our expectation.[9] We may sometimes be able to produce more value by discovering a new opportunity for action, such as a technological innovation, rather than by responding to felt problems; the fluoridation of water supplies was such an innovation, and was considered even in the absence of an increase in tooth decay. An opportunity may also be recognized first within an organization rather than by the general public (Brewer and deLeon 1983: 37–42). Conversely, there may be no available means for dealing with certain trends affecting a felt problem; an increase in incidence of the diseases of old age may result in part from the prolongation of life. Our expectations should be shaped, then, by some knowledge of what can be accomplished.[10]

Our definition of problems through comparison over time would seem to rest on a desire to restore the condition measured by previous points in the series. Zapf (1972: 249) has characterized the political orientation of the early social indicators movement as "liberal incrementalism." If the time series in question are relatively specific—e.g., income distribution, housing, educational achievement—an effort to restore them all to previous levels may well neglect possibilities of major change. If, however, we examine measures of more general well-being, or ends rather than means, we may be able to con-

9. Similarly, the current state of a general population with respect to some value, as revealed by indicator statistics, is of only indirect relevance to our policy choice and to the models that guide it. It is relevant only if the incremental value of the policy depends on it, that is, if there are nonlinear effects such as diminishing returns.

10. For example, both birth rates and crime rates are affected by the overall age distribution; if this were the only variable affecting them, we might not benefit greatly from knowing of changes in them. Bauer (1967: 191) pointed out that "The decision to observe a phenomenon implies a decision to be responsible for it, if such responsibility is within one's power." This may require a higher degree of responsibility for social scientists than we usually require of writers, artists, or free citizens when they point to problems.

sider a wider range of policies to restore or improve them; examining trends in health rather than in health services, for example, can thus enlarge our perspective (Parke and Sheldon 1973: 106).

Time series also have special relevance to problem definition if we are concerned with subjective well-being. If people's subjective well-being depends not simply on their present condition but on changes in that condition (Chap. 6), then the felt harm resulting from a decline in "objective" well-being will be suggested by trends in time series,[11] and the effectiveness of our actions in restoring subjective well-being may be correspondingly greater where a decline has occurred. The use of time series to define problems is thus connected with subjective well-being.

b. Choice of policies. Once a public concern has arisen, it is likely to be accompanied by the proposal of policies to deal with it.[12] If we act on a problem defined in terms of a measure of a given value, we may try to influence this same value through our action. This involves knowledge of causal relations, expressed in verbal or formal models, in which the dependent variable is presumably the same value measured by our policy indicator statistics. Thus benefit-cost analyses of particular projects use essentially the same value measure as the GNP or national income (see Chap. 5), and local energy interventions can fruitfully use the same energy-consumption measures as national energy accounts.

This correspondence of values between problem definition and causal models holds regardless of whether the models are derived from research or not. Much of the causal judgment on which policy choices are based rests on expert experience and public belief; some involves simple assumptions such as blaming the incumbent officials.[13] We try to improve our models of causation by research, but cannot usually expect policy choice to wait till research has produced definitive results.

11. For this purpose it would be preferable to have panel data so as to examine the distribution of individuals' changes in objective well-being.

12. Public problems often need to be redefined for analysis (MacRae and Wilde 1979: 17–21), and policy alternatives must be designed. The political feasibility of enactment, affected by the public's definition of problems, must also enter into our choice of policies. In addition, the choice as to what level of government or organization is appropriate for decision (Meehan 1982: 80–81) also depends on our knowledge of causal relations. Among the policies that may be considered are meta-policies, such as gathering better information, conducting research, or changing decision structures.

13. According to one theory of democracy, the task of citizens is to choose officials, and that of officials to choose in detail among policy alternatives. Citizens would then be more concerned with models of personal or group blame (as in the doctrine of responsible political parties) than with impersonal assessment of policy consequences. Experts, however, would run a special risk of political controversy if they dealt with models of blame. This theory of democracy would therefore lead experts to advise officials rather than citizens—an undesirable outcome from the standpoint of promoting use of policy analysis by citizens.

Insofar as our models are furnished by systematic research (policy analysis or practical social science), they may be general and similar in form to those of basic science, or they may be specific to particular programs. We often begin with previously established general models and use them (together with knowledge based on experience) for the design of a program; we may then move to more specific models as we try out experimental programs and evaluate them in a process of fine tuning (D. Campbell 1976). These general and specific models can share the same value concepts and measures while differing as regards the populations we study to derive and test them. When action is taken within a political community, it need not affect the entire citizenry; program evaluation and social experiments thus require study of special delimited populations. The data studied are likely to be for individuals rather than aggregates; and for fundamental model development, the populations studied can range widely in time and space, depending on the variables in question and the economies of research. Indicator statistics used for problem definition will have a relatively small part to play in general model construction, even when they are based on the same value measures as are used in the models.[14]

The variables we choose for public statistics are linked to those in our causal models not only because they are joined in public deliberation, but also because our knowledge of causation can affect our priorities for presenting statistics on particular dependent variables. If the important contributory variables in a given area are nonmanipulable, if the costs of changing them are excessive, or if we do not yet know the causal relations with assurance (Sheldon and Land 1972: 142), the value measures in that area may have lower priority for presentation.[15] Dissemination of policy indicator statistics may, however, stimulate research and policy analysis.

For inferring causation, we need information that aggregate national time series do not usually provide. One approach used for developing specific causal models of policy effects is program evaluation; but "there is increasing agreement that program evaluation requires the evaluator to demonstrate that government programs, not uncontrolled extraneous variables, determine the outcomes measured by indicators" (Sheldon and Parke 1975: 695). These causal inferences, and the construction of the corresponding models, require either experiments, quasi-experiments based on observation over time, or

14. Biderman (1966b) notes instances in which before-after studies of unique events can be useful for model construction. Land and Felson (1976) also show that "dynamic macro social indicator models" can be based on time series of aggregated indicator statistics.

15. The knowledge required for choice of variables combines public belief with the investigations of experts, however. As Bauer (1967: 187) pointed out, we must be able to choose indicators even in the absence of consensus on a model of society; he suggested that we must choose the parameters to be measured in terms of general agreement that they are "important." It is desirable to be more explicit, however, as to why they are important.

careful nonexperimental reasoning (Chaps. 3–4). Any of these methods requires attention to special affected populations.

We can nevertheless connect model construction with time series of indicator statistics in several ways: (1) the value concepts and measures used in policy indicator statistics can also be used in policy models, especially as dependent variables (though other variables in the models need not be used in indicator series); (2) at times the trend of a dependent variable in a general population can be used as a statistical control for the trend in a comparable treatment population; and (3) if the treatment is likely to have a significant effect on a general population (e.g., a state or locality), analyses of time series for that population can sometimes combine the development of policy-relevant causal models with the study of social change (Land 1982: 36).

The sense that a problem exists can sometimes lead society to support research or experimentation as well as to seek policies to deal with it. When research is supported for this reason, however, its results may come at a later time when concern for the problem has abated or a choice has already been made. The lack of timeliness of research findings can be exacerbated by this interrelation between research and public concern (Chap. 4).

c. Program monitoring. The causal models we use to choose policies and develop programs are incomplete; they cannot predict the effects of a policy in every detail and are often only rough approximations. The results of actually carrying out a program can add to our knowledge, especially for that particular program. We often monitor programs with time series, which can reveal time-dependent effects such as those that are delayed or that decay over time. They can also reveal deficiencies of our causal models, such as a lesser output than expected for a given input. The additional knowledge we thus gain does not usually come from general population time series, however, but requires attention to the population affected by the program. Like other causal inferences, it also requires careful research design to infer causation; for example, a disaggregated baseline "inventory" of the population can be useful in tracing differential effects of policies on various population categories (Meehan 1982: 74–75).

When we seek to monitor a program, we often need to know conditions promptly in order to act on them. When the effects of a program are likely to be long delayed, as for the long-run effects of education or of preventive health measures, we sometimes monitor intermediate or contributory variables that can be changed more rapidly, such as educational test scores or the delivery of health services. Thus while the variables we monitor should ideally be the same ones that gave rise to the problem definition and to the policy that dealt with it, we sometimes make use of surrogates for them.

Several examples illustrate the possibilities and limitations in our efforts to infer causation by program monitoring. In the effort to reduce pollution in the River Thames, British scientists traced the concentrations of pollutants and of oxygen over time. Because of their knowledge of the chemical and biological causal relations involved, they had considerable confidence that the reductions in pollution were due to the effluent treatment program (Harrison and Grant 1976: 28–36). When a garbagemen's strike produced a temporary increase in pollution, this additional cause was understood. Consequently, the monitoring served as a check and an encouragement to continue with the program until an acceptable level of pollution was reached.

A second example concerns the effort to equalize opportunities for upward social mobility in Hungary after it became a socialist society. Andorka (1980: 11), using a series of mobility studies, showed a considerable decline in inequality of opportunity for men between 1938 and 1949, which continued till 1962 but was replaced by stability thereafter. He suggested that monitoring of such changes may permit the introduction, if necessary, of "policy measures to counteract emerging negative phenomena" (12). This suggestion implies that the causal models are adequately known, presumably including the effects of new policies affecting education and employment. They are not, however, known with the precision of the natural-scientific relations involved in water pollution.

A third example is the monitoring of educational achievement by black and white students, at ages nine and thirteen, in the United States over the period 1970–1980 (Burton and Jones 1982). Data collected by the National Assessment of Educational Progress showed that the difference in achievement between the two racial groups decreased steadily over this period. In view of the variety of policies and other changes aimed at equalizing opportunity, however, no inferences could be drawn as to the contribution of particular policies. The results may encourage continued efforts to increase equality in numerous ways; but the time series alone do not tell us which of these efforts will be most effective. In summary, monitoring by time series can be useful, but its utility depends on our eventual knowledge of how to respond to them, and thus of causal relations.

d. Redefinition of the problem with information on the general population. It would be comforting (and welcome to program managers) if we could terminate this process with a summary evaluation once we showed that a program was furthering the general value that led to its establishment. We cannot do this, however, for three reasons: (1) the program may have dealt with only part of the initial problem (students in remedial programs were learning, but the outreach of these programs was inadequate); (2) the program may have negative effects outside the target population (increased police patrolling in one area may move crime elsewhere; new jobs for one group may be gained at the expense of another); (3) the program may have positive effects outside the target population (a new style of care is diffused spontaneously within a profession). In addition, unintended effects, desirable or undesirable, may occur with respect to other valued variables; these may be discovered through study of other statistical series, or from nonstatistical information. Land (1975b: 31) notes that there can be "inconsistencies between micro and macro effects" which require us to scrutinize effects at the macro level.

For all these reasons, we must continue seeking the effects of our policies more generally than target populations and program goals would suggest. This inquiry must be approximate and exploratory at the start; by searching in a broader population, with a wider range of dependent variables, we greatly complicate the task of causal inference. We may be guided by theory, or by complaints, to particular subpopulations and variables for which unintended effects have occurred. What is important to recognize, however, is that a quest for the general welfare, or for justice or equity in general, ultimately involves attention to general populations. Especially when policies involve social causation, for which our models are incomplete and fallible, we must look beyond narrow and short-run demonstrations that we have done our job. Valuable and difficult as such demonstrations may be, the ultimate subject of our public discussion must be general well-being, not simply the accomplishment of program objectives.

In summary, the value of the use of time series to guide action depends on the correct determination of the valued expectations against which the series are compared, in terms of the range of community values involved and of our capacity to deal with the problems thus defined. We may expect our schools, for example, to be educators, equalizers, socializers of a heterogeneous population, babysitters, and means of keeping delinquents off the streets; but unless we recognize that such goals compete with one another for resources, indicators reflecting these goals may only heighten political demand. The monitoring of series of measures of general values (such as aggregate subjective well-being) can sensitize us to general problems without clearly showing us their causes; the monitoring of specific series (such as the cleanliness of a city's streets) can direct our attention to more specific problems associated with more specific means. In the latter case we may know what to do about the problem, and may well be in tune with public opinion, but may neglect more general approaches.

When we use time series to define problems, in the absence of causal models of high reliability, we are using an approximate goal-setting procedure rather than an ideal. Comparison with the recent past is indeed a feature of incremental decision making, with its remedial aspect (Braybrooke and Lindblom 1963). In accepting this aspect we recognize the difficulty of seeking out innovative directions for policy—related to the difficulty of optimizing by scanning *all* alternatives. We are satisficing, not optimizing, in our decision strategy.

In a democracy this remedial strategy gains another possible advantage. If experts or policy analysts cannot often seek out innovative policy directions, the general public can do so even less often. Major political pressures will often support the restoration of a previous situation after citizens have suffered a loss of well-being; and indicator statistics can provide a public definition, fostering some degree of consensus on the magnitude of the problem, when this has occurred. Thus the definition of problems by fluctuation of indicator series is often in accord with the public's normal tendencies in definition of problems; the series may anticipate and influence that definition, or provide a firmer basis for public concerns than those based on personal experience. Governmental action on these problems, even if it does not optimize a welfare function, will meet with public support; and since policy analysts are minor participants in a large political system, they usually need this support. The definition of problems by changes in indicator series is thus a solid second-best strategy.

Concepts, Measures, and Values: Terminology

We must now suggest ways of choosing the variables used in the policy-related part of a public statistical system. For this purpose, the phrase *social indicators* contains four possible implications or ambiguities that we wish to avoid: (1) it has become associated with the analysis of social change, which often guides the choice of variables in a direction different from that of policy choice; (2) it has arisen from a criticism of, and opposition to, economic indicators, whereas a policy-relevant statistical system should include economic information and integrate it with information based on the other social sciences and natural science; (3) it unfortunately defines measures and statistics together,[16] causing difficulty if we wish to speak of the variables and measures used in social (or policy) indicator statistics when they are also used in the study of special populations for model construction or program evaluation; and (4) following the terminology of "economic indicators," it tends to confuse an indicator or measure with the concept that it measures (Meehan 1975: 36–37). Because this distinction is even more important in the less exact social sciences than in economics (and it should be heeded there as well), we need again to make this distinction.

In addition we shall make a distinction that has not been stressed in the policy-oriented approach to social indicators, between more inclusive and

16. Some economic indicator statistics (the Consumer Price Index or inflation rate) do not have corresponding disaggregated measures at the individual level. Others (the gross national product) *are* named distinctly from measures of the production of individual firms. Indicators of unemployment more nearly resemble those social indicators for which the same term designates both a national aggregate and an individual characteristic.

less inclusive values.[17] One value can include or subsume another by being the end to which the other is a means, or by being an inclusive class of which the other is a part. We focus first on the means-end relationship. Those values that are ends, and not means to other values, are known as *intrinsic values* (Lewis 1946: 397; Frankena 1963: 66–68, 71–75). Intrinsic values, especially those that are also general or inclusive, are especially important in relation to policy indicator systems for several reasons: (1) they may be of more lasting interest than the particular means to them; (2) they can play a part in the reconciliation of values in public discourse about policies (Chap. 2); (3) they can help to clarify and guide our measurement of contributory variables; and (4) they can be the basis of a type of practical research that is both general and policy-relevant at the same time (Chaps. 5, 6), connecting social research with a general aspect of public discourse about society's goals.

Let us first distinguish between a *concept* and a *measure* of it.[18] Social scientists conventionally distinguish measures from concepts (Duncan 1975: 130). This distinction is especially important for statistics used in policy choice, because such statistics may lack reliability or validity not only because of the usual sources of measurement error but also because of pressures from persons affected by their use (Chap. 9). A measure of the concept "subjective well-being" may be a respondent's answer to the survey question, "Taking all things together, how would you say things are these days—would you say you are very happy, pretty happy, or not too happy?" A measure of economic well-being would be the answer to "What was your family's income during the past year?" Measures of environmental conditions and facilities might include the concentration of a pollutant or the presence of a park in the respondent's neighborhood.

Most of the value concepts and measures we shall consider refer to individuals. This is because the principal intrinsic values we consider in this book (economic and subjective well-being) refer to individuals. Equity or justice, which we also take to be an intrinsic value, usually refers to relations between individuals, but can also refer to relations between groups. The substantive matters to which equity refers, even when they are comparisons between groups, are also usually based on the conditions of individuals. For any of these values, however, it is social rather than individual valuing that serves to define public problems or justify collective action.

17. The values we analyze here and in the next section refer in the first instance to individuals, and the judgment whether they are intrinsic is also made for individuals. When these values are measured and the measures aggregated for a general population, we assume that the aggregate of an individually intrinsic value is also of intrinsic value, and reflects the general welfare.

18. Although the actual design of statistical systems, and that of research on policy models, are centrally concerned with choosing the best measures of given concepts, we do not treat the choice of measures in detail in this book. Some of the organizational and political factors affecting this choice are discussed in Chap. 9.

Among the concepts useful for public statistics, however, are some that do not refer to individuals. The concentration of a pollutant is a concept of this kind; but the intrinsic (dis)value that results from it must be referred to its effects on individuals. Similarly, the integration of a social system is a concept referring to a supra-individual entity, even though we consider the intrinsic values it affects to refer only to individuals. We shall refer to concepts such as pollution and integration, designating states of systems other than individuals, as *contextual concepts*, and to measures of them as *contextual measures*. We are thus generalizing the sociological use of the term contextual, using it to refer not only to properties of social collectives (Lazarsfeld 1959: 69–73; Lazarsfeld and Menzel 1965) but also to characteristics of a person's physical or biological environment. An additional type of nonindividual concept refers to relations between individuals, such as we consider in connection with equity (Chap. 7).

Among national economic indicators we classify some of the corresponding concepts as aggregate end-values and others as contextual conditions affecting individual well-being. The national product and corresponding measures of net economic benefit are typically aggregated over goods and services, not people; but in taking these as examples of intrinsic values we may note that the equivalent national income can be aggregated from conditions of individuals (Chap. 5). Various contextual concepts measured by economic indicators (prices, inflation rate, investment, balance of payments) do not correspond to intrinsic values.

Aggregates of individual values (e.g., disease rates, average educational achievement) also correspond to concepts and measures that do not as such refer to individuals. They are aspects, positive or negative, of the general welfare. We shall refer to these measures of aggregate concepts, if the individual measures on which they are based qualify as indicators, as *indicator statistics* to distinguish them from individual-based indicators.[19]

We use the term *indicator* to denote a measure which is justifiably used as a basis for public statistics (i.e., a measure of an *indicator concept*); this parallels Carlisle's (1972: 25) definition of a social indicator. A *policy indicator* is a measure on which we can justifiably base statistics that are included in the policy-related part of a public statistical system; and a *social indicator* is a measure of social conditions that can be justifiably used in the system. We leave open several possible bases of justification for choosing social indicators, e.g., enlightenment, relevance to policy, or relevance to basic social-system models. In this terminology the aggregate or population data on social conditions that are published in a social report will be called *national social*

19. The concepts to which indicator statistics refer may be called "macro-concepts," by analogy with the terminology of Land and Felson (1976: 568–69). An example, though an imprecise one, is "the general welfare."

indicator statistics, and series of them *national social indicator series* rather than simply social indicators.[20] We can then meaningfully ask whether a policy indicator should be employed in studies of special samples for model construction or program evaluation.

Among the concepts and measures that are candidates for our information systems, we must make distinctions according to their valuative status as ends or means. Since measures refer to concepts, we shall discuss only the valuative status of the concepts in question.

We first designate the term *end-value* as an abbreviation for intrinsic value. An *end-value measure* is a measure of an end-value.[21] The designation of end-values will not be made identically by everyone; in the next section we shall suggest one way to make such designations, by analyzing the valuative connotations of some social indicator concepts. Chapter 2 will discuss further the ways in which means and ends are treated in the discourse of democratic political systems.

Among end-values, we are primarily concerned with those that are inclusive as well as intrinsic; by subsuming various specific values they aid us in comparing them. We noted above that one value can subsume another either by being an end to which the other is a means, or by being an inclusive class of which the other is a part. The means-end relation defines intrinsic values as ultimate ends, but not all such values are general; for example, the appreciation of music is generally regarded as of intrinsic value, but it does not include as many other values as does economic benefit or subjective well-being. Even health, though often considered an intrinsic value, is not as general as some others, as we shall see below. Because we are concerned with the potential role of end-value indicators in reconciling diverse particular values in public discourse (Chap. 2), we shall therefore stress *general end-values* (inclusive ones); it is these that we treat in Chapters 5–7.

Even before we choose a set of general end-values for discussion, we can

20. A similar terminology for social indicators, including multiple qualifiers, has been proposed by Michalos (1978). Our inclusion of the word "justifiably," in the definition of an indicator, calls for extensive discussion of issues treated throughout this book. Our definition of social indicators here reflects the social-reporting function more than the part played by indicators in scientific models of social change.

21. Two major components of a policy-related statistical system will be indicator statistics related to general end-values and to corresponding contributory variables (defined below). Value concepts include concepts of both these types. An end-value measure resembles what was called "a direct indicator of welfare" in an early definition of social indicators (U.S. HEW 1969: 97); it is a more general category, however, including equity as well as welfare. In view of the distinction between measure and concept, however, it is more accurate to say that it is a measure of a value considered intrinsically desirable (in this sense "direct"), even though it may not be a perfectly reliable or valid measure of that value concept (Sheldon and Freeman 1970: 98). The general end-values with which we shall be concerned resemble what Gross (1966b: 263–66) has called "grand abstractions" and what Dunn (1974: 107–10) has referred to as a "grand indicator."

use this definition to circumvent an earlier controversy in the social indicators literature. The claim (U.S. HEW 1969: 97) that social indicators were "statistics of direct normative interest" was challenged by Sheldon and Freeman (1970: 98) on the ground that other measures, which helped us to understand value concepts through their causal relations, should be included among social indicators. Policy indicator systems may well include such measures (called contributory measures below), though not always the same ones as would be used for tracing social change; these systems can include measures of both ends and means, but we can still distinguish between the two.

Our choice of general end-values as bases for some policy indicators (a choice in which citizens and experts both participate) derives from our judgment as to what is of intrinsic value, based on notions such as those of the good, of welfare, and of justice or equity. It does not derive from our notions as to what is useful as a means, because end-values define the very criteria of usefulness. Once we have chosen end-values, however, they can play a part in the choice of other policy indicators.

In seeking to influence end-values we make use of explicit or implicit causal models. We shall call the other variables in such a model, which affect a given end-value, *contributory variables*[22] to it. Measures of these variables are *contributory measures*, and some may be chosen as policy indicators. Contributory variables can be either manipulable or nonmanipulable—the former being possible policy instruments, or results of them, and the latter statistical controls or bases of subdivision for judgments of equity.[23] Thus if health is taken to be an end-value, then in relation to it conditions of sanitation or numbers and distribution of physicians may be manipulable contributory variables, i.e., means to health. For general subjective well-being as an end-value, satisfaction with health is a contributory variable.

A *value concept* will be defined as either an end-value or a valued means to

22. We use this term rather than "explanatory variables" to stress the decision-oriented character of the models that link them to value variables. Johnston and Carley note the distinction between discipline-based explanatory variables and variables more closely related to movement toward preferred future states (1981: 241, 250, 253). "Instrumental variables" can also have a wrong connotation. Land (1975a: 17) defines "analytic indicators" as components of models of social processes which result in "values of the output indicators," but contends that models developed for scientific purposes can be transformed into policy models by the simple addition of valuative criteria (p. 33). Basic researchers, however, are not properly motivated to include in their models just those variables most useful for policy purposes.

23. Land (1975a: 18) makes a parallel distinction among exogenous variables in social indicator models. Our terminology differs from his in distinguishing indicators from concepts, in not restricting contributory variables to exogenous variables, and in defining the final endogenous variables in terms of values rather than outputs. The question whether a variable is manipulable can depend on the situation; some variables become manipulable after relevant technology is developed (e.g., weather modification), and others are manipulable only in the long run or at considerable cost (e.g., a constitutional amendment).

an end-value. We can value a contributory variable as a means even though it is not valued for its own sake; it can be valued either positively or negatively (pollution reduces health). If the causal relation between the contributory variable and the corresponding end-value does not have the same sign under all conditions, then no one direction of valuation can be placed on the contributory variable relative to that value variable. For example, whether a divorce increases or decreases well-being depends on the circumstances of the marriage for the person in question; and whether increased social integration conduces to well-being may well depend on the current level of integration, if the relation is curvilinear (Chap. 8).

The new terms and definitions we have proposed are summarized in Table 1–2.

Each end-value can have its own set of contributory variables; these sets may overlap and at times a contributory variable may be a positive cause of one end-value and a negative cause of another. Suppose, for example, that we are concerned with two logically independent end-values—the GNP (Gross National Product or national income) and aggregate subjective well-being. We can imagine contributory variables that are causes of both. One such variable might be the total number of hours worked. If, as a matter of national policy, we went back to the ten-hour workday instead of the eight-hour day, presumably the GNP would increase; but overall subjective well-being might decrease in spite of the additional products and earnings available. In evaluating a ten-hour workday (or a six-hour day), we might have to decide on our trade-offs between the GNP and overall subjective well-being; the potential conflict between these two values could create a motive for reconciling them (Chap. 2). One end-value can also be contributory to another. The GNP, for example, might affect subjective well-being (Chap. 6). A given variable can be both an end-value in its own right and a contributory variable to another end-value.

Measures or indicators themselves cannot usually be taken as causes of other variables in the corresponding causal models (we cannot make time run backward by turning the clock back); they themselves are thus not usually contributory variables.[24] We may, however, wish to consider for our statistical system some variables that, like indicator items, are not simply causes of end-values but enter into causal models with them in other ways. Expenditures to alleviate a problem may be both results and causes: "An increase in expenditures for unemployment insurance signals both a weakness in the economy that should disturb us and a strength in the protection for individuals and families that marks an effective social welfare system" (Merriam 1978: 118). Levels of services rendered have similar properties. Biderman (1966a: 115) notes the possibility that "year-to-year increases in crime rates may be more indica-

24. The publication of indicator statistics can, of course, modify people's actions, and these altered actions can change the meaning of the statistics.

Table 1–2. Terms and Definitions Related to Indicators

Public statistics: Statistics available to the public, including statistics from nongovernmental sources.

Government statistics: Public statistics provided by government.

Concept: An underlying variable, not necessarily in operational form, entering into theory, models, or public discussion.

Measure: A variable measuring a concept. (Each type of concept below may have a corresponding type of measure.)

Value concept: An end-value or a valued means to an end-value.

End-value: An intrinsic value.

General end-value: An end-value that includes or subsumes numerous more specific values.

Contributory variable (to a given end-value): A variable (concept) that is a direct or indirect cause of that end-value.

Contributory value: A contributory variable that is a value concept.

General population: The population of a political community that defines problems and makes policy decisions, e.g., a national, state, or local political unit.

Indicator: A measure justifiably included in a public statistical system.

Indicator statistics: Statistics based on one or more indicators. (Corresponds to a widely used definition of "indicators.") Many indicator statistics are aggregate measures.

Indicator concept: The concept measured by an indicator.

Policy indicator: An indicator relevant to public policy choices made in view of ethical values. Many policy indicator statistics are aggregate value measures.

Social indicator: A measure of social conditions justifiably included in a public statistical system. (Possible bases of justification vary from enlightenment, to policy relevance, to relevance to basic social-system models.)

tive of social progress than of social decay." Factory smoke that would now be considered harmful in many industrial societies has also been considered a sign of progress at earlier stages in development. Leading indicators, such as housing starts as signs of future economic trends, may also be useful for early warning without being causes of the value that they predict (and thus without being, or even measuring, specific contributory variables). In the final section of this chapter we shall suggest some principles for choosing among contributory and other variables for policy indicator systems.

Value Analysis of Social Indicator Concepts

In our definitions we have distinguished end-values from contributory variables, and general end-values from more specific end-values. I shall now illustrate these distinctions by analyzing some frequently used social indicator concepts. This analysis will suggest that several general end-values (the ones to be discussed in Chapters 5–7) provide a more fundamental account of what we are seeking than do the initial indicator concepts; these end-values can thus be used as common denominators for comparison and as possible bases for indicator statistics. Yet even if I were to claim expertness in this analysis, the choice of variables in a policy indicator system would not be a matter to be decided by individual experts or even by expert communities. In the process of collective self-observation, "citizens, through their chosen deputies, can *impose* on one another an obligation to render information about their own situation" (Johansson 1976: 80). The choice of end-values is thus a choice by citizens, in which experts can participate but cannot decide alone (Chap. 2). Different political systems, or the same system at different times, may make different choices of end-values. In particular, their choices may differ as to the generality or inclusiveness of these values: one political system may choose a variety of relatively specific indicators rather than a few more general ones, and thus leave a relatively large part of the reconciliation of values to politics alone; while another may choose statistics on more general end-values to aid more in this reconciliation. I advocate the latter approach here for consideration by my fellow citizens (some of whom are experts).[25]

In this analysis I shall use as examples the areas chosen as chapter topics in *Toward a Social Report* (U.S. HEW 1969: ix), for statistics to guide strategic national decisions on social and economic policy (Olson 1970: 115). These areas have appeared in various lists of social indicators in publications on social trends or in social reports (Zapf 1974: 25) and are thus somewhat repre-

25. Similar efforts to subsume particular values under more general value variables are illustrated in MacRae and Wilde (1979: 45–46, 101) and in MacRae (1981: 108–26).

sentative of the field of social indicators; they are listed in the left-hand column of Table 1–3. They correspond largely to what Carley (1980: 185–90) has called "social goal areas," an example being the list of "areas of social concern" proposed by OECD (1976): health, development through learning, employment and quality of working life, time and leisure, personal economic situation, physical environment, social environment, personal safety, and social opportunity and participation. Some correspond to existing professions or to government departments. They also correspond roughly to a list proposed by Land (1975a: 23–24) based on a functional analysis of the "activities which define a human society." They therefore lie somewhere between the abstract needs of society and the particular political and institutional forces that predominate in a society at a given time. They may change over time, and we have some freedom to choose them (Johansson 1973).

In the analysis that follows, which gives rise to the center and right-hand columns of Table 1–3, we shall examine variables in each of these areas and ask whether they should be considered as intrinsic values or not; and if not, to what intrinsic values they contribute. The intrinsic values suggested, corresponding to each of these areas, are shown in the middle column as end-values; some are more general than others. Some corresponding contributory variables, which can include the variable designating the area itself, are shown in the right-hand column.

Health. Health, the first area listed in Table 1–3, might seem an obvious choice as an end-value. It has been listed as an intrinsic value in the philosophical sense (Frankena 1963: 72); and von Wright (1963: 9) considers it to exemplify a comprehensive ethical notion, "the good of a being or welfare."[26] Yet as we examine the meanings associated with various measures of health, we shall see that it cannot be considered an end-value without qualification.

Two concepts often used in the measurement of health are of special interest: incidence of diseases, and capacity to function. The first of these, disease, is a disvalued physiological condition defined by medical science and usually capable of being remedied by the health professions, especially physicians (Williams and Hadler 1983). A handicap, or an injury resulting from an accident or crime, may not come within the scope of a disease indicator, even though it can have the same effect as a disease on more general values such as a person's production, consumption, or subjective well-being. The loss of function of one arm from a stroke is more clearly relevant to a disease statistic than the severing of that arm through an industrial accident; injuries, however, are sometimes included in the categories of sickness or illness. The dysfunctions of old age, insofar as they are not easily curable and indeed seem

26. Following a Platonic line of argument, von Wright (1963: 12, 61) views health and the good of man as knowable by observers, independently of responses by the person in question.

Table 1–3. A Classification of Variables in Social Indicator Areas

*Area**	*End-Values*	*Contributory Variables*
Health and illness	Health as perfection Length of life Quality-adjusted life years Preference satisfaction Cost	Incidence of diseases Capacity to function Services rendered
Social mobility	Freedom Economic efficiency Subjective well-being Equity	Extent of upward and downward mobility Net upward mobility Mobility by groups
Physical environment	Quality-adjusted life years Subjective well-being Preference satisfaction Perfection in the environment Well-being of other species	Concentration of substances in air, water, workplace Housing Availability of recreational facilities
Income and poverty	Economic or material well-being Equity Needs, rights, minima	State of the economy Policies promoting equality of opportunity or redistribution
Public order and safety	Subjective quality of life, life years Economic well-being Equity (transfers to the poor) Punishment Quality and quantity of life of those punished	Extent of crime and victimization Expenditures on protective services Prison conditions (such as crowding)
Learning, science, and art	Production, effects on human capital Quality of life Equity Esthetic creation and appreciation, understanding of nature and society	Students' achievement School desegregation Extent of research (discoveries and publications) Availability of cultural facilities

Participation and alienation	Short-run well-being Long-run well-being related to policy or system change	Participation Alienation

* Adapted from HEW (1969: ix).

"normal," are not so likely to be counted in disease statistics as are the remediable or temporary malfunctions of younger persons. Thus, as Land (1975b: 28) points out, institutional outputs need not provide the ultimate definition of social goals or indicators.

The presence of a disease can simultaneously imply several conditions relevant to well-being. It can imply a lowered level of subjective well-being (pain), corresponding to an intrinsic value. It can imply a lowered level of functioning—with functional capacity being a means to the well-being of either the ill person or others, or an end in itself if normal functioning is considered an intrinsic value in a perfectionist sense. Disease can also contribute to reduced well-being in the future through effects such as impaired intellectual capacity or shortened life. All these aspects of disease are based on deficiencies relative to normality; excesses of valued conditions are not relevant to the concept of disease.

A disease statistic can therefore measure an end-value or a contributory variable, depending on the aspect of disease we consider and on our notions of intrinsic value. Even when we consider disease as a contributory variable, our judgments of the seriousness of one disease relative to another can vary depending on our notion of well-being; criteria of production and of happiness, for example, would lead to different classifications.[27]

Various cases of a single disease, and diseases collectively, are treated by using shared knowledge and technology provided by the health sciences and used by the health professions. This sharing of means of treatment adds to the significance of disease statistics as components of policy-related statistical systems. The alleviation of diseases corresponds to a common set of causal models affecting significant values; it thus represents a direction that policy can take. These causal models may deal with health generally if they involve variables such as health personnel, hospitals, and sanitation; or they may focus on specific remedies for particular diseases. Rates of incidence of par-

27. A concern for production might lead us to attribute low importance to schistosomiasis, according to the findings of Weisbrod et al. (1973). A concern for happiness could lead to emphasis on patients' moods and on means to change them.

ticular diseases can thus be policy indicators—justifiable variables for such a system—even if they are not highly general.

A second type of health statistic, which measures the individual's capacity to function, can include disabilities from whatever source they arise. Examples are statistics on days of bed-disability or restricted activity—and at the limit, death. An index combining death and disability rates into "expectancy of healthy life" has been proposed (HEW 1969: 99–100); but in counting death and permanent bed-disability equally, it ignores differences in our valuation of the two conditions. An index considering a larger number of levels of functioning has been proposed by Chen et al. (1975). Indicators of this type relate health (in a broad sense) to production—either on the job, or in "household production," which is closely related to consumption.

Neither disease indicators nor function-status indicators, however, place primary emphasis on the subjective aspect of quality of life. Some writers (e.g., Bradburn 1969: 7–8) have pointed out that mental health is closely related to subjective well-being. Subjective well-being (in the negative form of pain) is among the symptoms that physicians consider, together with capacity to function, in their diagnoses of disease. If, however, it does not fit into the diagnostic categories used, it will not be reflected in disease statistics. Pain and anxiety, aspects of subjective well-being, may be caused by environmental, work, or family conditions; and even if a manipulable cause cannot easily be found, reporting of the incidence of low levels of subjective well-being may stimulate a search for additional remedies.

In Table 1–3 we suggest that disease statistics should be related to end-values such as length of life,[28] quality-adjusted life years, or preference satisfaction. This is because these end-values are more general and permit trade-offs, or comparisons of seriousness or importance, among particular diseases as well as between disease and other disvalued conditions. Length of life might be measured by the mortality rate or the life expectancy for a population. A second possible general end-value might combine the length and subjective quality of life (Chap. 6) as has been proposed in measurement of "quality-adjusted life years" (Weinstein and Stason 1977).[29] Preference satisfaction might also be the basis of judgments of the relative seriousness of various diseases. Among the variables that might contribute to these end-values are aspects of health such as the incidence of diseases or capacity to function;

28. Mere survival in a painful or subhuman condition may seem undesirable to some, but to others it is intrinsically desirable.

29. This variable incorporates subjective judgments of quality of life by the person affected. We must recognize, of course, that a person may be blissfully unaware of his actual serious illness, when observers know that in the future his subjective well-being is likely to decrease or his life expectancy to be low. This does not necessarily mean that his present subjective well-being is not to be valued, but only that his well-being over time must also be considered.

and at a second remove, conditions of the physical and biological environment, number and distribution of health personnel, and health services rendered. Preference satisfaction as an end-value can also be related to patients' preferences for health services regardless of their effects on health. Cost, also listed in Table 1–3, is a relevant negative end-value for any policy; we consider economic benefit and cost as end-values because of their relation to preferences (Chap. 5).

Conventional indicator concepts can thus be usefully subjected to valuative critiques in view of the more general end-values that we seek. We must recognize, however, that costs of measurement and problems of reliability lead us to use indicators that are less than perfect measures of value concepts. In the health area, reliable statistics on disease are more easily available than statistics on people's capacity to function, and these, in turn, than data on subjective well-being. For similar reasons, a national commission assessing labor force data in the United States explicitly rejected the use of indicators such as unemployment statistics to measure welfare (Adams 1981: 128). In this tension between criteria for the choice of indicators, what is important is to recognize the direction in which value criteria would lead us in the longer run.

Social mobility. Social mobility, the second area in Table 1–3, has been the subject of extensive and sophisticated academic studies. Early research in this area (Lipset and Bendix 1960) was directed at cross-national comparisons, with mobility assuming significance both as a value (opportunity) and as a possible cause of the weakness of socialism in the United States. The notion of opportunity comes close to the end-value of freedom, widely held but seldom measured by social indicators.

More recent studies of status attainment have aimed at predicting the occupational status reached by persons in terms of their parents' education and occupation, their own education, their group membership, and other variables. The status attainment concept comes nearer to being a measure of well-being than does mobility itself. Mobility, in the literal sense, may be downward as well as upward; mobility between jobs may occur too rapidly, thus failing to make use of specific human capital (Becker 1975); and even upward mobility may bring psychic conflict and insecurity. Perhaps the hope of advancement is itself a source of well-being or of motivation to produce, and the hope of fair advancement a source of social cohesion; but such connections need to be verified. Thus, rather than considering social mobility as an end-value, from a valuative perspective we might better deal with the efficiency of occupational placement systems for overall economic benefit; the subjective well-being that people enjoy in their work, their consumption, and other activities in relation to mobility among roles; and the equity with which social statuses and their attendant rewards are distributed, as shown in Table 1–3. Social mobility as a contributory variable must relate to the end-values that individuals receive from changes in status; but we must also consider aggregate attainment when the advancement of some can entail losses for others (Hirsch 1976). By considering these more general end-values and treating aspects of social mobility as possibly contributory to them, we can clarify the relations between mobility and the values we seek. Equity is particularly relevant (Chap. 7), and policies affecting it often involve education (Chap. 4).

We therefore list various aspects of mobility as contributory variables in Table 1–3. The policy variables that affect mobility would be contributory variables at one further remove, and are not shown.

Physical environment. The area of physical environment deals with pollutants in air, water, and the land; outdoor recreation; housing; and population, urban space, and transportation (HEW 1969: Chap. III). None of these corresponds to an intrinsic value. The harm done by pollutants depends on their effects on life, especially on humans. Unspoiled natural environments are similarly of benefit for recreation only if people use them. Housing of given physical characteristics, above the minimum required for health, may be valued quite differently by people whose expectations differ. Population density and transportation also differ in their significance depending on expectations and on the activities people wish to carry out (Rodgers 1982a). We thus relate them to more general intrinsic values.

The area of physical environment involves similar value variables to that of health, such as quality-adjusted life years, subjective well-being, and preference satisfaction. Many environmental conditions affect health and the length of life. Other environmental conditions, such as urban sprawl, muddy water, and unpleasant odors, have effects that are more nearly esthetic, affecting subjective well-being or the satisfaction of preferences more than health. Again, we regard characteristics of the environment as contributory variables, grouped together because they share elements of technology that allow us to act on them.

Some "environmentalists," in the broad sense of the term, also have concerns that extend beyond the environment of human beings. They seek a state of natural perfection in ecological systems, independent of benefits to humans; or they consider the welfare of other species, especially of those similar to humans. In both these respects they deal with end-values not usually reflected in social indicators.

Income and poverty. The area of income and poverty (literally an economic rather than a social category) relates to economic well-being, which is often treated as an end-value (Chap. 5). Income is a means to buy goods and services, and prices provide a basis for placing relative valuations on them. If incomes are compared with a poverty level, or a level judged to be required in terms of needs or rights, they are in effect transformed into a dichotomy and statistics on this dichotomy (proportion below the poverty line) can be used to reflect equity as well as well-being. If the distribution of income is reported in terms of concentration, dispersion, or intergroup differences, the relevant end-value is again equity with respect to income.[30]

Public order and safety. Public order and safety is an area commonly understood to refer to crime, though numerous other hazards affect our safety. Several intrinsic values are associated with this area. Security or safety is presumably a source of subjective and economic well-being. For critics of the social order who consider that property is inequitably distributed, transfers to the poor may be considered social benefits even when they result from crime. For others, punishment of crime is itself of intrinsic value and is part of the notion of justice; it is valued for its own sake, and not as a means to public order and safety. For still others, the well-being and life of those who are punished are major considerations.

Learning, science, and art. The area of learning, science, and art involves diverse end-values which should perhaps not be grouped together. Learning, for example, is largely a contributory variable. Formal learning extends from the preschool years to postdoctoral and adult education, and informal learning occurs in many spheres of life. Learning is sometimes judged in terms of its contribution to production through the creation of human capital. It may also be judged in terms of more direct contributions to the learner's quality of life, either in the educational process itself or through the appreciation of uses of leisure. Aspects of equity enter into education as regards equality of educational opportunity. Education may also affect the quality of citizenship, which can, in turn, affect public decisions and thus people's lives. Finally, the cultivation of learning may be considered to be of intrinsic value, like the corresponding aspects of science and art.

Science, too, may be assessed with respect to various end-values. Economists have assessed the contribution of scientific research to national production (Griliches 1978). Science may also contribute to our understanding of nature and society in the sense of contemplation rather than action. These benefits, though, like those of art, seem to flow largely to a select part of the public. Because of the extensive effects of science on production and economic well-being, we classify it as largely a contributory variable. Thus a report such as *Science Indicators 1978* (National Science Board 1979), dealing largely with scientific activities rather than their outcomes, requires clarification as to its relation to end-values. This relation depends on the importance attached to science as an intrinsic value, and on our knowledge of the magnitude of causal relations between particular aspects of science and other end-values.

30. Equity can thus refer to the fairness or equality of distribution of some value; its precise meaning, then, depends on the particular value involved (see Chap. 7).

Art, if measured in terms of the creative or appreciative experience, may contribute directly to subjective well-being. Many art critics judge art in terms of its esthetic quality, however, and would object to the application of this Benthamite experiential standard to it. The domain of art is also one in which the value of equity may enter, if it is affected by public policy, since many artistic exhibitions and performances relate to the experiences of a small and affluent clientele.

The area of learning, science, and art takes us away from a narrowly economic notion of end-values in two ways. First, the values that they further go beyond preference satisfaction to consideration of a possible need to change preferences, e.g., through education intended to cultivate an appreciation of high culture or an understanding of diverse groups in society. Second, because the functioning of the political system may require support, understanding, and participation, we may seek types of education that are aimed not directly at the production of commodities or at personal well-being but at informed citizenship. Not only the cognitive conditions for citizenship, but the attitudes needed for democratic participation, may be a needed outcome of education (Merelman 1980).

Participation and alienation. The area of participation and alienation is another in which the relation of indicators to values is difficult to establish. At first glance, it might seem that a political system with a high level of alienation and low participation was in an undesirable state. Yet a counterargument might be that the authorities or the regime deserved criticism, and that the political alternatives currently offered to the public did not deserve confidence or participation. Alienation and withdrawal might then be better responses than identification with the regime or voting. Conversely, participation might be increased and alienation reduced by a carefully designed media campaign; even subjective well-being might therefore be increased in the short run, but at the cost of opportunities for more basic changes that might ultimately result in higher levels of well-being—or of equity. Thus, until we consider more carefully how statistics on participation and alienation would be used, we should perhaps not give them high priority as policy indicator variables. Participation and alienation may well be examples of a general type of contributory variable, measures of social integration (Chap. 8). Like marriage and divorce, they may contribute either positively or negatively to well-being, depending on other conditions.

In summary, our discussion of Table 1–3 has led us to a limited set of end-values that are more general than many of the "normative" social indicator areas of *Toward a Social Report*. By examining each area in turn we have proposed, in the center column of Table 1–3, a smaller set of variables that cannot easily be characterized as merely means to other values. For this reason we have treated social mobility, the physical environment, public order, learning, science, participation, and alienation, entirely or in large part as contributory variables.[31]

The principal end-values that emerge from our analysis (or that were introduced into it) are these:

*a. Economic efficiency, as a means of satisfying preferences and thus promoting well-being in one sense. This includes assessment of the economic benefits and costs of particular policies.

*b. The subjectively assessed quality of life (subjective well-being).

*c. The length of life, adjusted by its subjective quality.

d. "Perfectionist" values, which correspond to the development of preferences, tastes, or capacities that are considered desirable in themselves according to notions of human perfection. Examples are esthetic creation and appreciation, understanding of nature and society, and health in the sense of perfected functioning. In addition, perfection may be sought in nature apart from humans.

e. Freedom.

f. The well-being of other species.

*g. Equity in the distribution of the above values, including notions of needs, rights, and social minima. Equity, closely related to justice, is actually a multi-faceted concept including the alleged fairness of retribution for crime, the reciprocity between society and those who have served it or drawn resources from it, and relations between individuals or groups (Chap. 7).

Among these, the general end-values we shall stress in our further analysis are those marked with asterisks (*) above: *a* (Chap. 5); *b* and *c* (Chap. 6); and *g* (Chap. 7). These values include notions of the general level of well-being and of equity (Land 1975b: 20).[32] "Perfectionist" values will not be emphasized because of our concern with tradeoffs and value reconciliation, which do not seem easily applicable to them; nor shall we consider freedom or the welfare of other species, which have not been concerns of the social indicator movement. Conceivably these other types of values could be used as bases for policy indicators, but measures of them and models of their causes have not been extensively developed.

31. All those variables classified as contributory in Table 1–3 (except alienation) are objective variables, while of the end-values in the table, some are objective and some subjective. Some subjective variables such as domain satisfactions, perceptions, and expectations may also be contributory; but we do not usually think of them as directly manipulable by policy.

32. Land and Felson (1976: 566) suggest that a class of "welfare indicators" would refer to the conditions on which a social welfare function depends; but here we stress the direct measurement of social welfare and of equity, in the form of general end-value measures, as a major component of policy indicator systems. The inclusion of both life conditions and subjective quality of life as central types of indicators has also been advocated by Zapf (1979).

This valuative analysis of social indicator concepts, and the concepts that guide it, lead us to give a special status to a small number of general end-values. When we propose contributory variables for a public statistical system, it is because they are causes of general end-values. This approach differs from an approach that develops systems of indicators by considering a plurality of more specific values.

A pluralistic structure of values is exemplified by the notion of "needs." An observer of a human population who wishes to characterize its members' needs may begin from a notion of components of the good life, either in terms of "material or impersonal resources by which people presumably can master their living conditions," or types of basic needs such as having, loving, and being. Welfare, in this perspective, is "defined by the degree of need-satisfaction" (Allardt 1976: 228, 230). This differs from our approach in which welfare is defined directly in terms of end-values, and needs in terms of deficiencies in these values.[33] Lederer (1980: 3), for example, notes that an approach to needs as "objective" defines them as prerequisites to conditions such as human survival, development, or functioning as a human being. On this approach, we characterize these latter conditions as end-values, try to define and measure them more precisely, and seek the models of causation that connect them with needs and in turn with policies. We stress end-values because of their lasting character, the guidance they can provide for measurement and research, and the part they can play in reconciliation of values in public debate.

Choosing Contributory Policy Indicators

From our analysis of social indicator concepts we have suggested a set of general end-values that can guide our choice of contributory indicators and be considered for use as policy indicators. Much of the remainder of this book will be devoted to these end-values—their function in value tradeoffs (Chap. 2) and the part played by net economic benefit (Chap. 5), subjective well-being (Chap. 6), and equity (Chap. 7).

At the same time it is important to include measures of some contributory variables as policy indicators; indeed, as we have shown, nearly all the social goals commonly measured in indicator systems are contributory variables. We shall not treat these goals in full detail because of their complexity, their relation to the requirements of particular political systems, and the general

33. An alternative definition of needs is employed in needs assessment as practiced by professionals and evaluators in the United States. This definition also centers about particular social deficiencies (Rossi and Freeman 1982: 91), but emphasizes needs for particular types of services provided by professions, agencies, or programs.

agreement that seems to have been reached on a listing of social goal areas for national statistical systems among the nations in the Organisation for Economic Co-operation and Development (OECD 1976). We shall, however, list some criteria for choice of variables as policy indicators.

In choosing policy indicators we return to the purposes that they may serve: definition of the problem, choice of policies, and monitoring results of action (with a cyclical return to problem definition).[34]

a. Definition of the problem. In order for a concept (measured by an indicator) to be useful in aiding the public to define problems, there must be widespread agreement as to whether its presence is good or bad, so that it may be related to valued expectations. In other words, it should have *clear valuative polarity.*[35] The concepts that have priority for problem definition, therefore, should be value concepts, and the causal relations from them to end-values should be unambiguous in sign. A contributory variable that has both positively and negatively valued causal associations, such as expenditure for unemployment insurance (discussed above), is less useful in this respect for problem definition. A variable of this sort can still be useful, however, for monitoring the effects of a program; and the overall justification for a policy indicator depends on all three of the purposes we are now considering.

That a variable have clear valuative polarity is a necessary, but not sufficient, condition for its priority as a problem definer, and thus as a policy indicator. The variable (or statistics on it) must also direct the attention of the political community to useful action. The possibility of action depends on our knowledge of available means to influence this variable. Such means may be found in causal uniformities of nature and society, or in the existence of professions or organizations that can act and carry out relevant policies. For contributory variables to have priority as policy indicators, we must know not only how to affect them, but also how they in turn affect end-values. The nearer these contributory variables are to end-values in causal chains, the surer we are likely to be of their effects.

The indicators of highest priority for defining problems would thus seem to be measures of end-values.[36] This priority holds, however, only if statistics on end-values can suggest not only problems but also useful action. The national

34. We describe these activities first as a sequence of activities by "citizens." In actuality, however, we shall face the problem that researchers develop models, elected officials play a major part in policy choice, and administrators play a large part in monitoring programs, leading to a possible division of labor in the generation and use of information (Biderman 1970: 219).

35. We shall make an exception to the rule of clear valuative polarity in Chapter 8, where we deal with indicators of social integration, which can be desirable when it furthers desirable collective goals, but can have undesirable effects as well.

36. If we measure a general variable (e.g., health) rather than a specific means to it (e.g., health services), we may be able to agree more fully on the problem and search a wider range of policies to deal with it (D. Campbell 1969: 410).

product or income, as an end-value, can be acted on as a whole by macroeconomic policy, and in part by policies affecting such variables as education or productivity in particular industries. Thus a decline in the national income or its rate of growth, taken together with our causal knowledge, can suggest appropriate policies, and national income is a useful policy indicator. But the particular actions to be taken may also be suggested by tests of their incremental contribution to the national income (by benefit-cost analysis), without our using national income statistics themselves to define the problem. A general end-value can thus be used either as a problem definer (by means of policy indicator statistics) or as a basis of tradeoffs or value reconciliations, without requiring indicator statistics. We know less, however, about the effects of policies on aggregate subjective well-being, and its use as a problem definer may depend more on needed research (Chaps. 5, 6).

We thus give priority to end-values that call attention to useful action. We seek surrogates for them among contributory variables, however, when they cannot be measured reliably enough; thus objective variables are often substituted for subjective well-being. Moreover, the relation of our policy choices to end-values is often delayed and indirect. One feature, therefore, that "promotes" certain contributory variables to the status of indicator concepts, even when they are not intrinsic values, is the difficulty of measuring the changes in end-values that result from our actions. Through educational policies, for example, we may seek to affect the values of production, quality of life, equity, and appreciation and understanding; but these eventual effects are often so long delayed or difficult to measure that we often use achievement test scores as a presumed tangible "leading indicator" of them.

We may be tempted to go one step further toward practicality and immediacy, especially if we are members of a profession delivering services, and to measure inputs in the form of services delivered. College teachers who agree to measure their contributions in numbers of hours taught are as much at fault in this respect as physicians who accept statistics on services delivered as indicators of health effects. A useful contribution of the social indicators movement, however, is that it has tended not to accept input measures as satisfactory definers of problems.

The strength of the argument that we can make for including as a policy indicator a measure of a contributory variable depends, therefore, not only on the cost or difficulty of measuring its effects on corresponding end-values, but on the confidence we have in the strength of the causal relation from contributory variable to end-value.[37] This type of argument, including the perceived

37. We here approach Parke and Seidman's (1978: 11, 16) description of "the transformation of social statistics into social indicators." Sabatier and Mazmanian (1981: 8) illustrate the need for models by the choice of whether to measure the concentration of sulfur dioxide in the air. The paradox is that we can feel confident about causal connections when effects on the end-value are hard to measure; this may be true because our confidence results from commonsense knowledge,

strength of such a causal relation, might justify including in our statistical systems variables similar to those in the left column of Table 1–3, corresponding to "social goal areas" (Carley 1980: 185–90). Their place in the system, however, should ultimately depend on a demonstrated contribution to corresponding end-values.

In contrast, those contributory variables that are not manipulable means to a value variable (and thus not value concepts) are of less interest, presented by themselves, for defining problems. Thus the age distribution of the population, even though it affects rates of crime and disease, is not by itself a useful indicator statistic for crime or health policy. It may, however, enter into age-standardized crime or disease rates, or other analytical measures[38] (Parke and Sheldon 1973: 110), that can be meaningful to the public for problem definition. Similarly, the distribution of the population by race or geographical area, though useful as a base for equity-related tabulations, is not a useful policy indicator statistic by itself.[39]

Finally, public debate and public beliefs about causation are relevant to the choice of contributory variables for public statistics. The public can be educated to understand new indicator statistics; but as the ultimate political authority and the source of support and information for many public statistics, citizens also have a right to ask for information that will deal with questions of interest to them (Johansson 1976: 80).[40] They may ask for statistics on "crimes without victims" even when there is little evidence that these crimes do harm. They may ask for data on numbers of doctors, and on their accessibility, even when the effects of these variables on health are not clear. Policy indicators serve the citizens; and though the design of these systems cannot respond to every short-term public demand, it must be responsive in general.

The balance required between education of the public and responsiveness to it may be illustrated by problems in the presentation of input statistics. *Toward a Social Report* (U.S. HEW 1969: 97) claimed that "statistics on the number of doctors or policemen could not be social indicators, whereas fig-

---

with no measures of the end-value, or from longitudinal studies that measure effects on end-value only at great expense.

38. Winsborough (1975) suggests a method for distinguishing the effects on income of race, age, period, and cohort.

39. We noted earlier that means as well as ends could be policy indicators, even though we recognized the distinction between them. These non-valued contributory variables, however, are by themselves less useful than value variables for collective problem definition, and thus of less interest to the general public. Data on them may be in the public domain, but presented so as to be available to experts for model construction, or used for private motives such as market research.

40. An example of publicly available statistics relevant to public debate, produced outside government, is Stouffer's (1955) study of communism, conformity, and civil liberties in the United States.

ures on health or crime rates could be." In our classification, health and crime are not clearly end-values; and if they are not, numbers of doctors or policemen would be contributory variables at a second remove. Whether we should tabulate them depends in part on the guidance that they would give to problem definition, and therefore on the interpretation that people will give to them. The danger is that they will be inappropriately promoted to the status of policy indicators, with the encouragement of the occupational groups in question. This danger will be reduced if statistics on the corresponding outcome variables (health or crime) are also published and related to the inputs. Citizens will be aided, therefore, by reliable and intelligible models of causation. The models connecting health and crime to well-being are presumably much stronger and more reliable than those connecting numbers of doctors and policemen to well-being, and thus measures of health and crime deserve priority for problem definition. The more closely an indicator of "system performance" (U.S. Department of Commerce 1980: xx) is connected as a cause to an end-value, the higher the priority it deserves for this purpose.

b. Choice of policies. Indicator statistics are not related directly to our choice of policies, which requires causal models for comparing alternatives. They are connected indirectly, however: the same value concepts and measures can usefully be employed in both models and indicator statistics, and the best problem definers are those value concepts that we know to be affected by policy choice, in terms of valid models.[41]

A feature that makes some contributory variables particularly useful is thus their connection with a particular set of shared means that can be influenced by policy and in turn affect well-being in various ways; this joint-product feature increases their efficiency. Examples of such general means are the reduction of atmospheric pollution; health science and the health professions; educational institutions; and the crime and justice system.[42] Each of these means affects a variety of particular values or disvalues (types of pollution, disease, skill, and crime) in a "social goal area." Either because of shared manipulable processes and technological relations, or because of the organization of occupations and institutions that can further certain aspects of well-being, groups of possible activities and policies may be viewed as similar or interrelated. The technologies and groupings may change, however, and the importance of these contributory variables may change with them. Schools, for example, might conceivably assume the functions of reducing crime or pro-

41. A further connection may arise from the use of indicator data in the development of policy models. We have stressed the separateness of model construction from public statistics; but if the same data that are to be aggregated into public statistics can also be reanalyzed for model construction, and if we can make judgments of the prospective benefits from thus improving the models, these benefits may partly offset the costs of the data and increase the justifiability of collecting them.

42. We examine models related to education and crime in Chapter 4.

moting health. A list of social goal areas can thus depend on the political and institutional structure of society, which is susceptible to criticism and change.

The relevance of policy choice to our selection of policy indicators thus works through our knowledge of causal models. In addition, our selection of end-value indicators relates to their possible use in democratic policy debate to permit reconciliation of values through tradeoffs (Chap. 2).

c. Program monitoring. At the stage of monitoring of the results of public action, we move temporarily to more specific variables (D. Campbell 1976: 246–48), before returning to more general values at the end of the cycle. Monitoring, as we have indicated, requires careful research design if we wish to infer results of action from our observations.

Our list of contributory variables in Table 1–3 emphasized general values and causes rather than the specific policies and administrative choices that are intended to affect them. We have remained up to this point at the general level of "enlightenment" of the public (Biderman 1970) with respect to policy-relevant models. Specific administrative variables are also important, however, for monitoring programs. They are thus another type of contributory variable, less emphasized by the social indicators movement, and they also must be considered among policy indicators.

Legislators and citizens may wish to see whether a particular program is being carried out faithfully—whether its funds are being spent at the planned rate, whether skilled personnel are being hired, or whether services are reaching the intended population. If the program was initially based on a model relating its provisions to the general welfare, then we need to examine not only whether it furthered that welfare; but also, in case it should fall short of its goals, whether its actual implementation differed from that which was expected. Administrative variables, a specific type of contributory variable, are important for this task.

The monitoring of results, like the choice of policies, is often assumed to be more the task of special elite groups than of the public at large.[43] Administrators and evaluators are thus assumed to have a special interest in monitoring outcomes and inputs. Individual citizens' problems with government programs are often handled through legislative casework; but it is also important for the representative process to be concerned with more general conditions of implementation. These details may be followed up by the public more in mu-

43. The necessary information for program monitoring may be obtained from the program itself or from those affected by it. In the former case, it may lead to praise or blame of those in the program who produce the information; as public officials they can expect to be accountable in this way, to their superiors (Kaufman 1973) as well as to citizens. In the latter case, when the information comes from citizens, it is especially important that "the original sources of data have confidence that the information will not be used to sit in Olympian judgment on specific individuals or organizations" (Biderman 1970: 23), so that sources of this kind will not be impaired in the future.

nicipal administration than in national (Chap. 9); and in some issue areas, such as that of school desegregation, the presence of indicators together with a strong organized constituency for implementation can enhance monitoring (Levin 1981).

The indicators and models of interest to us for program monitoring thus differ from those of social goal areas in that they overlap less with those studied in academic social science. Some types of more specific contributory variables that may be of special relevance are:

(1) Direct transfers of money or goods to persons in need, whether through the private or the public sector. These include the aid of families to their dependent members, unemployment insurance, income maintenance, and restitution or insurance to victims of crime.

(2) The provision of services, or the availability of physical facilities, to individuals and groups (e.g., to promote health, education, or justice), and the amounts spent on them. The efficiency of these services and facilities is often uncertain, and precise measurement of their efficiency can be hindered by the resistance of those who provide them. Information on expenditures for government services is also relevant as part of a system of fiscal indicators (Chap. 9). Categories (1) and (2) together include program outputs (Land 1975b: 17), but also include the supply of the same outputs from nongovernmental sources.

(3) The prevalence of private cooperative behavior that furthers the purposes of public programs (implicit in [1] and [2]); Whitaker (1980) has referred to this as "coproduction." An example is protective behavior to reduce criminal victimization, transportation accidents, or lung cancer attributable to smoking. More generally, household production is a major contributor to welfare (Zapf 1982). This sort of behavior may be either modifiable by public policy or relevant as a statistical control. The resources spent on it are also relevant.

Among the contributory variables may be quite specific administrative characteristics such as pupil-teacher ratios in schools, details of the administration of food stamp programs (Rich 1981: 66), or measures of organizational performance. These variables may not seem general enough to include in national social reports (Johnston and Carley 1981: 243); but the transition from general to specific variables, and from citizen to administrative audiences, is so continuous that we would be well advised to consider all these sorts of variables as parts of a functioning indicator system (Garn et al. 1976). Indeed, the effective choice of policies includes following them through the stage of implementation, in which factored problems (Allison 1971: 80–81) and specific measures of process and outcome are more important than at the stage of enactment. But even while we focus attention on highly specific program goals, we must remember to link them to the more general values that the programs were designed to serve.

2 Citizens' Values and the Expert's Roles

When citizens use indicator statistics to define a problem, they contrast the state of affairs measured by those statistics with their own valued expectations. A policy indicator variable must therefore express a value that citizens are willing to apply to their expectations. If that variable is supposed to tell us "how well off we really are," for example, it must measure a condition valued by the public as an aspect of the general well-being.[1] If a policy is eventually chosen to deal with that problem, it should presumably further that same value.

Individual citizens can differ, however, in their values. Political communities differ in their degree of consensus, and within a given political community consensus will vary with the issue under consideration. Political communities differ also in their ways of resolving internal disagreements; some make more use of broadly supported general principles than others. In a political community that makes use of such general principles, indicator statistics measuring general end-values may have a special part to play, and expert communities may more easily propose such statistics. The task of this chapter is to show how such general values can link the experts with the political community.

In proposing policy indicator variables, we are not merely responding to the public's values, but are also advocating increased concern with some of them. We therefore face the question of what right any experts, such as social scientists, have to make such proposals. A claim to be expert in measurement, sampling, and assessment of causation does not usually extend to telling the public what its notions of well-being should be. Experts may, however, argue more justifiably for general end-values than for particular social goals. End-values such as the national income, happiness, and equity may be less partisan and command wider support than more specific values, which they may serve to reconcile. Economic indicators and benefit-cost analysis, for ex-

1. These values are further selected by a political community's capacity to act on them, both by the availability of manipulable causal processes and by the constitutional powers of its government.

ample, have encouraged the comparison of particular values in a monetary framework. They have also been criticized for doing so. Yet the use of other general end-values may serve as a counterbalance to economic values while encouraging a similar style of debate—seeking ultimate and general values and not simply specific goals.

The development of statistical systems that serve democratic decisions involves a two-way relationship between expert communities and political communities. The members of a community of experts, such as statisticians, social scientists, or policy analysts, review one another's work within a given specialty in order to insure its quality; in this respect such a community is autonomous and self-regulating. Yet when its members propose to change the list of policy indicators that guide public debate, they join with the political community in roles of service or participation and must thus cooperate with citizens and public officials.

If members of expert communities are to propose public statistical systems based on values, they must recognize that the political communities with which they work are constantly expressing values and choosing among them politically. They must ask how their proposed value concepts and indicator statistics should be related to these political processes. They must ask how technical information can be used democratically. Most important, they must choose among various possible roles defining their relations with the political community.

Let us assume that we have chosen to work with a democratic political community that is capable of using policy indicators.[2] We wish to recommend priorities for variables in a public statistical system related to policy choices made in view of the general welfare. Any notion of the general welfare involves combining the welfares of particular persons and groups, and usually also involves combining particular aspects of welfare.

An indicator statistic such as the incidence of a disease combines the well-being of many persons into a single figure. If this statistic improves, we cannot know that all individuals or groups have improved; some may have become worse off. We can, however, break down the statistic by groups or examine individuals' changes in health status so as to see more clearly how gains and losses are distributed. These latter statistics can relate to the political struggles between groups, or to notions of equity that may be advanced to justify distribution of health-related resources.

A more general indicator statistic often combines various particular notions or aspects of well-being: a general health index may combine incidence of particular diseases; or a measure of economic welfare may combine contributions from health, education, and crime reduction. If such a general concept were used by an individual to reconcile his own values, it might lead him to

2. We discuss the political conditions necessary for use of indicators further in Chap. 9.

set individual priorities in a more organized way. But when used by society, it can combine the values of particular groups as well as particular notions of value. The group that gains from reducing one disease may not be the same that would gain from reducing another; those whose health is improved may not be the same as those who gain in education. Thus general values may still involve the reconciliation of the values of particular individuals and groups, and not merely the reconciliation of notions of well-being. Policy indicators can play a part in the reconciliation of values in both these senses—among groups and among notions of well-being. Our main concern is with the second—that of measuring shared notions of the general welfare so as to permit shared problem definitions and policy choices. At the same time, however, politics and the use of public statistics involve conflicts among group interests and values, not all of which are reconcilable. If the statistics we recommend are available to the public, they will be used in controversies involving these interests, as well as in debate about the general welfare. We shall thus first examine the processes by which political communities go about reconciling values in both these senses, and then consider the roles open to experts and the uses of indicator variables in these processes.

Politics and the Reconciliation of Values

The formation of any public policy usually involves a decision with which some citizens disagree, or by which some are harmed. The invocation of the power of the state to enforce an enacted policy usually implies that in the absence of that power some citizens would continue to oppose the policy, or would not comply with it. They may have to pay taxes, undergo regulation, or see values contrary to their own expressed in policy, or they may fail to realize hopes or claims they believe are just. They may believe, even if these effects have not yet occurred, that the policy will produce them. Even when the state is simply a mediator or the calculator of an optimum, as in the ideal case of collective goods (Chap. 5), it may well leave some parties dissatisfied and seeking more than they receive. Policy formation can thus involve conflicting interests, conflicting perceptions and anticipations of the effects of policies, and conflicting notions of the public interest. Among those who accept a particular policy there may also be some who are harmed by it, but who do not protest because of ignorance of its effects, or respect for authority.

Thus, the political community usually begins any decision process with a set of disparate and conflicting values and interests on the part of individuals, and ends it with a collective decision or policy that accords with them in different degrees (Steiner 1983: 311–12). We shall say that policy B *reconciles* these values more than policy A if it accords more widely with them. One form that this wider accord may take is that of Pareto improvement. We may thus characterize the particular sets of values in one way, as individuals' preference rankings for the states of affairs that would result from various policies (Arrow 1963). These preference rankings may be said to conflict to the extent that they place two such states of affairs in opposite rank orders, so that not all individuals can have the policy they prefer. Preferences may then be said to be reconciled more by policy B than by policy A when policy B is a Pareto improvement over A.[3] Agreement with the policies chosen may usually be increased by moving from dictatorial to participatory decision processes, though a benevolent despot might also effect this sort of reconciliation. Such processes may, however, assure support only by a bare majority. Beyond that point, further reconciliation may require modification of initially held preferences, perhaps under the influence of more general preferences such as support for the political community, its processes, or its general welfare. When preferences change in the course of a decision process, we can say that reconciliation occurs if a given policy accords more widely with the preferences after they have changed; or if a new policy accords more widely with them than does a previously considered policy.

We may also characterize the disparate values in another way, as types of valued *variables* (e.g., health, education, safety, and other social goals; efficiency and equity) that people seek to realize through policies. Such variables, if used as ranking criteria, may also place policies or their consequences in opposite rank orders, and in this sense they too may conflict with one another.[4] Reconciliation of values, whether among individuals or among variables that they value, defines a broad class of decision processes that we shall explore with reference to politics and the use of indicator statistics.

Some policy decisions are not reconciliations in the sense of going beyond majority rule; they may simply allow one side to prevail and another to be

3. We have made several restrictive assumptions, which must be relaxed for a more general treatment of reconciliation. First, we have assumed that the results of policies are fully known. Second, we have assumed that a well-defined and fixed set of individuals are participants and are a properly chosen set of participants. Third, Pareto improvement is a restrictive form of reconciliation, since it cannot occur if one person prefers A to B. Another definition of reconciliation is possible if we can characterize the divergence or discord between each individual's values and a policy by a cardinal number, comparable among persons. The sum of absolute values, or squares, of these numbers could then be required to decrease in order for reconciliation to occur. A number of terms exist for processes of this kind: Pareto improvements and Kaldor compensation; Arrow's conditions of nondictatorship and responsiveness of social choices to individual values; distributive justice; interest aggregation, authoritative allocation of values, and conflict resolution.

4. The two problems of reconciliation are mathematically similar; the requirement of Pareto improvement for individuals corresponds to that of "dominance" for values or objectives (Cohon 1978: 69–70). The political processes of reconciliation in the two cases differ, however; conflict among persons or groups involves questions of equity and of "wider accord," while conflict among values does not press us to include all values that might be invoked unless they are advanced by persons.

neglected.[5] The winners, however, may try to justify their victory by arguing that they deserved to win. If the victory was a defense of an existing order, they may argue that that order was just and that benefits that the losers sought were undeserved. If the victory was a transfer of resources by government, they may argue that the losing side deserved to lose—that its previous gains were ill-gotten, that it needed the resources less than others did, or that the resources could be more effectively used by the recipients. The losers may then come to accept their losses, at least for a time, after a policy has been chosen.[6]

The means that are used, and can appropriately be used, for reconciling values vary somewhat depending on whether the values are self-interested or ethical. The reconciliation of self-interested values through politics can resemble market processes; bargaining or vote trading can thus be used to form a majority. Some have favored making this sort of political exchange easier (Coleman 1970), but it does not necessarily yield an optimum result (Buchanan and Tullock 1962: 139). Proponents of competing ethical principles must sometimes resort to vote trading; but this procedure rests on the balance of forces and not on the merits of the issue or even on moral compromise.

Reconciliation (for either type of values) can also work through the acquiescence given, at least temporarily, to decisions reached by accepted procedures. After the election of a president or a premier, a "honeymoon period" usually ensues in which it is considered inappropriate to oppose new governmental initiatives. Not only the legislature, but also the public, seem to express this feeling.[7] "Decision by interpretation" in political groups (Steiner and Dorff 1980) involves some political actors' refraining from challenging an interpretive decision, thus leaving an impression of unanimity.

Not only government but also other institutions and procedures may be accepted by the public because of general authoritative qualities, such as impartiality, attributed to them. For distribution of costs or benefits among a number of similarly situated persons when some but not all can receive them, a lottery can be accepted. Alternation, rotation, or sharing of public offices among groups or regions may satisfy the affected citizens' sense of equity. In earlier times, ordeals, oracles, or omens were taken as indications as to how

5. We assume here that sides exist, but they themselves may be formed by interest aggregation (Almond and Coleman 1960: 16–18). Although we often disapprove of the complete predominance of one side over another, there are values that we feel deserve to prevail, such as basic human rights or moral rules; and others that deserve to be ignored, such as evil or frivolous political positions. There remains a wide domain in which reconciliation seems desirable.

6. We exclude one form of reconciliation in which the policy is not substantially implemented, thus providing symbolic gains to one side but little material loss to the other; policy indicators could expose symbolic gains as unreal.

7. The "bandwagon effect" in which voters retrospectively report greater support for the winning candidate than they actually gave, suggests this sentiment.

public decisions should be made. At times, the judgments of experts may enjoy this same authority, as in the case of a blue-ribbon commission or an expert witness.[8]

Consensus can also be fostered through participation of conflicting parties in common discussion and decision procedures. Opportunity to be heard and to hear the other side, availability of intelligible explanations of predicted effects of policies, and the chance to play a part in the decision (as by voting) can contribute to the legitimacy of participatory mechanisms. This sort of agreement through interaction may take place at many levels of the political decision process, from citizen groups to party leaders (Lijphart 1977).

The arguments and processes that are used to change preferences and thus promote reconciliation suggest that the gain of one of the conflicting parties and the loss of another may represent something more than the effect of naked power. If they are effective in changing the losers' preferences, reconciliation has occurred, though it may not always be justifiable. When the arguments used derive from general principles, however, and are not simply chosen in an ad hoc fashion to support the decision at hand, they can reflect and sustain a normative order, itself defensible by argument, that links winners and losers. In terms of a reasoned system of values linked to such an order, reconciliation may become justifiable. The same is true of principles of empirical inference that enter into the factual aspects of justification. Arguments about justice, deservingness, need and effectiveness, and arguments connecting past conditions with present policies may later be used to set limits on the winner's claims and provide some reconciliation over time.

Most important for our purposes is the possibility that participants will agree because of shared notions of the general welfare or equity. The potential gainers may thus claim less, or the losers accept their losses more willingly, in view of principles that both support; or spectators, whose personal interests are less involved, may decide the issue on such principles.[9] This possibility can exist either at the stage of problem definition or at that of policy choice. Policy indicators that measure general values can play a part in reconciling values as well as in providing relevant information. The maximization of these values can provide an argument for reducing excessive claims and satisfying others.[10]

8. Differences or uncertainties about matters of fact may also be reconciled by information, including indicator statistics.

9. Schattschneider (1960: 3) sees the "crowd" as an initially uninvolved portion of the citizenry, whose influence can be decisive when it is drawn into a conflict. It is possible that they can be persuaded by general ethical principles. This process may seem antithetical to the findings of voting research and to modern campaign techniques, but perhaps it can be encouraged.

10. Maximization of the general welfare alone, however, is not synonymous with reconciliation of claims or demands. Preferences or political demands may not accurately express individuals' welfare (MacRae 1976a: 138–45). If some people were initially modest in their demands, a maximizing policy might give them more than they demanded and give others less in order to do

When maximization leaves a substantial number of losers, we might imagine that considering justice or equity would assist in further reconciliation of values. If the losers are poor or deprived, consideration of equity may cause us to reduce their losses. If the losses are deemed to be unfair, compensation may be provided. Equity in either of these senses may lead us back to a Pareto-like change by which more preferences are satisfied. It need not, however; the nonpoor may lose without being aided by notions of distributional equity (Chap. 7), and those who previously gained unfairly may also be omitted from this further reconciliation. Reconciliation, therefore, is a softer, less trenchant value, perhaps suitable as a latent value in social science or as a basis of political compromise, but not identical with either the general welfare or justice.

Citizens' Justifications and Their Reconciliation

In participating in political decisions, an individual may seek personal interest, group interest, or the public interest. The quest for personal interest does not itself provide a basis for persuading others; like "economic" action, it does not necessarily call for principled justification. Instead, a person may appeal to other actors privately in terms of their particular interests, making a bargain or providing inducements rather than presenting a general public appeal. He may also appeal to their particular values, including altruism toward the persuader. A citizen who is seeking to persuade a wide audience, however, is likely to appeal to a group interest or the public interest. Such an appeal must justify a particular proposal on the ground that it exemplifies one or more general values held by that audience.

An individual who enters into collective action with fellow members of a social system—a family, a firm, or a neighborhood—is thus drawn toward justification in terms that appeal to other members of that system. If that system—or a labor, business, or professional group—seeks to influence the decisions of a larger political community as a pressure group, it will be involved

---

so; this is not a reconciliation of demands. It might do the same in the name of efficient maximization of welfare, if some were highly efficient producers of their own welfare yet already enjoyed high levels of welfare and demanded little more.

A policy that maximizes welfare can therefore also reconcile preferences only under certain conditions. Sufficient conditions for the two to resemble one another would involve every participant's initially expressing an "excessive" preference, i.e., a preference for a situation in which he would receive more than his efficient share of the collective resources (the share that would correspond to maximization of overall welfare). Maximization of aggregate welfare would then lead each to receive less than his most preferred share of the resources, but with reductions shared by all; the result would be a reconciliation relative to any policy reflecting only one participant's initial preference.

with two types of justifications, internal and external. Its internal decision processes can involve discussion invoking particular group interests and values. Externally, it must formulate justifications of positions which are persuasive to its fellow citizens in the larger political community.

The move from individual to collective action involves a change not only in the type of values involved in justification, but also in the explicitness of the arguments called for. When an individual makes decisions affecting only himself, he may wish to do so impulsively or intuitively without systematic calculation.[11] Even when those decisions have far-reaching effects and careful consideration is warranted, he may still prefer to picture the expected results carefully in his imagination and again judge intuitively which choice is best.

If, however, this same person wishes to participate in the choice of public policies and persuade fellow citizens to agree with him in view of the public interest, he must be more explicit. Public justification requires reasons, and can call forth more explicit discussion of procedures and criteria for judgment.[12] As we move from individual action to group action, we must not only make the values behind our proposed action more explicit, but also spell out our factual claims of causation more fully.

Even though public argument derives much of its force from an interplay of private interests, those who speak for these interests often argue that their proposals promote the general good. Such arguments may be made by business ("What's good for General Motors is good for the country"), by supporters of the working class as embodying the general good, or by the professions contending that their autonomous practice benefits their clients or patients. The quality of public discourse can be improved if we take these arguments seriously, design information systems for measuring and monitoring such concepts of the general good, and conduct research to test whether such arguments are valid. We do not necessarily try to "unmask" these arguments (Mannheim 1949: 37) by tracing them to the social origins, personality characteristics, or interests of the speaker; rather, we examine their persuasiveness as reasoned arguments in their own right (Steinfels 1977: 5). In doing this we depart from political scientists' models of the "rational" (selfish) voter or economists' models of "public choice" based on maximization of self-interested individual "utility"; there may be truth in these models, but it is a partial truth.[13]

11. The use of decision theory for analysis of personal as well as public choices has been stressed, however, by Behn and Vaupel (1982).

12. The actual practice of public justification, however, and some philosophical treatments of it, do not proceed exclusively from general principles such as the end-values we discuss here; see Scott and Lyman (1968), Austin (1961), and Kress (1969).

13. Even the most public-spirited policy analysts, however, when seeking to maintain an organization and resources, will be driven to support that organization's narrower interests (Wildavsky 1979: Chap. 9).

Many of the value concepts that citizens use in their justifications, if we try to express them in measurable terms, resemble those in our policy indicator systems and models. They vary from the specific to the general (from a particular type of disease to health), and from means to ends (from school budgets to pupils' achievement, success, and productivity). Statistics on any one of these values can be used by a citizen to express an aspect of the general welfare. Taken together, the organized totality of such values that a citizen might use in justification may be called his *ethical system*. The component values in such a system, in addition to end-values and contributory variables, may include moral principles such as support for individual or group rights (e.g., the privacy of the family), and respect for established decision procedures or for public officials having certain traits. Within a citizen's ethical system, the values most closely related to possible policy indicators are those that correspond to the valued or disvalued consequences of policies.

The reconciliation of values depends particularly on the presence of general values in citizens' ethical systems, which may be organized in various ways. Such a system may consist of numerous specific values, each of which is considered desirable in itself; or it may include a smaller number of more general values that subsume the more specific ones. In a broader sense, it may lie somewhere on a continuum between the coexistence of numerous values and complete logical organization about a single value. We shall refer to these two extreme types of organization as *pluralistic* and *monistic*, respectively.[14] A relatively pluralistic ethical system might include basic values corresponding to various aspects of the general welfare such as some of the contributory variables (social goal areas) shown above in Table 1–3: health, occupational opportunity, environmental quality, income, safety, learning, science, art, and participation. A more highly organized ethical system might emphasize a smaller number of the general end-values shown in that table, such as economic efficiency, quality-adjusted life years, and equity. A fully monistic system would emphasize only a single principle, such as the maximization of happiness.

14. The various valued variables in an individual's ethical system (or in a set of policy indicators) may be related to one another in several ways. They may be independent of one another, as in a pluralistic ethical system; they may be equivalent to one another (as possibly on examination "the public interest" and "the general welfare" may be); one may be logically included in another (as "security from crime" may be included in "security" generally); or one may take precedence over another in hierarchical fashion, as in the case of a moral constraint that takes precedence over teleological principles. In addition, contributory variables may be means to end-values.

A monistic ethical system (or ethic) is one type of logically consistent ethic. In the economic literature, a frequently mentioned form of monistic ethic is the maximization of a social welfare function. Another type of consistent ethic that does not derive from a single principle can involve a clearly specified hierarchy or lexical ordering of principles (Rawls 1971: 42–44). I have called an ethic that is clear, consistent, and general in application an "ethical hypothesis" (MacRae 1976a: Chap. 4).

A person who subscribes to a pluralistic ethical system may do so either for intellectual reasons or because of the structure of values and interests in the surrounding society. Among the intellectual reasons for doing so are that one feels it inappropriate to subsume important, particular moral values under some single value such as economic production or happiness. Examples of such moral values are the prohibition against murder, the value that some place on family stability, and the disvalue of crime as such. In addition, one may lack the information necessary to relate various particular goals to a more general one through causal models or a common valuative measure. Thus an economist may disvalue both unemployment and inflation, or value efficiency and equity, without being able to specify how much of one is worth how much of the other.

Pluralistic ethical systems can also reflect the presence of a plurality of interests in the surrounding society or political community, giving rise to a corresponding type of public debate. This debate can involve interests and goals ranging from particular interests to the general welfare.[15] At the former extreme of this dimension, numerous persons or groups interact in quest of their goals,[16] testing the reactions of others periodically; at the latter, some members of the political community attempt to calculate the effects of various policies on all those affected and to evaluate them in terms of a general system of values. An example of such a general system is benefit-cost analysis, which is used to rank particular policy alternatives in terms of a general economic value.

The valuative aspect of this dimension appears to run from private, nonethical value systems, expressing particular interests, to ethical systems. Yet, at the "incremental" or "interaction" extreme, diverse notions of the general welfare may be advanced by various groups—health by doctors, equity by labor, efficiency by business. These and other participants can thus develop relatively pluralistic ethics involving respect for one another's values and thus incorporating these diverse principles. This mutual acceptance of specific group values can be carried even farther. Health services, for example, may be treated as an end-value through respect for the health professions, even though an outside observer might consider them valuable only as a means to health. Indicators such as numbers of health personnel per capita and their distribution, used to measure this variable, can themselves become such cen-

15. This dimension corresponds roughly to that from incremental to synoptic (Braybrooke and Lindblom 1963), or from interaction to cogitation (Wildavsky 1979: Part 2). The diversity and divisions among interests can also give greater weight to private as against ethical values in an individual's motivation.

16. Lindblom (1965) thus refers to the process as "partisan mutual adjustment." The goals in question require reconciliation when we consider public policies that compete for common resources or have multiple outcomes.

tral focuses of discussion that the question of their efficacy in producing health is neglected.

Members of a political community may thus develop a respect for one another's values so that in effect each individual supports a variety of groups' notions of the general welfare without incorporating them in a single logical system. Members committed to the overall community and its decision procedures may agree on this set of values; but rather than forming them into a single logical system, they may simply pursue them through group interaction.[17] An individual with a pluralistic ethical system may then make a personal judgment of the relative importance of two incommensurable values (such as health and education), aided by indicator statistics for social goal areas (Johansson 1976: 85). He cannot, however, easily persuade others of this judgment by explicit arguments. Instead, the public reconciliation of such values takes the form of political pressure, bargaining, and compromise, or appeal to vague notions of the general welfare.

Political processes that make use of multiple incommensurable values are likely to be different from those guided by more nearly monistic ethics. An example is the politics of needs assessment. *Needs assessment* is a procedure that is often applied to estimate the extent of a problem or the amount of some social service or commodity that should be supplied to a human population (Rossi and Freeman 1982: 90). It has been used in fields as diverse as social work (Carter, 1966), health, education, nutrition, and housing. Its aim is usually to specify a statistic indicating the amount of a corresponding good or service needed by the population in question—each of these numbers often being specified by a different service agency or profession and an associated client group. This amount is often related to standards that are judged by a community, or by experts, to be minimal requirements. It may be considered to be a right to which citizens are entitled. But such standards also involve an element of social convention, varying among communities and over time (Bulmer 1982: 58). They are implicitly related to the available resources.

Competing claims as regards different sorts of needs cannot, however, be reconciled within this framework (MacRae 1981: 112–14). Each assessment of needs provides a separate statistic for the corresponding service or commodity, but the procedure does not provide comparisons between the values associated with different services. If these claims for community resources should conflict, one means of reconciliation would be a form of pluralistic politics in which various service bureaucracies, each allied with its own clien-

17. An important body of literature on social pluralism was developed in the 1950s advocating multiple groups with overlapping memberships as a protection against the threat of extremist political movements (Shils 1956; Kornhauser 1959; Selznick 1960). A plurality of interests can, however, be fragmented into intransigent single-issue groups when social integration is low.

tele, compete with one another (Gates 1975a). Such competition would eventually allocate the available resources, but not necessarily in a desirable way. The greater the diversity of values to be reconciled, the greater we may expect the role of political processes to be in the determination of tradeoffs among those values.

The difficulty with such a pluralistic political process, for our present purpose, is that it provides no firm basis for reasoned assessment of the adequacy of the process itself. When professionals, service bureaucracies, and their clients bargain about needs, there are risks that each set of professionals and bureaucrats will unconsciously exaggerate the efficacy of their own services and the extent of their clientele's needs, that client representation will be biased toward persons associated with the professions, and that taxpayer interests will not be adequately represented. For these reasons among others, a systematic analysis of the values expected to result from serving these needs, if introduced into the political process, may lead to better social results than political bargaining alone.

In contrast to a politics of multiple groups and pluralistic ethics, we may try to reconcile particular values through reasoned argument about the general welfare, as in the section of Chapter 1 on value analysis of social indicator concepts. We there analyzed concepts in a number of social goal areas and argued that most of them should be considered means to a smaller number of general end-values. These more general variables can then be used for reconciling the more specific claims. We might engage in similar argument in the political community, trying to persuade fellow citizens that their particular ethical values should be regarded as means to, or special instances of, more general values. With the aid of general policy indicators, public debate might be focused on more general values.

When general variables are used in this way, however, some critics consider them a threat to democratic processes. Benefit-cost analysis, for example, implies a different intellectual approach from that of a politics driven by the demands of interest groups. It implies a greater degree of "fine tuning"—the assumption of a social welfare function and the allocation of scarce resources so that at the margin, each unit of resources is used so as to produce the greatest contribution to total social welfare. This fine tuning approach is at odds not only with the pressure and bargaining of interest groups (though this too has been considered at times to produce a fine-tuned "resultant"), but also with ideas that support large coalitions of groups and that unite social movements in more dramatic types of politics. It seems to imply an "end of ideology"—pronounced prematurely to have occurred in the early 1960s. On the one hand there seems to be a politics of experts, of calculation, of incremental tradeoffs at the margin; on the other a politics of drama, of meaningful symbols, of struggle, and of public participation (Gustafsson and Richardson

1979: 434). Some critics would claim that the politics of benefit-cost analysis—and of abstract tradeoff indicators generally—is bloodless and elitist.[18]

Yet a politics of abstract values such as net economic benefit (or subjective well-being) can never exist by itself; it can at most be an ingredient in a larger political process. Our uncertainty about causal relations provides ample grounds for disagreement. Moreover, as we consider various general definitions of the public interest, the conflicts among these definitions can themselves contribute to democratic politics even while the definitions express aggregations of interests. This difference among definitions arises especially for justice or equity. Even if one such definition should prevail, its use would at most have a damping influence on the normal oscillations of democratic politics, by introducing tradeoffs and optimization more into public debate so that public policies would swing less far between extremes.

Tradeoffs and Indicators

An important potential function of policy indicators is the reconciliation of diverse values. An indicator statistic such as the gross national product, or the measure of economic welfare developed by Nordhaus and Tobin (1972), directs our attention to an aggregate to which particular policies can contribute. At a lower level of aggregation, the net economic benefit (benefit minus cost) from a particular project is the contribution of that project to the national product or income. The net benefits expected from efforts to improve health, education, safety, and conditions in other social goal areas can thus be compared if their benefits and costs can be estimated in monetary terms.[19]

Tradeoff procedures need not be limited to monetary valuation, however. If we were interested in equality of educational achievement between races, we might use a measure of this equality as a criterion for comparing two policies aimed at equality, such as school desegregation and redistribution of funds among schools. If we were interested in safety as a criterion, we could use it to compare effects of using a given amount of resources for fire prevention and for automobile inspection. In either case if we could predict the effects of alternative policies through knowledge of causal relations, and if we could

18. Altshuler (1965: 338, 343) argues that "the profession which consolidates all the measures of its success into one immeasurably enhances its claim to be called expert in its sphere"; but notes also that "it is obviously impossible to specify technical qualifications for the *most* general evaluators." Purely political means of reconciliation may risk excessive consensus (Steiner and Dorff 1980: 187–94); but reasoned consensus based on policy indicators cannot go quite so far.

19. A budget constraint, if it could be varied, could be entered into the tradeoff calculations similarly; but a fixed moral constraint could not.

measure the effects in comparable terms, we could compare the benefits of the two policies.[20]

Indicator variables can play two roles in these tradeoffs. One is encouraging the public to define problems in inclusive terms, by calling their attention through indicator statistics to general values. The other, more directly relevant to tradeoffs, is relating a common general value to each of the more specific variables (such as program outputs) under comparison; this does not necessarily depend on using the aggregated public statistics, but is joined with them through the cycle of policy choice (Chap. 1).

This function of comparison or reconciliation is suggested in an assessment of indicator validity by OECD (1976: 25–26), which proposes that "the most valid indicator is the one that is able to compare the effects of as many instrumental variables as possible on the concern under consideration. In the health area, for example, an indicator that could be taken as a criterion for the effects of both a cancer detection program and a new technology for treating heart attacks would be more valid than indicators which could not compare these two influencing factors with respect to a common unit."

Particular values may be reconciled not only by applying an explicit external standard (such as economic benefit, equality of educational achievement, or safety) but also by asking people to make judgments or ratings as to which of a number of conditions, or outcomes of programs, is more serious, more important, or more deserving of priority.[21] In the study of the seriousness of crime, this approach has often been used to devise indexes by which various crimes could be compared (Sellin and Wolfgang 1964; Rossi et al. 1974). Physicians' degrees of concern for patients with various symptoms have also been expressed in scale form (Holloway 1973). Syntheses of component values into general value indices have also been proposed in the areas of environment (Odum et al. 1976: 167–77) and health (Culyer et al. 1972: 108–18). In each of these areas, there has been an effort to increase the reliability and validity of the indexes by asking for judgments on component aspects of the condition rather than for a single global judgment. The resulting combinations or indexes reflect the organization of the respondents' value systems, and suggest possible consensual weightings that others may use, but they lack the conceptual integration of the economic approach (in which price represents more than an intuitive basis of comparison) or the inclusiveness of general

20. There remains an implied criterion of cost, involved in monetary valuation of input resources. It is also possible that citizens would prefer to compare two risk-related programs in terms of subjectively perceived risk rather than in terms of probabilities of harm based on past occurrences.

21. These judgments are analogous to consumers' judgments of willingness to pay (Chap. 5). They are comparable with one another, and thus may be analyzed statistically; but they lack the collective deliberation of public debate.

indicators of well-being that measure value variables to which other values are means. Judgments of the seriousness of particular conditions thus need to be examined critically to see whether the judges' criteria conform to our end-values (Bursik and Meade 1982).

A more formal and mathematical approach to this type of problem, which still does not rely on the use of a single explicit external criterion, is multi-attribute decision analysis (Keeney and Raiffa 1976). In this procedure, a decision maker is required to reconcile a set of disparate criteria to be used in valuing alternative courses of action. He does so by making explicit his own judgments of relative value, and by following systematic procedures to combine them in a multiattribute utility function. The methods of constrained maximization can also be extended to deal with multiobjective choices (Cohon 1978: Chap. 6). If more than one decision maker is involved, some reconciliation of valuations among them is also required.

It is thus possible to devise tradeoff indexes at various levels of generality—within an area such as health or crime, or more generally so as to compare such areas with one another. These indexes can also involve various degrees of explicitness in the criteria used in the tradeoff—from judgments of undefined qualities such as "importance" or "seriousness" to quantitative causal links connecting various specific conditions with an explicit end-value. Although all these methods have been used, we are here proposing the more explicit procedures that work through the measurement of end-values and estimation of the effects of particular conditions on them.

General or inclusive indexes of this sort have been criticized on the ground that they lack clear meaning for scientific or policy purposes. Rossi and Henry (1980: 498, 501) point out, for example, that indexes of the seriousness of crime may not be appropriate bases for making criminal law, as they might stress retribution at the expense of deterrence or rehabilitation. They also note that such indexes might be inadequate for program evaluation, and that offense-specific crime rates would be more useful since particular programs would be oriented to corresponding types of crime. This criticism—that various component parts of such indexes might be more relevant to policy because they are manipulable through distinct causal models—is true of components of the mortality rate as well (Land and McMillen 1980: 20–39). The criticism is relevant to many general indicators, including the GNP; but it reflects the needs of science and of evaluation at the expense of tradeoffs among policies. If the general index is a linear combination of component indexes for particular crimes (diseases, etc.), weighted in terms of social valuations, we do indeed need to see what means are appropriate to dealing with each separate component. Having done so, however, we must still allocate our resources to these means in terms of the overall value that is to be maximized; specific indicators alone provide no guidance for these tradeoffs. This maximization must take into account both the effectiveness of allocating resources to particular components and the relative value of the corresponding outcomes.

The combination of component values in a general value measure of this sort must always rest on judgments of relative ethical value, consistent with prevalent values in the political system in question. For this reason we do not here propose the construction of composite[22] indexes that combine variables by methods such as arbitrary weights, the addition of standard scores, or factor analysis; such indexes cannot provide the value weighting of their components that we need. Duncan (1969: 4), for example, has argued against indexes of health formed by adding together the numerical values of aspects of health (mortality and bed-disability), and against similar indicators of the status of minority groups. Similarly, a number of indexes of the presumed material well-being of various localities have been computed by adding variables such as median income, years of education, and the presence of municipal facilities; several of these indexes have been described and criticized by Carley (1981: 150–65). Not only is their valuative basis unclear, but they lack intelligibility for public discussion.

In contrast, the Gross National Product (GNP) and the Consumer Price Index are aggregates for which a clear conceptual and valuative definition specifies the universe of component measures, and an external criterion variable—price—defines their weighting. This advantage suggests that we might seek further general criterion variables, other than price, in terms of which the contributions of more specific values can be compared.

The degree of generality of the variable that is used for tradeoffs depends on the generality of the comparison being made. In public debate this might vary with the level of abstraction chosen in "mixed-scanning" (Etzioni 1968: Chap. 12). Within government it is likely to vary with the organizational level at which the decision is being made. For example, the director of a public health agency charged with controlling a particular disease might examine an indicator of the incidence of that disease and models of the conditions that affect it in order to compare policies for dealing with it. This indicator would permit the comparison of various programs such as preventive measures, personnel training, and the therapeutic use of drugs. But a government agency concerned with overall health policies might make use of mortality rates or patients' capacity to function in order to compare the effects of different diseases and choose among policies affecting them. At a higher level, a general budgetary agency might be concerned with the relative value of health policies and crime policies, and would require a still more general value measure to compare them.

22. Duncan (1969) refers to "aggregation" over items or indexes. In Chapter 7, however, I shall use the term aggregation to refer only to combination over persons.

The public can be concerned with choices at any of these levels; but its distinctive role is that of the supreme authority in a democracy. It thus has the right to be concerned with the most general tradeoffs in a way that intermediate administrators cannot be. Those who speak in public debate in the name of the general welfare thus have a reason for measuring the general welfare in general terms; but they must make those terms intelligible.

When we recommend general end-values as major components of policy indicator systems, we encounter two distinct problems concerning the roles of experts: (1) Are we not proposing a technocratic role that is inappropriate for social scientists? and (2) How can indicators as sophisticated as these be understood and used at all in a democracy? One question asks whether social scientists might not exercise too much influence by proposing new variables for public discussion; the other raises the doubt whether they will make any contribution. Both these questions are significant. We shall deal with them briefly in this chapter; again in analyzing economists' use of the value variable of net economic benefit (Chap. 5); and finally in Chap. 9 in connection with political feasibility.

In examining these questions it is useful to look again at the role of economics. Benefit-cost analysis has grown greatly in importance in the last three decades (Haveman and Margolis 1977: Introduction) but is far from the dominant criterion for governmental policy choice even in areas where its calculations can be made validly. It is open to criticism, but broad and intelligent criticism requires clear presentation and widespread citizen education. The necessary criticism and review require clarity and openness to examination of the entire process by which data are gathered, analyzed, and presented (Hanke and Walker 1977: 352). A central problem in the use of benefit-cost analysis is in fact not simply that it transfers power to experts, but that political interests distort the numbers that purport to result from its reasoned calculation.

The acceptance of the GNP as a value indicator, and of economic benefit-cost analysis as a corresponding basis for tradeoffs (Chap. 5), has not depended on their wide understanding by the public. Consensus on the part of economists as to the theoretical meaning of the GNP and the technical procedures for estimating it, together with acceptance by other elite groups, have aided this acceptance. The prestige of a discipline can sometimes forestall criticism and support the use of a policy indicator that is not fully intelligible to the public.[23] The success of economics in proposing this indicator therefore provides a useful example for both emulation and caution.

23. As regards macroeconomics, more serious criticisms have been posed by Lindberg (1982), who claims that the antiinflation policies of the Federal Reserve Board under Paul Volcker in 1980–82 unjustifiably focused on this one value at the expense of others (equality and the well-

The capacity of citizens to understand technical procedures for both trade-offs and causal inference depends on their education, formal and informal. I shall speak repeatedly of citizens who engage in reasoning similar to that of professional policy analysis; yet the proportion of such citizens in the electorate is likely to be small even in countries where education is widespread. Such a small group might be suspect as an elite, or as an extension of a profession with its particular interests and biases; but they may also provide reasoned voices for the general welfare. Some may be among the leaders of the citizenry, occupying significant positions in government and in private associations. In these positions they cannot hide behind claims of professional autonomy, but must engage in discourse with other citizens. If they achieve leadership, we can hope that it will be because they have argued rationally as well as persuasively. Moreover, as citizens rather than mere experts, they will continually make value judgments in public debate. Their claims as to the meaning of the general welfare will be tested in that debate. In this perspective, it is not wrong but highly desirable to try to educate more such citizens; they may be an elite in the earlier, valued sense of this term rather than in the prevalent disparaging sense.[24]

The Expert and the Political Community: Values and Roles

Policy indicators can play a part in reconciling particular values because they measure more general values. In proposing these general values, expert communities can exercise some influence on public debate. To see how they can and should do this—and the limits beyond which they should not go—we must examine the roles that experts can assume when they design statistical systems or develop corresponding policy models. We shall first describe two alternative roles, those of outsider and citizen, that may be assumed by social scientists engaged in practical work. In the following section we then recommend a third role, intermediate betwen the two, for those who design indicator systems.

Let us assume initially that we have chosen—as a first of two polar types of roles—to serve a political community as our ultimate client, recommending

---

being of the poor). This valuative bias, he contends, resulted from a monopoly by economists of policy recommendations in an area which is identified with macroeconomics but which also affects the equity of distribution of reward and suffering.

24. Some preparation for informed citizenship is given by undergraduate education in public policy analysis (MacRae and Wilde 1979); the expert technical communities we are about to discuss, and to which we return in Chapter 9, can also contribute to it but correspond more to groups of researchers.

policy indicators or related causal models.[25] We face questions about the community's collective values, and eventually, our own roles. If we intend simply to serve this client we have taken an *outsider* role, in which we provide expertise (means) and the community provides values (ends). We often deal with parts of a community, however, thereby making ethical choices similar to those we make in choosing to work with the community in the first place. In either case we forego the right to participate, in that role, as a citizen.

This outsider role, which takes the community as client, may be justified by social scientists on the basis of either philosophy (positivism) or prudence. Prudence may dictate their keeping a certain distance from current controversies in order to sustain longer-run values concerning method and theory, or to further the viability of the institutions that generate information. The outsider role is occupied of necessity by foreigners (Brewer and Brunner 1975: 10–11);[26] it is usually also the stance of paid consultants or employees, who by taking pay commit themselves to serve the client's or employer's goals. Some elected public officials also come close to assuming this role, seeking to serve their constituencies but not to persuade or even necessarily to lead them.

In taking the community as client, we still face a difficulty in identifying community values, as we can illustrate by contrasting this task with that of the designer of management information systems. In the management field, there is an extensive literature on the technology and uses of information systems. This field of application is typically one in which the users occupy roles in an organizational hierarchy, and in which managers can presumably be more effective by using better information about their organizations. The designer of such a system serves a manager or a small group of managers.[27]

In a democracy, however, it is the departure from hierarchy that is essential; this remains true for a wide range of political systems with varying degrees of democracy. Even though government officials and citizen leaders have a special need for accurate information on the state of their society and economy, in

25. The relation of expert to client is taken as central by Lazarsfeld et al. (1967: x–xi). Even though these authors mention "the stance of general social critic," the general connotation of working with a client is that of service rather than participation. When we choose the political community as client, this role limits us to those aspects of the public statistical system that are affected by government, either in its statistical agencies or in its aid to private statistical work. The notion of an "ultimate client" stresses the members of the political community in distinction to its officials (Alterman and MacRae 1983).

26. Carley (1981: 125) points out that "government statistical agencies are duty bound to remain politically neutral." This does not mean, however, that their managers can afford to ignore politics, either partisan or organizational. Scientists also have a special claim to enter public debate on narrow issues such as the correction of technical errors. In contrast, universities face complex questions of value and fact in judging the bearing of human rights issues on research conducted abroad (Hoffmann 1984).

27. The use of social indicators for providing feedback on an organization's environment was analyzed by Rosenthal and Weiss (1966) with respect to NASA.

a democracy it is vital that this information and the policies based on it be open to use and criticism by the general public. The task of developing information systems that are accessible to the public, and that contribute to reasoned analysis by the public of issues and policies (*democratic information systems*),is far more difficult than the design of management information systems, both in the identification of values and in the need for intelligibility.[28]

If we are serving a political community as we would serve a client and designing a democratic information system for it, we must identify the values that we shall serve among the multiple sets of values, some explicit and others scarcely articulated, in the community. Just as evaluation researchers often have problems in discovering the "true" goals of the programs that they seek to evaluate (Weiss 1972: 26–30), the consultant to a community has problems that are similar but greater. A political community, unlike an organization or program, is rarely founded to serve a specific purpose. Its culture often embodies dominant values, but in a free society a respect for diversity is also encouraged. If we are recommending value guided information systems for a diverse or pluralistic political community, we cannot easily discover the particular values that are "official."

In serving the political community, we work through government. A major statistical activity of government is to generate and disseminate information itself. Private organizations, perhaps with aid from government, are more suitable for experimental and temporary statistics; government statistical reporting tends to be more regular and stable. If government is to sponsor the collection of information over the long term, then the various groups in the community must have sufficient agreement to support the gathering of this information as government incumbents change. This agreement, if it does not exist on every particular value, may be built through bargaining or logrolling so that multiple clienteles, each seeking its own values, nevertheless form a coalition to sustain the whole enterprise.[29] Such a coalition requires cultivation if it is to last. It may also require the limitation of public support for information that can exacerbate major conflicts within the community.

The values implied in statistical series developed in this way are likely to be least-common-denominator values: either subjects of broad consensus (the GNP); matters of interest to particular organized constituencies and not strongly opposed by others, supported by coalitions (statistics on science or

28. We do not propose simply to measure and report opinions as a means of guiding policy; statistics on variables of this kind can detract from reasoned analysis by citizens, by fostering premature consensus rather than contributing to it. Certain variables often proposed as indicators, such as satisfaction and alienation, may also have this effect.

29. The values supported by the members of such a coalition do not necessarily represent aspects of the public interest. In addition to reflecting various concepts of the public interest, they may also be values accruing to distinct groups and more appropriate to an economic view of politics.

on the handicapped); or measures that may be interpreted diversely by various groups and used for their own purposes (statistics on marriage and divorce).

I have spoken of the political community as an ultimate client; but I have also acknowledged (by speaking of groups) that we may not be dealing with its citizens directly. We cannot easily do so in a large community; typically we deal with government officials or group leaders as immediate clients. We may think ideally of the citizens as the primary users of indicator information; but in actuality information systems for urban or national planning are used primarily by public officials.[30] An information system in Cologne in the early 1970s, for example, required permission from the responsible department before someone outside that department could use it (Beresford et al. 1976: 303). The acceptance and support of indicator systems can also depend importantly on bureaucratic politics (Rich 1981). The public's participation may then depend on the initiative of active group members or of the media, seeking out statistics, selecting and publicizing them.

The government officials with whom we deal may be moved by private motives or by specific, factored aspects of public problems, as well as by concern for the public good in general. So, too, may be the academic or professional who advocates the production of public statistics to advance his career or favor the activities of experts from his own specialty. If we regard the political community, rather than the present government officials, as our client, we may wish to deal with persons who are dedicated to its long-run well-being. We may also wish to promote wider participation in the decision as to the choice of indicator variables. Even if we (as professionals) could choose variables that reflected the true interests of the community, without participation we could still risk lack of support for the indicator system.[31]

Even if we could bypass government officials and deal with the public, the problem of representativeness would not be automatically solved. The groups that can most easily pay us as consultants or understand our technical language, and those that use statistics most, are not necessarily those most in need of help. Political systems that are officially democratic can never represent all groups with exact equality, and the group systems in those polities often reflect the same bias as the officialdom. One approach to evenhandedness has been proposed by Coleman (1980: 346–47), who suggests a style of "pluralistic policy research" serving not a single client but a variety of interested parties in society. Presumably a designer of indicator systems

30. Sheldon and Parke (1975: 695–96) refer to "goal-oriented analyses" as lying within the scope of administrators and their agencies, and only general "enlightenment" as being appropriate for the public; but citizens may also engage in analysis.

31. At the same time we must recognize that technical information not easily understood by the public (e.g., economic or environmental) can benefit the public; and that information about a nation's environment (foreign intelligence) can sometimes be of greater benefit to the public if it is not made public.

might do likewise, seeking to diversify their use; this role would maintain some of the detachment characteristic of academic research. Alternatively, we might consider setting ourselves against the bias of the system, taking underrepresented groups as our clients, but at the same time risking departure from a broadly shared notion of the public interest.

The option of choosing and advocating particular values and policies, however, suggests a second general type of role we may occupy besides that of the outsider—that of an active member of the community. For citizens and leaders of the political community there is an opportunity to take the initiative, to try to persuade others, and not simply to seek the composite preference of the other citizens. This opportunity exists by right for citizens of a democracy, and is not merely insinuated into the relationship as it is for a consultant with a client.

I have addressed the reader so far as though "we" are social scientists—whatever other roles we may occupy. Those of us who are also active members of the political community may be elective officials or candidates for office, seeking to lead a constituency and not simply to follow it. We may be government officials, trying to facilitate our work, coordinate it with that of other agencies, and respond to the public directly or through its representatives. We may be citizens, organized in groups, seeking to increase consciousness of an issue by publicizing conditions such as unemployment, illness, illiteracy, or inequality and to monitor possible progress toward reducing them.

The goals and resources of these various groups and roles may differ. But regardless of which of these roles we ourselves may occupy, all of us must be involved ultimately in the governmental choices that affect any successful policy indicator system. Public officials, or continuing private organizations aided by government, usually administer the information-gathering system, and other officials and organizations must be among its users. Citizen groups must provide continuing political support and criticism. Together with these groups, outsider experts are also necessary to monitor the quality of the information presented and to relate it continually to their theoretical understanding of causal relations. The interrelations among such groups have been an essential condition for the institutionalization of indicator series (de Neufville 1975: Chap. 10).

These active roles in the community may simply happen to be occupied by persons who are also social scientists. A person who occupies both roles may compartmentalize his or her activities into those undertaken as citizen and as scholar. Yet it is also possible for the two types of roles to be combined. One relatively active role is that of "public interest science" (Nelkin 1979: 114–16). Some natural scientists, who feel social responsibility for the uses of their specialized knowledge in relation to their notions of the public interest, have joined in groups to analyze public issues and take stands on them. Social scientists have done likewise (MacRae 1976a: 75); and groups of them, as in Britain, have had close connections with political parties (Bulmer 1982: 35). Such roles are not, however, always consistent with the "neutral expert" role that is more characteristic of scientific communities. In practice, those who take this role and become involved in controversies risk belittlement by their fellow citizens of their claim to scientific expertise, especially if they are social scientists. Such public judgments may extend to their disciplines as a whole.[32]

The combination of scientist and citizen roles by social scientists thus faces practical limits. The scientist role seems to require detachment from political controversy, but values are embedded in the discourse of the social sciences (MacRae 1976a) and invite controversy. In economics, however, an entire discipline has been able to associate itself with a value position which (at least in the United States) does not seem controversial to the same degree. The values of benefit-cost analysis, or of the GNP, have been used within the discipline and in public discussion as bases for policy recommendation.

The Proposal of Consensual End-Value Indicators

We have described two possible policy-relevant roles for experts in a democratic political community: neutral service to the community and citizen involvement. Citizen involvement in a limited degree, however, may still allow us to preserve the values and the autonomy of expert communities, and may be appropriate for the designers of policy indicator systems. This type of limited involvement seems to have contributed to the success of economics in combining policy relevance with scientific quality.

The relation of economics to public policy suggests a third role open to experts: concern with general values without direct involvement of the expert community in partisan politics. One form this may take is persuasion that does not work through government. Many nongovernmental decisions affect our public statistical systems. Some indicator statistics have been gathered by private organizations, such as the National Bureau of Economic Research, the Institute for Social Research at the University of Michigan, and the National Opinion Research Center at the University of Chicago. The directions taken in data collection, analysis, and presentation are influenced in important ways by the decisions of expert communities as to what work is significant; they are

32. These limitations bear primarily on the experts engaged in designing and administering indicator systems and developing policy models. They bear less on persons who are educated in these same skills but who function as citizens rather than members of a profession or discipline. The distinction is clearest if the two roles are filled by distinct persons rather than by the same person wearing different "hats."

also aided by decisions of universities and private foundations as well as of government.

The relative autonomy of these private decisions has been sustained in part by the devotion of university based organizations to basic research rather than to partisan controversy. Yet modest ventures into value-related types of research have succeeded and have drawn public attention to related general values. In addition to economic indicators, the study of subjective well-being has provided a potential alternative focus of attention. Happiness, like the GNP, does not seem to be an issue of partisan disagreement.

Research on the measurement of general end-values can provide a firmer foundation for collection of statistics on these variables. The quality of this research, and university support for it, can be enhanced by its connection with problems of basic scientific interest. We shall suggest in Chapters 5 and 6 that there are significant scientific problems related to end-values, and indirectly to policy choices.

Economics, as we have pointed out, provides a political community with a criterion for policy choice: maximization of the GNP. Taken by itself, this would constitute a monistic ethics of maximization. At the heart of an ethics of maximization is the measurement of the end-value that is to be maximized. Net economic benefit, however, though its use is instructive, is not the only end-value that might be used; others might provide similar general bases for choice. The social indicator movement has represented, in part, a protest against the values involved in economic indicators. It has resulted in a proliferation of measures, some noneconomic and others (such as data on the work force) bordering on the economic. It has not, however, sufficiently emphasized alternative general end-values in terms of which more specific social goal areas can be compared. A significant remaining task, therefore, is to retain the general tradeoff capacity of the GNP but to exemplify it in other types of values—stressing, for example, well-being that is not associated with money and possessions, and providing for greater equality among persons in assessing well-being.

One such general variable would combine two components already familiar in the literature on social indicators: subjective quality of life and duration in time. In principle, each interval of time for each person could be assigned a weight in terms of the quality of life experienced during that interval. Formally analogous weightings of time by the value of time have been proposed by K. Fox (1974: 54) and by Juster et al. (1981: 76) as components of social accounting systems; but the use of this variable need not await the detailed development of such systems.

In connection with the evaluation of health policies, Weinstein and Stason (1976, 1977) have proposed that we not only consider the aggregate length of life that various policies will produce, but also include in that aggregate a weighting factor for the quality of those life years. The result would be a total of "quality-adjusted life years" (QALYs). In particular choices of health policies, patients would be asked to compare their life experiences when under medication with those of "full health," and judge what fraction of a year of full health would be equivalent to a year under medication. More generally, techniques that have been developed for the measurement of subjective well-being (Campbell et al. 1976) might be adapted to weighting of life years by their quality.

If this type of measurement should become a central concern for those developing policy indicators, it might provide an additional basis for comparison and criticism of more specific indicators. Unlike economic measures, it places value on experience rather than on the satisfaction of preferences in the market. In addition, it is based on time, which is more nearly equally available to all than money, and might thus circumvent some of the inequity of weighting the demands of rich and poor in dollar terms. We shall discuss this indicator further in Chapter 6; but no matter what aggregate end-value we use, it will have to be supplemented by indicators of equity (Chap. 7).[33]

The judgment of the rightness of acts according to their contribution to a total of subjective well-being goes back to Bentham's ethic of greatest happiness. Bentham's utilitarianism left as a legacy the employment of "utility" as a central concept in the early development of economics (MacRae 1976a: Chap. 5). The subsequent development of the discipline led to the identification of utility—initially a subjective variable—with the satisfaction of preferences by consumers and producers.

Preferences have been inferred by economists largely from observation of market behavior. Subjective well-being, on the other hand, is almost always measured by questioning individuals in sample surveys. The two approaches are nevertheless potentially closely interrelated. Survey methods have been used to some extent to measure nonmarket economic valuations (Chap. 5). But equally important is the fact that these two approaches, though of common origin, can yield different answers as to the conditions for human well-being. The comparison and contrast between these approaches thus gives rise to important questions of value as well as of method.

These two divergent approaches can therefore benefit from comparison in both academic research and public debate. Such comparisons raise questions about the sorts of lives we should be living, and the social arrangements that will promote one or another version of the good life. The study of subjective well-being can provide a contrast to economic valuation; but it can also benefit from the systematic style of optimization for policy choice developed by economics (Chap. 5).

33. Hampshire (1973: 27) also argues that "large-scale computations in modern politics and social planning bring with them a coarseness and grossness of moral feeling." Thus we need not only to counter one general end-value indicator with another, but also to be aware of the nonquantified consequences of public policies for individual human beings.

Part II.
Causal Models for Policy Choice

3 Successful Models and Their Limits

The ultimate test of the usefulness of policy indicators is their contribution to improving our choice of public policies. Statistics based on them must direct citizens' and leaders' attention to important issues on which public action can be taken and to the consequences of that action.[1] If we measure health, for example, we must eventually be able to predict the effects of preventive and remedial policies on health, so as to compare those policies. Such prediction requires models of causation. The models that are used often derive from everyday experience; but we assume they can eventually be better formulated and improved by scientific research.

Models of causation are needed not only for our choice of policies, but also for our choice of indicators based on contributory variables. They can give priority to contributory variables that have a clear and strong effect on end-values (Chap. 1); thus air pollution is measured regularly now that we know some of its effects on health. They can also give priority to contributory variables that affect multiple valued outcomes; our presumed capacity to cure a wide variety of diseases through health professions and hospitals has led to emphasis on these inputs and to the consideration of health as a social goal. Similarly, police and correctional institutions deal with numerous types of crime, leading us to choose crime reduction as a social goal.

We must ask how causal models can best be developed to aid our choices of policies and indicators. The causal models we shall discuss in Part II will relate mainly to social goal areas such as transportation, health, education, and crime reduction, and will have as their dependent variables possible policy indicator concepts in these areas; they will range widely over policy-related science, especially social science. In Part III we shall return to the causes of end-values that connect these more specific social goals to one another.

In this chapter and the next we shall compare and contrast a variety of policy models in terms of their practical success and the types of knowledge on

1. The criteria of choice need not be purely consequential or teleological; moral constraints may enter. But when policy indicator statistics are useful, there is presumably some causal connection from policies to values.

which they are based. We shall show that although the most successful models seem to have drawn on the exact sciences, policy models cannot escape drawing substantial components from less exact fields. In this chapter we treat some of the more successful approaches and their limits; in Chapter 4 we deal with some important areas in which research has not yielded similar success, and make suggestions for their improvement.

I shall make the following proposals in this chapter as to priorities for policy model construction:

1. Since policy indicator statistics measure ethical values, we should use the corresponding indicator concepts as the dependent variables in policy models; the other variables (causes) in these models need not, however, be policy indicator concepts, and the models need not be tested on the general populations used for indicator statistics.

2. For practical use in the shorter run, we should assign the highest priority to development of models in particular social goal areas and lower priority to overall social system models.

3. We should free policy models somewhat from the criteria of basic natural science as regards its stress on elegant, pure research problems; its high standard of theoretical development of models before they are tested in practice; its separation of more exact from less exact science; and its separation of research problems from issues of public debate.

At the same time I shall argue (in this chapter and the next) that we can develop and use policy models successfully even when they contain important components based on social theories that are somewhat inexact. The ways of doing so include successive trials and experiments that permit us to revise our models under realistic conditions related to practice. They also include a style of policy model construction that is continuous, guided by general goals analogous to those of basic theory development, and distinct both from pure disciplinary research and from mere service to the immediate requirements of policymakers.

Policy Models

A *policy model* shows how, and to what extent, public policy alternatives cause outcomes that can be described by one or more value concepts.[2] The findings of basic science, such as Newton's laws, are not pol-

2. We shall use the term policy model to refer to a causal statement that a policy of a given type, in comparison with a specified alternative, is likely to produce specified results. This can include the statement that a policy is likely to be ineffective. In this chapter we are concerned mainly with policy models that have been developed by expert communities. There are also, however, many commonsense and intuitive causal judgments (Polanyi 1958; Oakeshott 1962: 8)

icy models but may be components or building blocks within them. These laws are used, for example, in the design of publicly supported highways or supersonic jets; but the full policy models include other components as well. Policy models are linked with policy indicator statistics when the same measures (indicators) used in the statistics are their dependent variables, making the models relevant to the problems that the indicator statistics help to define. Thus a macroeconomic model that shows effects of public policies on unemployment is linked to unemployment statistics; and a model of the effects of pollution control on health, to health statistics. Since policy indicators can encompass a wide range of value concepts, from general end-values to specific administrative variables, a similarly wide range of models can be linked to statistics measuring these values.[3]

A useful policy model must be characterized not only by the variables it contains but also by other features that conduce to its use. These include its availability to the right persons at the right time (Coleman 1972b; Wildavsky 1979: 219–20), since many policy decisions are open for choice only at limited times and essentially closed at others.[4] Moreover, if the findings favor certain policy alternatives over others, enactment of these alternatives depends on the alignment of political forces at a given time. It is thus important that expert communities maintain continuity in the study of policy problems that are likely to recur, so as to reduce the time needed for translating findings into practical terms. These communities should also try to anticipate new problems and to deal with nonrepetitive needs for policy models (Chaps. 4, 10).

A useful model must be intelligible; this depends in part on its expression in the users' language and on repeated explanation, but also on the content of the model. Some models are harder to explain to users than others. The users' capacity to understand, which may depend on their familiarity with scientific principles or with other models, can be increased by prior education, but users must also be continually brought up to date.

Research-based policy models are more often focused on social goal areas than on end-values alone.[5] They are thus likely to be concerned with predict-

---

that, although stated approximately and not subjected to scientific testing, play an important part in the judgment of policy alternatives.

3. Although the vast majority of policy models will be related to indicators in this way, the connection is not logically necessary. Policy indicators are chosen in part because of their breadth of application including nonpolicy uses, their cost, and their intelligibility. The priority of value concepts for use in policy-related public statistical systems might not, therefore, be the same as that for policy-related research.

4. This concern for timing need not imply subservience to an existing power structure; information may be made available to the public or to social critics.

5. Policy models are defined and developed most systematically in the field of public policy analysis. In social science they overlap with policy-relevant research, especially when that re-

ing whether one policy or another will provide more health, education, transportation, or public safety, improve the physical environment, or the like, and how much it will cost, rather than simply seeking net economic benefit, happiness, or equity in general. Research-based policy models may involve economic benefit as an end-value if we place monetary valuation on the dependent variables; but they need not be general social system models predicting net economic or social benefit from all sources.[6] Our models assume this form for the same reasons that lead us to choose certain contributory variables as policy indicators: policy problems are conveniently factored into parts in which separate institutions (Land 1975b: 25) use separate means to suboptimize these social goals. Moreover, some end-values, especially subjective well-being, are harder to measure reliably and to predict than are the objective social goals that we take as surrogates for them. We thus deal here with models of causal processes affecting the attainment of social goals, and return to models of the causation of more general end-values in Part III.

Policy models in the various social goal areas are nevertheless linked together in several ways. First, they may contain similar components, as generalizations of natural or social science may figure in them, and they also share methods of analysis from statistics, decision science, management science, and policy analysis. Second, they are connected by causal relations working through policies: a policy centering about one social goal, such as education, may generate side effects on other goals, such as crime reduction; or one social goal variable (the reduction of poverty) may affect another (health) directly. Third, they are potentially connected by end-values that, if we can show how particular social goals contribute to them, allow us to compare those goals in value. Fourth, they are connected by common exogenous variables: demographic trends such as migration or changes in the age structure of the population affect social goal areas such as education, health, and crime.

This separation of policy models according to social goal areas differs importantly from the approach of economic indicators and from much that has been proposed in the area of social indicators as well. When unemployment increases or the growth rate of the GNP declines, the remedies we often con-

---

search is centered on value concepts and policy alternatives rather than on nonmanipulable variables or the theoretical criteria of particular disciplines. Evaluation research is also concerned with the development of policy models, but has been largely devoted to a restricted range of alternatives that are closely related to preexisting programs (MacRae and Wilde 1979: Chap. 7). Models related to particular policies (Larkey and Sproull 1981: 240) rather than goal areas have the advantage of including consequences in multiple goal areas; but if policies shift more rapidly than goal areas, research communities organized to study them may have less continuity.

6. Social system models, or models combining social goal areas, would be useful for predicting side effects (Land 1975a: 18) from one goal area to another; but we propose to approach this problem through causal models of general end-values (Chaps. 5–7). Macroeconomic models of the causes of economic growth are, however, an example of both types at once.

sider first are macroeconomic policies that affect the economy as a whole rather than (say) technical education or productivity in particular industries. This overall approach has been imitated within the social indicators movement by a quest for models that will reveal the causes of national aggregates[7] through understanding of the social system as a whole.

Citing the interdependencies among the various goal areas as well as the example of macroeconomics, Land (1975b: 32) has argued that "to be minimally adequate to the types of tasks which Bauer and his colleagues intended for social indicators, we need entire equation systems capable of capturing the structure of institutional interdependencies of American society in the same way that econometric models have attempted to measure the structure of the American economy." The development of such models will be very difficult and will require extensive cooperative work. It may be attempted in two ways: by basing causal statements about aggregates on models of individual behavior, or by examining the relations among aggregate variables independently of individual data, as in macroeconomics.

If we wish to build up estimates of national aggregates from individual units, we might attempt to construct complex socioeconomic simulation models (Orcutt et al. 1976) involving processes such as family formation, fertility, education, employment, migration, and production. To aid this process we might follow the approach of social accounting, which seeks to build national accounts from elementary data on individuals. Like national economic accounting, this would be an ambitious enterprise involving systematic interrelation of large bodies of data. Two major approaches to social accounting are time-based (tracing individuals' use of time) and demographic (tracing individuals through various statuses over a period of years). While providing empirical grounding for the construction of social-system models, social accounts are not causal statements themselves (Land and Juster 1981: 9–10).

An alternative approach, that of "dynamic macro social indicator models" (Land and Felson 1976; Land 1979; Land and McMillen 1980), deals directly with aggregates in time series models that predict national social indicator statistics. A model of this sort "deals exclusively with determining the aggregate level of [an] end-product social indicator" (Land 1975b: 31).[8] Such an aggregate indicator statistic might be the national continuation rate of school pupils from the eleventh to the twelfth grade (Land 1975b: 29–31), or a national crime rate (Land and Felson 1976). One virtue of such a model is that it can attempt to deal with overall effects of policies rather than the localized effects of particular programs. It thus abbreviates the sequence discussed in

7. Land (1975a: 30–32) also includes individual-based ("micro") models among social indicator models.

8. Macro variables need not always be aggregates: examples are environmental conditions and the level of prices, which can also be contextual variables (Chap. 1) in models predicting value concepts for individuals.

Chapter 1, from problem definition to specific model to program monitoring to redefinition of the problem; rather, if applied to policy it goes directly from problem definition to modeling for a general population in terms of the same aggregate indicator statistic that was used to define the problem. An anticrime policy that reduced crime in one neighborhood by displacing it elsewhere, for example, might seem successful locally but would not appear successful if a model for the entire city were developed. If dynamic macro models, formulated for particular social goal areas, can be applied to units smaller than the nation, they can supplement our inferences from models based on individuals or on the target populations of particular programs.

Models of entire national social systems might be desirable if the difficulties of building them could be surmounted, but they also seem to require considerable simplification of a complex reality. A special opportunity for overall social system modeling may arise when a local social system such as a new town is being planned, or the social impact of a new industry on a locality is assessed. At the local level, we might also be better able to assess the effects of policies on particular groups by some degree of disaggregation of the model.

Instead of seeking national social system models, we propose simplification of the task by concentrating on the effects of particular policies in social goal areas. Models of these subsystems can probably be developed more rapidly. Land (1975b: 25, 27) also recognizes that social goals are centered about particular institutional structures and that social system models need to be classified in this way as well; he suggests as an example of a social indicator model, a representation of processes affecting educational outcomes, including policy and organizational inputs, nonmanipulable background characteristics, and side effects (1975a: 20). Such a model resembles the policy models we shall discuss in Chapter 4 for education, and need not be based on national aggregates. We may welcome knowledge of overarching connections among such particular models; but while waiting for it, we accept a degree of incrementalism by making policies within social goal areas. We can also look at relations among pairs of goal areas rather than seeking to model the entire social system.

Our factoring of the modeling process into social goal areas at this point and our acceptance of this degree of "disjointed incrementalism" (Braybrooke and Lindblom 1963: Chap. 5) imply, however, that the models in particular areas are liable to change. Not only because human groups can alter their perceptions of "objective" conditions (MacRae 1983), but also because of the effects on a particular goal area of overall social change and of policies in other goal areas, our models of particular goal areas will not be immutable. This possible "decay" requires that our knowledge be continually brought up to date if it is to be useful.

The Limitations of Policy Models with Exact Components

The fields that contribute to policy models differ in their apparent success. In policy analysis, the fields that are drawn on most often are the natural sciences, operations research, and economics. Contributions to models are sought much less often from political science, social psychology, or sociology. We must ask, then, why some fields seem more successful than others and whether this judgment is justified.

A policy model is a causal statement relating policies to valued consequences, and we are here especially concerned with those models that can be tested by scientific procedures. To be *successful* it must achieve acceptance and use outside the expert community that has formulated and tested it.[9] This acceptance may be accorded by policymakers, by the public, or by other groups of experts who assess its practical value; *use* means placing the model in operation through enactment of a policy.[10] We look for models that have succeeded in this sense in order to see what features are necessary in policy models that stand the test of practice. This is an exacting test of success, because it disallows numerous valuable techniques such as surveys, statistical analysis, program evaluation, and experimental methods, that can be used in developing models. It disallows "research utilization" that merely refers to or praises models, as well as the valuable finding that a proposed model is false. It also deemphasizes principles that are adapted to diverse local situations, in ways that make it difficult to say a single model has been used in all cases.

The relative attention given to the natural sciences, operations research, and economics in policy analysis suggests that the most successful models are

9. The success of a causal model does not depend on its being initially proposed by scientists, or the process embodying it being designed by them. Numerous social institutions that have evolved without clear invention or choice, such as family and bureaucracy, give rise to predictable consequences without having been designed by scientists. Some deliberately chosen institutions, such as the United States Constitution or the council-manager form of municipal government, have resulted from study and design by persons not certified in scientific disciplines. We are here concerned, however, with policy models originating in groups of experts.

10. We do not specify exactly who must accept and use a model in order for it to be considered successful, except that the users must be outside the scientific group that tested it. Nor do we require that the use be altogether for desirable purposes; it is only the practicality of the model that is our concern at this point. The acceptance and use of a model, moreover, may depend on somewhat irrelevant conditions such as the prestige accorded to a technical field of work, or on the noncontroversial character of the results, such as the accomplishment of goals together with budgetary savings. Apparent success may also result from symbolic aspects of perception and placebo affects. Possibly some of these effects can justify a perception of success (Krieger 1981a), especially if they affect subjective well-being. A procedure that involves interaction or negotiation with those affected may succeed, but the result is less likely to be seen as the success of a scientific model.

those that make use of the most exact fields of science. But in order to explore this hypothesis further, we must say what makes some scientific fields *exact*. These fields, and especially the natural sciences, are characterized by elegance of expression, capacity to predict the results of our actions, and durability of their findings. We shall refer to a field as exact to the extent that it possesses these qualities.[11] The contributions of exact fields are characteristically expressed in mathematical equations that are more precise than linear approximations. The statistical techniques of linear regression, involved in some of the models discussed in Chapter 4, are most often used for models that are less exact. Qualitative verbal expressions are usually still less exact, but can nevertheless be given relatively precise meanings that set them apart from everyday language.

Those scientific fields that we shall refer to as *less exact* tend to share several features that lead them to be less elegant and predictive and their findings less durable. Their subject matter is not easily amenable to manipulation; they involve complex systems; and the human participants in these systems can modify the meanings of aspects of the systems by altering their symbolic representations and perceptions of them. In addition, these fields often suffer from a lack of the precise measurement that would facilitate correction of errors in models.

We may imagine a rough continuum from physics, an example of an exact field; to astronomy, which is exact though its subject matter is nonmanipulable; to the theory of evolution, also nonmanipulable in subject matter, somewhat less exact, but with extensive empirical support; to experimental social psychology, in which subjects' interpretation of the situation is controlled by instructions (sometimes deceptive) given to subjects; and finally, to the study of "natural" social systems in which participants can collectively interpret events, including policies, and react to them.[12]

This contrast has led some in the field of policy analysis to regard the less exact fields as less worthy of attention. Together with parallel successes in basic research, it has also led some in the less exact areas of social science to emulate natural science[13]—seeking elegant, lasting generalizations through basic research at the expense of practical research. Yet we may question these

11. The contrast is also complicated by differences within disciplines. Demography, for example, often considered a subfield of sociology, is relatively exact. Several of these features of exactness are mentioned by Machlup (1961: 177–78). In classifying operations research and economics as exact we are emphasizing their mathematical formulation. An additional advantage of economics in policy analysis is that it provides a valuative calculus of benefits and costs.

12. This capacity to interpret and react goes hand in hand with a considerable overlap between the subject matter of commonsense discourse and social science discourse. In economic policy models some reactions of affected persons are included and anticipated, but they are not due to revised perceptions of policies.

13. We assume that some areas, less exact in the past because of their subject matter, may be studied in part by more exact methods.

inferences as they affect policy models. I shall contend, first, that the contributions of natural science cannot stand alone in the development of policy models, but must almost always be combined with less exact social components; and second, that the less exact areas of social science require somewhat different approaches from that of basic natural science if they are to contribute most fully.

We begin by examining models based on the natural sciences, from which the most successful policy models seem to come; yet viewed more closely, this success can be less convincing. For example, in the light of experience what seemed successful at one time may later appear less so. Our success in exploiting fossil fuels as sources of energy promises to be temporary, though it has satisfied many preferences while supplies seemed plentiful. The success of physicists in harnessing nuclear energy, though it attained a wartime goal and provides an alternative to fossil fuels, has been increasingly questioned. Simon (1982: 7) thus judges "that there is much less disparity than we have sometimes imagined between our abilities to produce results with technologies based on the natural sciences and the social sciences respectively."

The fact that a science can have widely recognized findings while the policy models that incorporate such findings are less clearly successful results from a central feature of policy models: their dependent variables are value concepts. It is always the degree of fulfillment of these value concepts, which may lie outside the conceptual schemes of science, that we must predict in order to serve policy choice. Policy models may contain components from the exact sciences, but these must be supplemented by components that link them to value concepts. As Morgan and McMichael (1981: 346) point out, an increase in sulfur dioxide in the atmosphere may have numerous effects: "crops raised in low-sulfate soil may show increased yields, paint may discolor, sunsets may become prettier, and people's health status may be affected." A still more complex set of social components is needed to assess the human effects of possible climatic changes related to increased carbon dioxide in the atmosphere, and of policy options for dealing with them (Schneider and Chen 1980). Many scientists would like to limit their responsibility to the more clearly defined task of prediction or explanation of value-free scientific concepts; but if scientists do not propose policy models, someone else must. It is policy models, not simply scientific models, that are relevant to the use of indicators and to the outcome values realized through our choice of policies.

Policy models are inescapably involved with less exact sources of information and judgment, not only in the realm of values, but in factual inferences as well.[14] These sources include the less exact sciences and intuitive judgments based on commonsense knowledge and experience. These types of judgment

14. The contrast between more and less exact methods in policy analysis is discussed in MacRae (1980a).

are involved because policy models link policies carried out by human beings to values experienced by other human beings; because they are often complex and, in their most useful form, very specific; and because the systematic knowledge that enters into them is always incomplete (Lindblom 1979: 518).

A special advantage of applied physical science is its capacity to construct and manipulate artificial objects. Policies that make use of these sciences deal with objects (such as automobiles, CAT scanners, or machinery for production) that can be tested in controlled environments and improved in design. These objects are closed systems, the properties of which we can understand not only through the principles used in design, but also through manipulating them in whole or in part. When these systems are used, however, human beings must usually set them in action, read their meters or displays, and control their operation. For this reason (over and above "errors of design" if they are indeed distinct from human errors), we observe some failures of "well-engineered" technical devices such as nuclear power plants, and automobile and aircraft accidents.

We might imagine that the construction of a bridge (often financed by government) would involve one of the most exact and best-understood policy models. Occasionally the technical knowledge involved fails to anticipate some hazard to a bridge, such as the wind-induced oscillations that caused the collapse of the Tacoma Narrows bridge in 1940. But more important are the human effects involved in carrying out the policy and in realizing its values. The construction of a bridge may be assigned by a mayor to a contractor who is his friend, and whose concern is more with profit than with solid construction; thus the ideal conditions of the engineering portion of the model may not be implemented fully at the start. Eventually, the conditions needed for maintenance and repair of the bridge may be slighted for political reasons, as numerous aspects of urban infrastructure in the United States have been, and the predicted useful life of the bridge may not be realized. Finally, the demand for the bridge may change as areas of residence and work change or alternative means of transportation come to compete with it; it may suffer from disuse or congestion. The physical properties of the bridge may change; or it may retain its physical properties while the human values realized from it decline. The policy model should include all these aspects and not merely the engineering technology. Such a model would relate the initial decision to build the bridge, made perhaps by the city council, to the final human values realized from the bridge.[15] These values might have earlier been measured as problem definers:

15. Engineers do consider these human factors, sometimes drawing on social science. Some of their judgments of causation result from experience rather than social research; but because these are not drawn from their central area of expertise and responsibility, such judgments may not be tested rigorously. They may still compete successfully with social science models, however; specific knowledge can be as useful as general, and deterministic models may be inappropriate for personal relations.

in the social goal area of transportation, by indicators of travel time; or at a more contributory level, by administrative indicators of congestion or of the bridge's current state of repair.

Certain components of policy models are also made less exact by the time constraints on decisions; such models must often be used rapidly when we cannot wait to improve them. Many of these components involve special knowledge related to the problem at hand, which is not continuously available from a research community. The knowledge required may either be too specific to command the continuing interest of this community, or be subject to decay in accuracy and in relevance over time. Thus special ad hoc research is often required for policy recommendations. The projection of future traffic across a particular proposed bridge, for example, might not be part of the general stock of knowledge of economics or regional science. Given time, planners or analysts could develop projections that were more reliable than the initial rough estimates. If such problems recur, a community of experts might develop a repertoire of solutions that could speed their response to a new problem of a known type (Chaps. 4, 10).

Models based on applied physical science can also encounter limitations from sheer complexity—the combination or concatenation of the errors of particular components. Even the artificial objects mentioned above have this property, but their parts can be examined and improved separately (Simon 1969: 21). When we deal with complex systems that cannot be taken apart and modified, such as the atmosphere or the economy, these difficulties are exacerbated. The time horizon over which predictions are possible is a measure of the adequacy of models of these systems (Ascher 1981: 258). An assessment of weather prediction suggests that "detailed forecasts for more than a few days are currently beyond the capacity of the National Weather Service" (Lutgens 1979: 259–60). Stephen H. Schneider (in a letter to the author, 1982) judges that approximate weather forecasts can be made by physical models for periods of 6–10 days, and that "there is a theoretical limit to weather predictability (2–4 weeks)," which researchers are still seeking to attain (Kerr 1983).

In applied biological science we also encounter complex systems, and the development of policy models in this area often requires carefully designed randomized experiments rather than controlled laboratory experiments. If the number of variables to be controlled is large and unknown, and treatment effects do not dominate the effects of other variables (Gilbert et al. 1975: 138), rather than trying to manipulate each relevant variable we protect ourselves against their variation by randomizing. Thus statistical procedures for social experiments have been based in considerable measure on the earlier use of randomized plots in agricultural experiments.

Similarly, in testing drugs both efficacy and side effects need to be examined in statistical terms. The proposal of new drugs is in part a matter of chemical analogy, and their detailed effects require trials first on animals similar to humans and then, where justifiable, in clinical trials on humans. Successive experiments of this kind may lead to more reliable estimates of effects (Gilbert et al. 1975). Moreover, the part played by our wishes and expectations as they affect our actual treatments and our observations of results seems greater than for policies that use physical science technology; hence we sometimes insist that such experiments be double-blind.

Models Based on Simpler Aspects of Human Behavior

Policy models that extend from human actions to the realization of human values must presumably include statements about human action and reaction. Some aspects of human behavior may, however, be treated in more exact terms than others; and if these models are appropriate to our purposes, the enterprise of model building may be simpler.

Perhaps the simplest example of treating a human being in terms of an exact science would concern that person as a falling body. When falling from a height to the ground the body would obey Newton's laws, including the effect of air resistance. The corresponding prediction of time of fall and velocity at impact would be accurate but would not depend on whether the person was alive or dead at the start. The person's subjective well-being might differ greatly depending on whether he were falling from ten feet above the earth feet or head first, but the conditions of fall would be much the same. From this purely physical perspective we are dealing not with human behavior or action (once the fall begins), but only with an occurrence affecting a human body.[16]

This is not the only instance, however, in which human beings' action or behavior, or what happens to them, is reasonably predictable. Numerous applications of operations research involve humans, but in roles or capacities that permit predictability. Queuing theory, for example, applies equally to humans and to nonhuman objects as long as they arrive at a queue with a known probability distribution over time and remain in the queue until dealt with; in either case we can estimate the expected waiting time. Administrative indicators of waiting time or queue length can be combined with other information, such as the estimated cost of waiting, to define the relevant problem. We customarily treat the social cost of waiting in terms of a monetary cost per unit of waiting time.

Models based on queuing theory can appear less successful, however, if

16. Dunn (1974: 65–66) contends that the imitation of physical science has restricted social science to areas "where the issues of human purpose can be treated as *a priori* forms arising outside science, . . . where social entities and their transactions are somewhat stable in time," and to those areas where conflict can be resolved by markets, rules, or games.

some aspects of human valuation and action are introduced. Our customary assessment of the value of waiting time is made in terms of a figure commensurable with an average wage rate—a certain number of dollars per hour. If, however, we were to treat different people differently according to their wage rates, we might move a manager to the head of the line past the manual workers who were waiting. Differential treatment of this sort might lead to resentment and to redefinition of the situation so as to alter the process being modeled.

The perceived success of a policy model depends not only on its accuracy of prediction, but on participants' and citizens' agreement with the values it serves. Thus even if a queuing model were accurate in predicting the different costs of waiting to different income groups, observers might disagree on the desirability of treating these groups differently. The model may be seen as successful when it can be used in noncontroversial ways, or when the responsibility for avoiding controversial uses can be placed on the client rather than the expert consultant.

The validity of a simple queuing model can thus be modified by social behavior—by the interaction of people in the queue with one another. The existence of a queue depends on social norms that differ somewhat among nations and occasions. Even setting aside the possibility of changes in arrival probabilities, we may find that persons in a queue behave differently after communicating with one another. The queues for registration at the University of California may have contributed to the students' protest there in the 1960s, in which they asked not to be treated as mere IBM cards. Members of a queue may also consult with one another, seek information, decide that the queue is not moving, and leave it. Norms may develop in a long queue, such that one place can be saved by a group of people who take turns holding it, and thus reduce the effective cost to each one (Mann 1969).[17]

Humans also drive cars, and their driving behavior enters into traffic models. Parameters such as traffic density and distribution of destinations can be estimated from past experience (a narrowly empirical model) or from the distribution of locations of residences and workplaces together with the availability of alternative modes of transportation (a more general planning model). The corresponding models provide estimates of the large scale properties of transportation systems (also measured by indicator statistics), and of the effects on them of policies such as the pattern of streets and their connections. The simple model of motivation used as a component in traffic models requires only that we postulate a desire to go from one place to another, perhaps with some consideration of cost, whatever the intended activity at the destination may be.

17. The real cost to individuals depends not simply on total waiting time, but also on its scheduling, on the length of waiting intervals (if cost is a nonlinear function of interval length) and on activities carried on while waiting.

Both of these types of models—of queuing and driving—make use of statements of human motives that are simple and manageable. One cause of their simplicity can be the existence of social norms, inculcated for drivers by training, and enforced by licensing and police action. The models seem to work best if the participants do not communicate with one another so as to redefine the situation for themselves. Communication can, however, reinforce the expected motivation through social control. The models also seem to work best if the participants are seeking goals of their own—service or transportation—that are not forced on them by public policy, and thus have less reason to redefine their goals in protest than if they were coerced. Under these conditions, queuing or traffic models are not complicated by participants' elaborate conceptual structures or group formation—though conceivably such complications might occur and invalidate the models.

These models are also subject to trial and modification. We may alter the number of queues and the type of service, or the traffic regulations and street patterns, in order to examine their actual effects. Thus although humans are involved, we can single out some aspects of their behavior that can be treated more simply than others, and in which the conditions surrounding that behavior can be changed in successive trials.[18]

There are also successful models of human interaction that are formulated in relatively simple terms of antagonism or exchange. A model that was extremely successful in producing major changes in U.S. strategic policy was developed at the Rand Corporation in the 1950s to deal with the question of where strategic aircraft should be based. Up to that time, bases had been assumed to be distributed throughout the world near potential theaters of operation. A team led by Albert Wohlstetter, given the task of analyzing optimal base location, radically altered the definition of the problem. Considering the possibility of war initiated by the enemy, they made the key assumption that in case of war the enemy would seek to win by attacking our forces rather than our cities.[19] From this postulate they concluded that bases far from the United States and near the enemy were especially vulnerable to attack. This led to the doctrine that our forces should be prepared to deliver a second strike—that even after an enemy attack they should be able to survive and counterattack. This posture was expected to deter the enemy from making a first strike. This

18. Models of this sort need not be imposed on participants unilaterally as in the common notion of social engineering. Citizens who find long waiting lines at the post office, after a budget reduction, or additional traffic in a residential area due to a rerouting aimed at reducing travel time, can protest in the same way as if the conditions had not been predicted by technical models.

19. The actor here is a nation, or its armed forces, rather than an individual. Whether this type of "rational actor" model is adequate has been analyzed by Allison (1971). For accounts of this analysis see Wohlstetter et al. (1954) and Dickson (1971: 59–60). The recognized success of this model may have resulted in part from its substantial cost savings.

model was conceptually simple, and quite influential after numerous briefings persuaded Air Force officials of its correctness. Wohlstetter (1968: 26–27) has argued, however, that detailed strategic models are not likely to last more than fifteen years because in such "opposed systems" there is a strong motivation for each side to try to change the technical parameters of the contest.

There is thus a class of models of human behavior that can be expressed in relatively exact terms. In the examples we have given, the parameters of these models may vary from case to case and over time. For models of traffic and international strategy, technological parameters may change; the models are thus mathematical frameworks that can be fitted to various particular situations but that do not give rise to universal empirical constants like those of gravitation or the speed of light. If we were to include psychological models of simple perceptions and responses, we might find more fixed parameters related to human physiology.

These models will be overrepresented in technical policy analysis relative to those produced by the less exact social sciences, if only because their precise formulation allows them to compete more successfully against the verbal models supplied by common sense.[20] To succeed in this competition, however, they must also predict behavior more successfully than these verbal models. The most successful of these models assume rational maximization; and among these, the most prominent are those of microeconomics.

Economic Models

The discipline of economics has succeeded in formulating elegant models of complex systems of human interaction by assuming maximization on the part of participants within a given institutional framework. Thus the general microeconomic model of a market can tell us some of the long-term effects of controlling the price of a commodity or a factor of production. It predicts that rent controls will lead to lesser investment in the production and maintenance of rental housing, and that a minimum wage law will lead producers to seek automation or cheaper foreign labor in place of domestic labor. In either case the model shows that a policy of regulation will

20. "Common sense" here includes not only citizens' policy models, but also the models of behavior used by professions such as law and engineering. These professions deal with human behavior and action, but do not necessarily attempt to formulate the same sorts of generalizations about them as do social scientists.

Models based on simpler aspects of behavior, including some discussed in Chapter 4, imply that the analyst occupies a separate and superior role in relation to the persons studied. In many instances, however, the persons studied may also act as citizens and can thus influence the policy choices that are made.

decrease the equilibrium efficiency of an otherwise perfect market in allocating factors of production, and will induce reactions by those affected that run counter to the initial goal of the policy—provision of low cost housing or adequately paying jobs.

The actual application of these models is complicated, however, by the presence of values other than that of efficient allocation of productive resources. Tenants and wage earners are concerned in the short run with their living conditions based on rent and wages, and regulation of these payments will give them short-run benefits that are not counted when we consider only long-run equilibria. Considerations of equity between landlords and tenants, or between employers and employees, lead to concerns that one party will benefit from deregulation while the other suffers. Thus, as we have seen, a model that successfully predicts one value concept (long-run allocative efficiency) may not tell us enough about others.

With respect to factual "exactness," microeconomic models attempt a more ambitious task than that of the queuing or traffic models we have described. They deal with a complex system in which humans interact, not simply by heading toward a single goal, but by adjusting numerous decisions about production and exchange. Within this system, however, the motivational assumptions used are simple and limited.

The field of macroeconomics has also contributed significantly to policy modeling. The resulting models are more complex in form than those of microeconomics; make less use of the assumption of rationality; and when used in forecasting, involve more empirically estimated parameters. A particularly interesting class of policy models are thus the macroeconomic simulation models that have come to be used for forecasting general economic conditions such as business activity, employment, and price levels. These models are valued "by businessmen for detailed forecasts relating to individual industries and markets" (Zarnowitz 1980), and can be used by government for adapting to anticipated changes as well as for predicting the effects of possible policies.

Since the 1930s economists have developed models of increasing complexity for the simulation of a national economy. A functioning computer model of this sort can involve over a thousand individual equations. The prediction of aggregate variables for the economy at a given time depends, in the model, on both prior (exogenous) variables and simultaneous reciprocal relations among current values of variables. The solution of such a model can involve subdivision of the larger model into subsystems or "blocks" that are solved in sequence: first those blocks dependent only on exogenous variables and then those dependent on the first. Instances of mutual dependence of variables (nonrecursiveness) and of nonlinearity are dealt with by iteration (Fromm and Taubman 1968: 18–19; Klein 1975: 22–23). The components of such a model are drawn from macroeconomic theory: for example, components of personal and disposable income; financial market variables; prices, production, employment, and the labor force; and demand behavior.

The predictive accuracy of these models seems to have made them useful only for an interval of about two years. Kuh, summarizing a stochastic simulation of nine years of earlier economic data (1953–62) by Nagar (1969) using a Brookings model, writes: "Tentatively, I would conclude that this model can serve as a valid rough approximation of the real economy for a two-year period, perhaps for a complete [business] cycle, but that mis-specification plus 'true' noise make it a not totally reliable tool for cyclical simulations extending beyond that period" (Kuh 1969: 5). Christ (1976: 333) compares a number of policy simulations, where various models of the U.S. economy are run with and without a particular policy change, by noting that "after two years, the agreement [among the models] begins to evaporate."

It is difficult to say just what we should expect of economic models in the nonexperimental, real-world conditions of policy intervention. Zarnowitz (1980), writing of the prestige accorded to econometric models in recent years, notes that "objective analytical methods have come to play a major role in economic forecasting. Informed judgment is as important as ever, but the purely subjective elements are reduced." He goes on to note, however, that "forecasts of econometric model builders often have been no more accurate than the forecasts of those who analyze business conditions using less formal methods." Ascher (1981: 249) goes further, in summarizing an ongoing survey of such models, to state that "econometric models have not, and still do not, forecast quite as well as the judgmental approach relying on no explicit routines." Burns (1966: 59) observed earlier that "the 'new economics' provides no assurance of continuing prosperity. . . . What history discloses is a succession of business cycles no two of which have ever been alike." Ascher (1981: 258, 260) notes that the continual development of large-scale econometric models has not led to any improvement in their accuracy; perhaps their context, especially that provided by relevant governmental policies, is changing in ways that are hard to predict. The subject matter may be limited in its exactness even if the models are elegantly expressed.

As Zarnowitz (1980) points out: "In econometrics itself, the process of building larger and more complex models of national economies seems to have been pressed forward too far in recent years. Certainly, it left behind efforts to test each model's analytical foundations and evaluate its performance." A simulation model that combines many individual relationships may cumulate their errors, and its overall performance cannot tell us which relationships need correction. Such a combination may be useful, however, as a policy model, and tests of prediction and postdiction can suggest its future degree of accuracy; indeed, policy-related simulation models can be useful in domains where their theoretical components are less exact than those of macroeconomics (Chap. 4). Fromm and Taubman (1968), for example, used such a macroeconomic simulation model to analyze the expected effects of excise tax reductions on variables such as GNP and income tax receipts. Klein (1969) subsequently analyzed effects of the tax cut of 1964, using two alternative econometric models. Nagar (1969) included in his stochastic simulation estimates of the expected effect of a reduction in income taxes.[21] More recently, econometric models have been used to guide more complex decisions, as in French national planning (Gauron et al. 1982). Nevertheless, Hargrove and Morley (1984) judge that economists' hopes "to give policymakers the ability to fine tune the economy" with macroeconomic models "have not been realized."

Economics resembles the natural sciences in viewing the individual's action as determined by the "objective" characteristics of his environment, unmediated by variations in perception or symbolism (Rosenberg 1980). In microeconomics the individual is characterized by a set of preferences—usually fixed for the period under analysis—and by endowments such as "human capital"; but these individual differences do not introduce the problem of new collective perceptions of the same objective world, or of the formation of social norms (Stinchcombe and Wendt 1975: 66–73; MacRae 1983) through collective development of new categories and goals. Similarly, the predictor variables of macroeconomic models are usually objective aggregates.[22]

We shall distinguish economics from those social sciences that explicitly include these subjective variables (perceptions and norms) in their models;[23] the latter sciences, which we shall call *subjectively social,* include much of sociology, social psychology, social anthropology, and political science. Thus economics tends to rule out, by its mode of formulation, some of the most important sources of indeterminacy and change in the models entertained by other social sciences. A deliberate reluctance to investigate the empirical details of individual human choices makes the economic paradigm more nearly self contained. As a result, economic models are more elegant in form than most models in other social sciences; the discipline's attention is drawn to

21. There is an important difference, as we have noted, between using models to forecast the state of the economy in order to adapt to it, and predicting the difference that a proposed policy will make. In the latter case we need not know the exact values of the outcome variables. The differences made by nonpolicy variables that simply introduce an additive effect can be ignored, as they do not affect the value added by a new policy.

22. Models can be constructed that use subjective variables, such as consumers' expectations, to predict economic actions and conditions (Adams and Duggal 1976); but these variables are more often taken to be endogenous and omitted rather than measured. Changes in preferences, which might also be an exogenous subjective variable in microeconomics, have been subsumed by Stigler and Becker (1977) under "consumption capital."

23. Similar to economics in this respect are operations research, demography, and human ecology, as well as some applications of economic approaches to topics in sociology and political science (Becker 1976).

problems that do not involve major changes in social perception of the world; but some errors of prediction occur because of this omission.

The dependent as well as the independent variables of economics are usually objective rather than subjective; utility is rarely measured directly, but is taken to be a function of variables such as prices and quantities of goods. When we are interested in well-being as an outcome, however, we sometimes wish to express it in other terms. Thus Garn et al. (1976: 6–7) contend that the assumptions of "the core model of economics" require modification when we are studying the effects of service delivery on human welfare. "This is not due solely to the [fact] that they are frequently unrealistic empirically. The customary use of such assumptions is to derive analytic solutions which have the appearance of greater precision than is justified, given the wide range of probable influences on welfare outcomes." These authors stress the continual interrelation of production and consumption in policy areas such as housing and education. They go beyond the economic approach to household production by emphasizing that some of the resources acquired and used by individuals are normative and symbolic rather than tangible "goods." Especially in the production of services, the person being served "is inevitably part of the production process" (1976: 14), and "the output is always a jointly produced output" (1976: 15). Routinized patterns of service delivery may be counterproductive, as they can ignore needed communication with the client. Thus in addition to assessing asset bundles and preferences we must assess user *perceptions* and the diversity of their consumption technologies (1976: 8–9, 14, 15, 19).

We group economic models here with those based on natural science, not because they are equally accurate in prediction, but because they share with natural science models a degree of formal development, public confidence, and policy use that has few parallels among policy models in the other social sciences.[24] The success of economics does not rest entirely on demonstrations by intervention such as natural science provides, but is judged by other criteria as well. The values that the discipline serves, such as preference satisfaction, freedom, competition, and production, have substantial support. The expertise of economists has been recognized and they have been employed widely in government and industry; in the private sector, their success has been tested by the marketplace. As for their advice to government, we do not always know just what would have happened to the economy in the absence of economic advice—or even with advice from another school of economists under the same conditions. My classification of economics as a relatively exact science thus rests on a style of theorizing as much as on proven effects.

24. Skinnerian models of psychological reinforcement have been used extensively in practice, but few of these uses work through public policy.

The Use of Successive Trials: The Fairweather Lodge

We have dealt so far with policy models that are regarded as successful because they make use of natural science or because they combine mathematics with precise and objective characterizations of human behavior. In spite of their desirable qualities, however, these models face intrinsic limits because of the human inputs and values that policy models necessarily involve. They are also limited by complexity and cumulation of error.

Some policy models are successful, however, not simply because of the exact or objective character of their components but because they have been tested and refined repeatedly through successive trials.[25] Such testing is important for models (and devices) based on natural science; but it can also be used with models based on less exact social science, as we shall show. After describing one such example, we shall then examine the more general conditions necessary for the use of successive trials.

We have suggested that the ability of human groups to reconceptualize their situations collectively rather than simply responding in a predictable way to their objective environment, hinders our development of exact models of their behavior—and therefore presumably of successful ones. We have taken this feature to define the "subjectively social" sciences, which we expect to be less exact. Yet in a series of randomized experiments[26] reported by Fairweather (1980) and by Fairweather and Tornatzky (1977), a successful method for promoting the self-care of groups of former chronic mental patients has been developed through repeated trials. The method in this case involves developing a certain form of human organization, and the corresponding causal model has a general part (statements that a certain type of human interaction will produce desired results) and a specific part (that the embodiment of the model in an organization with specific procedures of formation, and specific rules, will produce desired results in particular instances).[27]

These authors describe a series of investigations and experiments, conducted over a period of more than two decades, aimed at reducing the chronic hospitalization of mental patients. After assessing the problem, they first experimentally compared several existing forms of therapy over an eighteen

25. We refer here to what Deutsch (1963: 71–79) has called "learning" in contrast to "goal-seeking," which applies a given model repeatedly so as to adjust results toward desired values. Meehan (1982: 110) points out that even weak theories can be useful if they are corrigible and thus are improved.

26. We shall subsequently use the term "experiments" to refer to social interventions using randomized treatment and control groups.

27. In this case successive trials seem to have converged on a fixed institutional design. But in complex subjectively social systems, the successive improvements may not lead to a permanently fixed model or policy but may require continuing adjustment to changing conditions and perceptions, over and above changes due to scientific advances.

month period and found that they were ineffective in producing community adjustment outside the hospital. The research group then turned to innovative approaches centered about the idea that "if small cohesive groups of mental patients could be developed in the hospital they might function to maintain each other in the community and thus lead to a reduction in recidivism and an improved adjustment in the community" (Fairweather and Tornatzky 1977: 34–35). They concluded that "in the community new social situations needed to be created so that people previously hospitalized could live and work and at the same time be protected from the negative reactions of those who considered their behavior deviant" (Fairweather 1980: 5).

An experiment with this new type of intervention, carefully designed and administratively coordinated, showed that in hospital work situations, groups of chronic mental patients who were unsupervised developed more cohesion than groups who were staff supervised. Next, a five year experiment examined the effects of organizing a hospital ward "along small-group principles in which the patients made the major decisions about their lives in the hospital and for the future." This experiment showed that the innovative ward, in comparison with traditional treatment, "enhanced all aspects of within-hospital adjustment." When these patients were separated from their groups and returned to the community individually, however, the effect disappeared and rates of return to the hospital became similar to those of the control patients. The need for group support was brought to researchers' attention by follow-up interviews with the patients; thus patients contributed in a way to the reformulation of the model (Fairweather and Tornatzky 1977: 36–37).

A new institution external to the hospital, the Community Lodge (later known as the Fairweather Lodge), was then established and compared experimentally with the external use of traditional community services. This experiment, again lasting five years, was carefully planned from an administrative standpoint and, as before, randomized. Because "the immediate move to the community as a group was chaotic" (Fairweather and Tornatzky 1977: 38), the in-hospital preparatory training procedure was redesigned so as to make the patient groups more independent of the hospital social structure. Among the provisions of the Lodge was the organization of economic enterprises to be conducted in the community, similar to that of a sheltered workshop. Jobs such as janitorial service and gardening were undertaken with help at the start from staff and later from temporary outside consultants. Work crews were organized so as to consist of three or more men so that patients with more difficulty could be helped by their coworkers. "The Lodge did not become self-sufficient until its fourth year of operation" (Fairweather 1980: 29–30).

The results, in terms of selected administrative indicators, were then higher employment rates, lower rates of return by patients to the hospital, and lower cost for the Lodge than for the control group. The intervention was successfully replicated twice in quasi-experiments.

Theoretical notions from social science were centrally involved even though they were not of the exact sort. Fairweather and Tornatzky (1977: 38–39) concluded that "the major variable manipulated in this study, of course, was the role and status structure that historically exists in mental health organizations between patients and mental health workers. Rather than mental health professionals supervising patients' behaviors, they helped them attain and sharpen their own decision-making processes and thus participated in liberating rather than controlling them." Fairweather (1980: 5–6) reports a study of the literature on small groups in social psychology and sociology made prior to the planning of the Lodge. This literature "suggested that small groups might be able to establish norms that were quite different from the typical societal norms and that such groups might also provide roles within their organizations in which behaviors that were socially deviant in the larger society might be accepted."

The institution created in this way has not only been a result of successive trials; internally it has also exemplified a process of successive trials by its members. "Through constant trial and error, organizational change, and role change, members began taking more responsibility for governing themselves" (Fairweather 1980: 20).

The analysis underlying this program was based at several points on careful choice of administrative indicators. By concentrating on patients' adjustment in the community rather than on psychiatric symptoms or test scores, the investigators addressed more directly the question of patients' well-being in relation to various policies. At the same time the investigators considered the cost of various possible interventions. Moreover, their attention to group cohesion made use of a key contributory variable that we shall discuss in Chapter 8. Some of these indicators were related to values that were of concern to the cities where Lodges were started, and thus might have been used for problem definition, but citizens of these cities often learned about the working of the Lodge from nonstatistical sources.

From the reports of this work we conclude that it is possible to develop successful causal models in the subjectively social sciences. The program, designed initially on the basis of verbal causal models and refined through trials, seems to work reliably. By 1978, several dozen Lodges had been established throughout the United States (Fergus 1980). The demonstration of workability of such a model gives additional significance to indicator statistics on mental health, by adding to the policies available for dealing with it; and it points to particular administrative indicators that can be used for monitoring the success of our efforts.

Before we take this policy model as representative of what could be done in other fields, however, we should ask whether Fairweather and his colleagues faced particular advantages or hindrances in developing it. Their continuing efforts have taken place in a changing national environment as regards policy priorities and research funding. In the 1950s an interest in evaluation of various existing modes of therapy created an opportunity for the initial research that showed a disappearance of post-hospital effects after eighteen months. Even though research outside the hospital then seemed desirable, "the interest in community treatment was very low at that time." Consequently, the next step was to study effects of patient organization within the hospital. By 1964 the Lodge concept was ready for trial, and opinion was more favorable to community treatment; conceivably this general change in opinion was related to indicator statistics on numbers of chronic patients in mental hospitals. Care was nevertheless required in the choice of a location for the Lodge in view of possible neighborhood reactions (Fairweather 1980: 4, 6, 26).

Dissemination of this innovation was by no means automatic even after the positive results were first reported. Rather, systematic dissemination procedures had to be devised with the aid of further randomized experiments. These problems may be general ones for many social innovations; but in this case they were exacerbated by the need for role changes on the part of health professionals. Larch (1980: 60) reports, for example, that nurses had to learn to omit "tender loving care" as a central part of their professional role. Kiepke (1980: 68), an eventual adopter, reports an initial doubt whether "these people could really live on the outside without a live-in supervisor." She acknowledges having been "proven very, very, wrong"; but these attitudes, related to professional authority and responsibility and patient dependence, are widespread.

Conditions for Improving Models by Successive Trials

The success of this series of experiments is consistent with the results of a review of social and medical experiments by Gilbert et al. (1975). These authors review a large number of randomized and nonrandomized trials, and recommend randomized field trials for assessing complex social innovations. They point out that "we often do not know which is the best of several policies," and that in spite of "economic reasoning, sophisticated analysis, sample surveys, and observational studies . . . we still will not know how things will work in practice until we try them *in practice*" (46). From their review they conclude that large effects are rare even from well-designed programs; we require carefully designed experiments to estimate these effects, since poor designs or lack of randomization make estimation unreliable. They suggest that we should then expect "evolutionary improvements" (40) from the understanding and cumulation of small positive effects that come from particular trials.

The work of Fairweather and Tornatzky and the review by Gilbert et al.

provide good examples of the randomized field experiment (i.e., an experiment in conditions of practice) and its repeated use. These experiments represent the most rigorous form that successive trials may take.[28] In actual investigation, however, careful use of quasi-experimental and nonexperimental methods can be valuable (Cronbach 1982: Chap. 2). Moreover, in practice we conduct a wide variety of types of successive trials, ranging from randomized experiments to the incremental testing of types of policies through repeated policymaking. Although Gilbert et al. (1975: 150) consider the latter types of trials, without systematic controls, to be of little value ("fooling around with people" rather than experimenting), such trials are widely used in policy processes. We shall therefore examine them in more general terms.

To see more clearly when and how successive trials can succeed, we need to examine the various ways in which such trials can be conducted. The diagram in Figure 3–1 will allow us to identify some general features of the process. We start at the left and follow the arrows (which indicate sequence rather than causation) from block to block.

First the process in which we are interested (our own or others' activity) leads to the design and conduct of a trial (Simon 1969: 69–73); in Fairweather and Tornatzky's work, this might be a trial of the Community Lodge or of one of its precursors. This trial gives rise to results such that the model can be tested[29] by application of a selection criterion, which in their case involved patients' rate of return to the hospital, supplemented by the degree of organization or "chaos" in the Lodge, employment rates, and cost. Conceivably such an application of selection criteria would have led to the conclusion that the approach was unworkable and should be discontinued (upper right arrow); but in spite of initial failures, these authors did not reach this conclusion. A trial might have been judged promising but not fully satisfactory (center right arrow), in which case the process would lead us back to the left (lower continuation of center right arrow) to modify the model and design another trial. This process might continue until the selection criterion led to the conclusion that the model was satisfactory (lower right arrow), in which case it would be continued without change; dissemination might then be tried, and success in a broader sense would depend on its acceptance. Changes in the context (types of mental illness, public regulations, community attitudes) might nevertheless later change the results when the selection criterion was again applied, leading to further cycles.

28. Randomization eliminates many sources of interval invalidity of our inferences; but because it cannot usually be performed on representative samples of human populations, it is less effective in removing external invalidity (Langbein 1980). The requirement that an innovation be eventually tested in conditions of practice is aimed, however, at some degree of external validity.

29. When repeated trials of a given model give rise to statistical variation, a replicated set of trials may be necessary before a reliable assessment of the model can be made. The cycle of Fig. 3–1 then enters only after such an assessment has been made.

Fig. 3-1. Cyclical Procedure for Improving a Causal Model

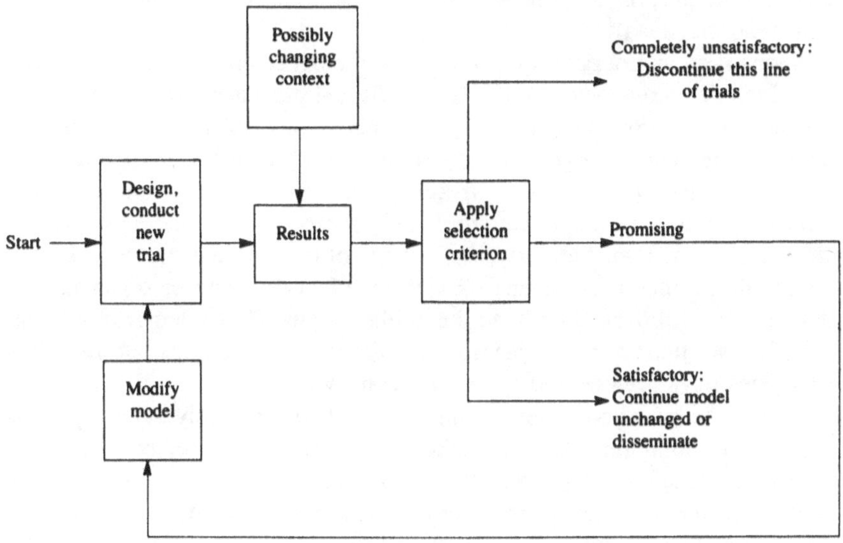

The diagram in Fig. 3–1 can be applied, however, to a much wider range of types of trials. We sometimes refer to such a succession of trials and improvements as evolutionary, reminding us of the process of evolution of species. In organic evolution new trials are random mutations, and the criterion is survival. We also often compare governmental policy design with the processes of a competitive market or a marketlike competitive situation; in the market the source of new trials is the entrepreneur's models, often aided by applied science and management science; and the motivation, his desire for profit. In policy analysis (including Fairweather and Tornatzky's work) the source of new trials is the analyst's technical and theoretical knowledge (aided by intuition), and the criteria are often indicator statistics such as rates of patients' return to the hospital or their employment rates. Finally, in the case of actual policy decisions, the source of new trials is the political process with its changing focus of attention and the necessity for interest aggregation to form majorities. The criterion of selection is then political approval, which combines political and technical aspects.

Each stage in the diagram can be used to illustrate conditions necessary for the successful application of this broader selection process.

First, new trials must be produced. In the evolution of species, this requires the production of progeny with mutations that are the source of innovation, and the propagation of these mutations in the gene pool. In a competitive market, new trials require freedom of entry by new firms. In policy-related social research, new trials require a willingness to innovate, or tolerance of innovation, on the part of program designers, decision makers, and influential persons who may be affected by the results.

Second, an appropriate selection criterion must be chosen; such a criterion may include an outcome variable and statistical procedures for making judgments about it. From the perspective of public policy, such a criterion variable should be related to the general welfare or equity. The mere survival of a firm or a productive process in a competitive economy need not imply a positive contribution to the general welfare, if only because of external effects not reflected in the market. Similarly, the survival of policies in democratic and bureaucratic politics may be more a sign of effective group or organizational action than of proven benefits to the public in general. To demonstrate those benefits, we must not only measure the appropriate variables but also show that gains have been caused by use of the model.

Third, the selection criterion must be applied repeatedly. The trial must generate information—in Fairweather and Tornatzky's case, on rates of return to the hospital, employment rates, and costs. To generate this information, someone must devote resources to gathering it and ensuring its quality. Managers of government programs often give low priority to this sort of information, and subordinate it to the evaluation of their own programs in their own terms or to the quest for constituency support. We must strive to maintain the use of well-measured criteria related to the general welfare while recognizing the part played by democratic processes. When experts do the testing, they must often continue this work over long periods, and the final results may not be known for years; the work cannot easily be done in doctoral dissertations or by junior faculty members seeking tenure.[30]

Fourth, if the cycle is to be continued, the designer must be able to modify the model so as to improve it. Fairweather and Tornatzky made use of observation and interviews, in addition to measurement of output criteria, to see where early versions of the model were inadequate. Both general theory and practical experience can play a part at this stage. In the evaluation of programs, this part of the cycle resembles "formative evaluation" (Scriven 1967), but it may be unduly limiting to screen modifications for acceptability to an existing organization. The new trial need not maintain the existing program. If the managers and members of the organization that implements the program are committed to it, and are not merely conducting an experiment, they may resist changes and thus constitute an important part of the political environment.

30. Even tenured faculty members often lack Fairweather's steadiness of purpose; sometimes an organization rather than an individual is required to provide continuity (Riecken and Boruch 1974: Chap. 7). Other locations than a university may be preferable as bases for long-term social policy experimentation (Fairweather and Tornatzky 1977: 384–93).

We must also consider the benefits and costs of the entire cyclical process. The generation of an improved model is a potential source of benefit, but there are also costs associated with each cycle and with the changes that it brings. In the evolution of species, the vast majority of the new organisms created by mutation die; we do not usually think of this as a cost. The survival of business firms in a competitive market has also been likened to evolutionary survival. New firms are not produced at random, however; the design of new businesses, even though many fail, is guided by entrepreneurs' models of causation. From the valuative perspective of the general welfare, their calculations reduce the aggregate social costs of failures. The net social benefit from innovation and competition derives not only from the eventual survival of efficient firms that can meet new conditions,[31] but also from the human cost when firms fail. The contraction or failure of firms produces short-run costs for their employees in the form of unemployment, job search, and retraining. This cost can be reduced if existing firms can adapt to a new market environment while retaining their employees, or if the founders of new firms can calculate the conditions for survival accurately.

Trials can therefore be costly. Their costs can be reduced by reducing the number of trials required to reach a given goal; this in turn requires the careful planning of each trial, including efficient research design. Total costs can also be reduced by reducing the cost of each trial; the most straightforward way to do this is to conduct a small-scale trial rather than instituting a full-scale policy in order to learn what works.

Technical and Political Trials: Types of Incrementalism

For a policy model to be successful, it must meet the tests of both technical examination and public acceptance. These two types of tests confront it with different environments. We must therefore ask when, and under what conditions, a model should be submitted to one or the other. Though the political environment is essential for public support, premature submission of a model to it may impair its success.

The environments in which trials are evaluated and selected can vary, and this can result in the application of one set of selection criteria or another. We may illustrate this point for the survival of firms in a competitive market, where consumers are a major part of their environment. The net social benefit

31. In organic evolution, the survival of a species depends not only on its adaptation to the nonliving environment but also on its competition and cooperation with the changing local population of other species and its use of or by them for food. In an ideal market competition occurs only through multiple offers of exchange between producers and consumers.

from competition depends in part on the quality of judgment by the consumers in the firm's environment, and on the information they receive. Uninformed and unthinking consumers can demand (or be persuaded to buy) goods that they may later dislike (e.g., because of problems of safety or durability).[32] The result may be either an increased failure rate of firms, or continuing purchases of demerit goods that do not yield consumer benefits commensurate with demand.

Similarly, policy models face various environments as they are developed and tested. In the earliest stages of design, trials are tested against criteria set by an environment of scientists, analysts, and evaluators. These designers and testers must be technically skilled, practically oriented, and continuous in their work as an expert community (MacRae 1976b).[33]

Later (or perhaps continually), the trials will face a political environment made up not only of citizens but also of leaders, political groups, legislators, and administrators. Experts can also be part of this environment (Chap. 2). This political aspect of the environment alters the conditions for the survival of trials. Some models judged to be good by experts may pass unnoticed in the political environment because of failures of communication, lack of priority, or opposition of interests; some models that experts consider less adequate may be favored instead.

This transition from the technical to the political realm for testing trials poses two types of questions: first, when should the transition be made? Second, how can the discourse in the two environments be coordinated?

The actual transition from the technical to the political environment can depend on the political temper of the times; proposed policies can wait and be further perfected if the mood of the public or its representatives is unfavorable, or can be rushed into legislation if the mood is favorable. There can also be reasons why we should or should not encourage this transition from technical to policy trials, depending on the nature of the policy and our state of knowledge. Those eventual trials that can take place on a large scale involve special costs. The degree of exactness needed in our models before we intervene thus depends on our capacity to intervene repeatedly, on the speed with which we can learn the effects of our action, and on the costs of error. It also depends on our capacity for improving the models after we learn their results. If a policy requires a large capital investment, will not realize its full effects

32. The importance of the quality of consumer choices is often slighted in economic analysis because of the assumption of rationality, either in choice of products or in choice of the amount of resources devoted to information seeking.

33. This expert community corresponds approximately to Coleman's (1972b) "world of the discipline"; but it can also involve standards of a more practically oriented "technical community" such as we advocate in Chaps. 4 and 10. The political environment corresponds to his "world of action," which we here regard as modifiable rather than as a given source of constraints.

for decades or longer, and has high potential error costs (e.g., a nuclear reactor), then we need to know a great deal about its potential effects. If, however, we are "steering" a system by repeated and relatively inexpensive interventions, then all we may need to know initially is the direction of association between intervention and our criterion.[34]

A political environment in a broad sense figures in any model's survival as soon as it is tried at all, even experimentally, when people and organizations are affected. So far we have written as though a trial can be carried out at the early stages of model development and used for learning. This is true in the research laboratory, in the design of prototypes of material products, and sometimes in the design of social policies. But in most policy areas where people's actions and well-being are liable to change, it is not so simple to undertake a trial. Randomized experiments usually have to be conducted with volunteers. Proposed interventions, if they are not merely matters of administrative discretion but require public assent, have to be justified as beneficial; potential experimental subjects are often reluctant to face risk or inconvenience without thorough justification, or when great uncertainty exists as to the value of the intervention. Often the amount of support required to persuade people that a trial is desirable is so great that it can be obtained only when there is enough support to enact a national policy.[35]

We might therefore wish to subdivide the block in Fig. 3–1 labeled "design, conduct new trial." Trials in a broad sense can range in their degree of realism and intervention from full-scale trials on the full intended population, to trials on samples, to trials on volunteers, to nonexperimental studies of ex-

34. More precisely, this is true if the system adjusts instantaneously as a result of our interventions, and if the results of our steering are sufficiently visible that we can learn to steer. If there is a lag in the response, we may either waste time getting to the desired condition or produce oscillation through mistimed corrections. We may also reduce oscillation by the use of leading indicators. Unclear results may lead us back to theory as a guide, however, rather than sheer learning "on the job."

The question also arises whether we can learn in time, with an adequate model, what combination of changes in input variables will improve the outcome. Arie Lewin has argued (in a lecture at the University of North Carolina, 3 November 1980) that in trying to improve the performance of public organizations we need more than mere knowledge of the outputs from, and budgetary inputs to, these organizations; that a more precise model is necessary.

In this discussion of steering we assume there is only one system from which we can learn through trials. For many policies, however, there are numerous real programs or units (such as states or localities) that can generate information for one another, as well as experimental systems for trials of models.

35. Gilbert et al. (1975: 180) contend that even though randomized trials might well precede the adoption of large scale programs, "political pressures often make this position unrealistic." Thus they urge "embedding the trials in the program." Ravetz (1971: 341–44) also points out difficulties in maintaining a "tentative and exploratory approach" to external reality in major projects for dealing with practical problems.

isting programs, to basic research on underlying principles. Real world trials are the ultimate test of our models; and if policy analysts and social scientists do not conduct them, they will be done by enactment of public policies in ways that provide less information (MacRae and Wilde 1979: Chap. 7) and thus reduce our capacity to learn.

The real world in which models are tried is thus not simply a world of realistic causal processes but also a world of politics. To try a policy involves persuading people that it is desirable. Their definitions of the problem, including the effects of policy indicators, can influence the success of our persuasion. In this process we may find ourselves, willy-nilly, trying to alter the priority of one policy relative to other policies that are aimed at the same problem or that compete for the same resources. Those who favor the other policies or wish to use the resources in other ways can oppose us. Even a comparative trial of two policies may be less than one side wants.

Recognizing that politics is involved in trials, we are led to consider a fully incremental approach (Braybrooke and Lindblom 1963). On this approach, each trial may be a quest for knowledge, not only of the effects of our proposed policy, but also of the political terrain. Some trials will not surmount the political hurdles; but discovering this, we may be able to revise our policy before wasting further time in analyzing an approach that is not politically viable.

Those who propose this sort of trial recognize the inadequacy of our knowledge of both our model and the political situation. In a series of writings, Lindblom has explicated and advocated a system of policy formation by which participants seek specific organizational goals, plan ahead for only a limited span, and depend heavily on repeated observation of results (including political results) and on a series of remedial interventions. On this approach we learn to do by doing, not by armchair reflection, laboratory research, or synoptic planning.

These trials can be more or less successful, however, depending on the style of analysis that goes into each one (Goodin 1982b: Chap. 2). The speed and regularity with which each single intervention improves on the preceding one, and the reduction of side effects, are affected by the quality of our analysis at each cycle. If each cycle passes through the political system, then this analysis must be political as well as substantive, and selection and improvement will be judged partly in political terms.

Braybrooke and Lindblom (1963: 3–4) correctly argue that the standards of academic science often impose an impossible burden on policy analysis, because we need to know things that cannot be known with the degree of certainty that science requires. At the same time, however, we must not relax our standards excessively; faith in the working of an incremental system cannot be an excuse for sloppy thinking or for failure to use information when it is likely to produce a net benefit. In this respect, the dichotomy between incremental and synoptic approaches is an artificial contrast of extremes and may distract us from doing the most useful analysis we can.

Lindblom (1979: 517) has distinguished more recently between incremental politics and incremental analysis. The process of successive trials that we have described corresponds to this distinction by starting more nearly in the realm of analysis before moving more into that of politics. Analysis may deal with small changes—especially at the stage of administrative fine tuning of a proposed policy—but it can make use of all the analytic methods that our resources allow. In policy analysis we are thus engaged in constrained optimization, with a shorter time horizon and fewer policy alternatives than the hypothetical procedure of pure synoptic choice would imply; but we are still trying to do our best.

Having separated analysis somewhat from full involvement in politics, we may ask again how the two types of trials should be related. A condition for the survival of desirable trials in the political environment is a certain degree of cooperation between technical and political communities. The relevant expert community, for its part, must recognize the public and private values and value conflicts embodied in politics, and not pursue its own criteria blindly. Some of its members may even become advocates of particular interests, and others will represent diverse notions of the public interest. The expert community must also recognize that the political community defines "real time" (Coleman 1972b). It must thus anticipate problems rather than simply respond to them; and it must maintain continuing interest in recurrent policy-relevant problems, so as to respond to questions that arise so rapidly that judgments on them preclude the design of special research (Chap. 4).

The rules and styles of action of political communities may require change if this cooperation is to be achieved. The environment that a political community, such as a city, presents to policy trials is often insufficiently receptive to the models proposed by experts (Szanton 1981). We have noted that in the market, consumers' models of product effects play an important part in shaping the environment for businesses. Citizens' and leaders' models of the effects of policies play a similar part in the process of political testing of trial policies.

To speak of citizens' models of the detailed effects of policies may seem to fly in the face of contemporary findings concerning political behavior. Sample surveys show that few citizens are involved in politics in the intervals between elections, when policies rather than candidates come to the fore. At elections most citizens simplify their choices by making use of the symbols of party, ideology, and candidate image to group many choices into a few. Their retrospective judgments on past policies are not the same as comparisons of future alternatives.

We nevertheless speak of citizens and leaders together—not necessarily of a representative sample, but of representative institutions. We ask, then, whether legislators and their staffs will understand the models involved, and whether the representatives of interest groups and of the press will judge policies in terms of reasonable inferences about their effects.

We may thus speak of "the fitness of the environment" (Henderson 1913) for the assessment of policy models. Democratic political institutions are supposed to perform many functions, including forestalling tyranny, protecting rights, assuring a degree of equity among citizens, and promoting the general welfare. Elections and the media focus the public's attention on issues from time to time. Legislative calendars and chief executives' timing define limited intervals in which major policy proposals are placed on the agenda for public discussion.[36] At these times the public and its representatives are most involved in choice among the models that support these proposals. The quality of the models chosen is limited by the capacity of the political system to judge them. Their quality in turn affects the valid use of indicator statistics; if we have valid knowledge as to how to act on the values such statistics measure, we will be more fully justified in using these statistics to define public problems.

In the very long run, we may wish to ask whether a democratic political system can be modified so as to permit more continuing attention to policies and scrutiny of repeated trials, perhaps through some form of collective memory. Education may play a part in such changes. But in the shorter run—a matter of decades—we must consider those changes in academic and scientific communities that would help these communities to cooperate with the political system as it is. Among them are a practical focus (though not to the neglect of general questions); a continuing concern with persisting policy-relevant questions; and an increased skill in conducting controlled field trials of policies. We shall ask in Chapter 4 how these improvements could be made in nonexperimental research in two important policy areas—education and crime control.

36. The attention that the public gives to an issue may also die away in time (Downs 1972). This may be due in part to a habituation effect that occurs after the presentation of new stimuli (Chap. 6).

4 Nonexperimental Social Policy Research

The models on which policy indicators depend are drawn from diverse sources. Many come from common sense, but we wish to test and improve these judgments. Some come from the natural sciences, but are limited by the need of all policy models to encompass human actions and values. Others achieve formal exactness by treating simpler aspects of human behavior. Still others come from the less exact sciences, like Fairweather's Lodge model, but are refined by repeated experiments. There remains, however, a broad domain of less exact and subjectively social science, not easily amenable to experiments, on which we must depend. This domain—which draws on social psychology, sociology, and political science—has given us fewer clearly successful policy models or components such as Newton's laws, Wohlstetter's basing proposals, or Fairweather's Lodge.[1] We shall refer to relevant research in it, now excluding economics, as *nonexperimental social policy research*. The fields involved are essential to the reasoned choice of policies, but they study aspects of human activity in which people communicate and form collective perceptions and norms.

In this domain we must deal with nonexperimental statistical generalizations as to the effects of policies in subjectively social fields. In the absence of experimentation we must consider numerous variables other than the policy treatment that may affect our valuative dependent variable. A major concern in model building of this kind must therefore be with the statistical control of these effects, so that we can make valid inferences about the effects of the treatment.[2] We shall ask here how these inferences can best be made—not

1. Numerous instances of policy application have been reported for these fields (e.g., Lazarsfeld et al. 1967). They include specific researches using techniques from these disciplines; consultation by individual social scientists; and utilization of research reports and analyses by policymakers. Relatively few of these contributions, however, have attained the status of a recognizable entity—a model or a component of a model—that outsiders can take and use.

2. Some fields of social science, such as history and social anthropology, are less concerned with statistical approaches to problems of causation. In more statistical fields such as sociology, much of the literature deals with nonpolicy effects; in comparison with these complex models, a model developed by experimentation can appear barren of detail. The more complex models of

simply through the application of textbook methods to individual researches, but also through the general organization of research communities and their goals. The goal of this organization is to focus the work of expert communities on continuing policy-relevant problems; it is thus relevant to policy models using exact and experimental approaches as well.

Many models in this domain are either formulated verbally, or applicable only in specific contexts such as evaluation of a particular public program. The most general theories in disciplines such as sociology and political science are expressed in words rather than equations; these may enter into specific models as the laws of natural science do, but they afford a less exact basis for formulating these models. There are efforts at mathematical modeling in these disciplines, but these efforts do not seem to have yet given rise to successful policy models.[3]

The policy models we shall examine in this chapter are based primarily on nonexperimental studies of phenomena in the social goal areas of education and crime reduction, such as the effects of public schools on their pupils (through cross-sectional surveys) or of rehabilitation programs on prisoners (largely through quasi-experiments). These are areas of persisting concern. Research in them utilizes statistical models such as multiple regression, which are not based on mathematical expression of scientific laws, but are chosen for their simplicity (e.g., linearity) in view of the extensive uncertainties involved.[4] Models of this type play a major part in nonexperimental studies of policies in subjectively social fields.

Models in specific social goal areas are likely to be an important ingredient of practical social science for the indefinite future. We should not, therefore, postpone the use of indicator concepts in policy models while waiting to develop models as exact as those of natural science; in the interim we can still improve on commonsense models. Although it is conceivable that our models of subjectively social phenomena will eventually become more general, more mathematical, and more rigorously based in empirical research, this is by no means certain. Such a state of affairs may be long in coming; the actual and perceived conditions of social life may change and our research may not cumulate; and thus at least some parts of the social sciences should be guided by practical concerns rather than simply by the quest for elegant and long-lasting theories.

---

basic and nonexperimental research seek to account for variance in general, while experimental policy models deal with the effects of action (Rossi and Berk 1981: 292).

3. The same observation may be made about Becker's (1976) application of the elegant mathematical techniques of microeconomics to sociological problems.

4. It is true that linear regression can be used to test some nonlinear physical laws. If for a falling body the theoretical distance fallen in time t is $s = gt^2/2$, then a regression equation of $\sqrt{s}$ on t, with the intercept set to zero, or of log s on log t, will show how well the predicted relation is approximated and permit an estimate of g.

Experimental studies such as those of Fairweather, aimed at testing the workability of programs rather than general propositions of social science, are also highly desirable. Some have been conducted in education and in the treatment of criminal offenders. In spite of their virtues, however, we cannot rely on experiment alone for our policy models. Experiments may be difficult in some policy areas, or may need to be supplemented with nonexperimental information on more representative samples.

We have seen (Chap. 3) that even those policy models that use natural science must be supplemented by empirical generalizations and verbal models of human action, since general and mathematical theories alone cannot supply all the necessary components of policy models. In the subjectively social realm, various types of models (e.g., statistical and verbal) and of research may also usefully coexist. Our task is to do as well as we can, to set our level of scientific aspiration realistically for policy guidance, and to specify as well as we can the sorts of expertise that may be claimed by various groups of experts contributing to policy formation. These groups of experts, in turn, must define their relations with one another and with informed citizens in their groups' discourse, which contributes to both scientific and public debate.

The Specificity of Practical Social Research

The work done in the subjectively social fields—especially in sociology—that is most directly relevant to policy seems to involve the construction of highly specific models through ad hoc studies of particular problems. These studies may draw on general theories and cumulative bodies of empirical research as did the Fairweather study, but they do not always do so. In developing policy models we are much more concerned with useful findings, whatever their source may be, that are related to specific policy choices than with adapting knowledge already developed in some academic field of study. In program evaluation the researcher is usually interested in knowing whether a treatment was effective as a whole, and "is relatively less concerned with determining the causally efficacious components of a complex treatment package" (Cook and Campbell 1979: 83). The findings that are of most use in the planning of a new public program may in fact derive from the evaluation of similar programs elsewhere, by studies that do not lie exclusively in any single academic discipline (as the field of evaluation research shows). A growing dissatisfaction in Washington with the results of grants for policy-relevant academic research, even before the Reagan administration, may well have resulted in part from the actual irrelevance of much disciplinary research to policy. Coleman (1972b: 22) has referred to the earlier relation between allegedly policy-relevant researchers and granting agencies as an "unhealthy symbiosis."

The contrast between use of general theories and of specific findings is exemplified in the fact that natural scientists and economists are often employed by government and industry for advice that they can give without conducting additional research (Zetterberg 1962: 16–17). Political scientists have served similarly as governmental advisers, for example, on foreign policy, reorganization of the executive branch, or municipal government. Sociologists, on the other hand (like evaluation researchers), most often seem to come to a policy problem bearing only a tool kit of research methods that will allow them to collect new information on the problem at hand. Theories aid them in interpreting the results, but seem to play a secondary part in their advice.

Lazarsfeld and Reitz (1975: 10), in describing applied sociology, point to a similar distinction between natural and social science: "It is misleading to draw an analogy between the natural and social sciences. Nowhere in the social realm are there unconditional laws and basic theories already well established. Quite to the contrary, it is the study of concrete and circumscribed practical problem-areas that has contributed a good part of the present-day general sociological knowledge." Thus in the organization of the volume, *The Uses of Sociology* (Lazarsfeld et al. 1967: x), a central question concerned the difficulties of "translating practical issues *into research problems*" (emphasis mine). This assumes that the major task of applying sociological knowledge is to gather new information through research, rather than to apply existing theories.

Lazarsfeld and Reitz (1975: 52, 56) do cite instances in which a sociologist, unaided by sample surveys or other such research data, made perceptive observations that helped to solve a client's problem. Nevertheless, they consistently write of "research" rather than the broader category of "knowledge" as the contribution of the expert social scientist. We need to ask under what conditions this is true; and in contrast, under what conditions observation short of a research project, aided by perspectives and theories of social science, can lead to useful policy suggestions.[5] Perhaps if practical social research were organized so as to generate its own models, rather than as a mere appendage to the basic disciplines, it could bring the general knowledge of these models, and not merely ad hoc research, to practical tasks.

One approach to policy models that is more general than ad hoc research is the development of simulation models such as those used for the prediction of future prison populations (H. Miller 1981) or of urban land use. These models are concrete entities, more easily perceived by a client than general knowledge in a consultant's mind, and they can be tried with modifications in more than one setting and therefore be "successful." The assembling of particular

5. Conceivably the methods and data provided by research are not merely means of developing relevant knowledge but are also means of differentiating expert knowledge from that of laymen or of casual observation.

parts of such models is distinct from the research that tests and verifies the correctness of those parts. So, too, is the testing of the overall predictive accuracy of the models, by prediction or by postdiction from past data to known past outcomes. These specific simulation models thus face problems similar to those of macroeconomic models (Chap. 3); yet there is a difference at least of degree between the two. The theories underlying the best macroeconomic simulations are more general and have been more thoroughly tested than some of the components of these specific simulations. The systematic cumulation of uncertain and empirical generalizations in simulation models may, however, aid policy choice if the alternative is to cumulate these parts unsystematically. If persons who understand these models can examine their properties in public debate, and if the models are not merely used as symbols to impress the audience, they may help to clarify our public discussion.

Among these various kinds of practical advice are a set of common themes. A specific problem is often defined in "the world of action" (Coleman 1972b).[6] The adviser brings to a specific task a body of theory and methods from a discipline or profession, which contribute part of what is needed to deal with the problem. An additional part is then drawn from specific observations at the site of the problem (or on its specific population); these observations may be qualitative or quantitative, and may deal either with particular parameters in a preexisting general model or with exploratory efforts to discover the nature of the relevant model itself. The adviser's recommendations are then based on a combination of prior knowledge with the additional information gained from specific observations.

This process is similar to a Bayesian decision procedure. Prior probabilities are modified by sample information, leading to posterior probabilities that can then enter into a decision. The prior information is likely to weigh more heavily when it derives from the theories and findings of a relatively exact science, and less when it comes from a less exact science. This is one reason why sociologists have to do more research in order to give advice than do chemists. At the same time, however, a field possessing knowledge relevant to the advising task is also less likely to require extensive new research; chemical engineers are likely to be able to give better advice than pure chemists when processes of production are concerned. We shall suggest later in this chapter that even in the less exact social sciences, a community of experts in a domain related to practical questions (such as education or crime reduction), if they share a common literature and procedures for its quality control, can also give more timely advice on policy questions, with less extensive research, than basic social scientists.

6. Such a problem may receive social scientists' attention, but not always in such a way as to contribute to cumulative knowledge of a class of similar problems. Occasionally a practical problem can be reinterpreted so that the results of studying it can be generalized and made theoretically relevant to a basic discipline (Merton 1968: Chap. 12).

In the following sections we shall examine the development and use of policy models in the subjectively social policy areas of education and crime reduction. These areas involve considerable research in which the economic approach is not dominant.[7] They are chosen to illustrate the problems of model construction when the data are nonexperimental and there is a lack of general, quantitatively verified theory; they are not a representative sample, and we thus need to seek out and explain exceptions to any generalizations drawn from them. They are useful, however, for posing a set of important problems concerning the sorts of models we should seek. We shall use them to examine how problems are defined and how the various expert and political communities review the results of research.

Cycles of Model Development for Educational Policy: Three "Coleman Reports"

Educational achievement appears on many lists of social indicators. In public discussion it is often treated as a value concept or even an end-value, though social scientists have not shown definitively that education furthers all the values in later life that it purports to serve. These values include the cultivation of the mind, the enjoyment of leisure, productivity, the capacity to function effectively as a democratic citizen or leader, earnings and occupational success,[8] and fairness in the distribution of these results. Studies of status attainment have thrown light on the contributions that years of education make to occupational status; economists have analyzed the contributions that they make to income.

Education within schools occurs by clearly subjectively social processes. Schools develop particular social climates (Coleman 1961); their staffs' morale may vary, in part in response to leadership; and the community's image of what happens in the schools may also change over time, not necessarily in strict relation to events there as seen by participants. This policy area is thus one where we expect models to be less exact.

The principal educational research we shall discuss is nonexperimental, based on sample surveys, administrative records, and census data. This is not because educational experiments are impossible; experiments have been done on the effects of numerous aspects of education such as curriculum and class size. Experimentation is nevertheless difficult, since parents are concerned

7. They involve sociological research, which though linked to indicators is not sufficiently linked to policy analysis.

8. The relation of education to occupational success depends on the state of the economy and on the connection between learned skills and job requirements. In planned economies, education is closely geared to expected occupational needs. In a free economy, the educational system responds to some extent to the market, but often with insufficient anticipation.

with possible harm to their children, and uniformity of education in many respects is guaranteed to children who attend public schools. Moreover, experiments are likely to be limited to special local populations; and if valid causal inferences can be made, national samples can sometimes tell us more about the generalizability of our findings.

Not only because of social research, but also because of prevalent commonsense models of causation, it is widely believed that access to schooling is a gateway to success,[9] and therefore that equity in the educational system is a major component of the equality of opportunity that an open society should provide.

Equality of educational opportunity in the public schools has gained attention in the United States in the years following 1954, when the Supreme Court decided in *Brown* vs. *Board of Education* that separation of educational facilities by race was an intrinsic denial of "the equal protection of the laws" guaranteed by the Constitution. Although some of the values listed above, such as citizen competence, may be judged in terms of aggregate educational effects as well as their distribution, special emphasis has been placed on distributional equity as regards the opportunity for success. In the discussion that follows we shall see an emphasis not only on overall educational accomplishment, as measured by tests of cognitive achievement, but also on its distribution.

Within this context, a major style of policy-related research in sociology is illustrated by three successive research reports in which James S. Coleman played a major part, each of which gave an important impetus to policy debates. The first of these reports (Coleman et al. 1966) concluded that the mix of students in the classrooms in American public schools was an important factor in students' achievement. The second (Coleman et al. 1975) studied "white flight" and concluded that it was increased by school desegregation. The third (Coleman et al. 1981; 1982) compared public and private schools and concluded that private schools encouraged greater learning because of their greater discipline and stress on homework. Even though no individual can fully represent a line of research for an entire discipline, Coleman is a distinguished sociologist and these reports illustrate important features of sociological policy research that distinguish it from the approaches discussed in Chapter 3.

Coleman's researches illustrate several features shared by much practical social research:

9. An alternative interpretation is given, however, by Gintis (1970), who sees education as a means of disciplining the labor force, that is, of inducing an unquestioning conformity to the goals of the owners and managers of industry. This same attitudinal component of educational experience has been recognized by Crain and Weisman (1972), however, as contributing to individual success. Thus neither the outcomes of education we choose to study nor the value placed upon them can be assumed to be given automatically.

1. They are based on specific empirical studies rather than on cumulative bodies of theory and research—somewhat in the style of evaluation research, but examining a wider range of programs than are usually studied in program evaluations and placing less emphasis on the effects of specific interventions.

2. They are relatively atheoretical. The reports, prepared for a general audience, make little reference to previous theories and do not attempt the full bibliographic citation that usually accompanies basic scholarly research.

3. They carefully study events that have occurred in the past, and then draw implicit inferences regarding future policies from the results. These inferences, like those of evaluation research, are often less explicit than those that policy analysts use for direct comparison of the merits of alternative policies.

4. The focus of these researches changes over time. The very situation studied in later researches may have been brought about by public action relating to prior phases of the research. Busing, white flight, and lower test scores in the public schools are phenomena that are partially responses to the public's and the courts' concern for equity and students' rights. Heraclitus's observation that we never step twice into the same river seems to pertain to the American educational system, continually modified as it is by changing urban spatial processes, markets for schooling, and governmental decisions.

These features of the three studies may not, however, all be necessary characteristics of social policy research. To some degree they may result from the training and values of James Coleman, whose role we have emphasized by choosing these studies for attention. They may also derive from an approach to applied social research developed by Paul F. Lazarsfeld, one of Coleman's teachers and an author or editor of two major volumes on that field (Lazarsfeld et al. 1967; Lazarsfeld and Reitz 1975). Some of them, however, apply more generally to educational research, which Averch et al. (1971: 157–58) judge to be "seriously deficient . . . in the integration of research results" and "lacking in focus." These writers also judge that this field is typically "not founded on a wealth of previous knowledge and understanding nor is it directed toward the needs of the educational policymaker."

I shall therefore enlarge our discussion by considering criticisms of these three reports from the perspectives of other disciplines and professions as well as sociology. Major criticisms of the first report came from economics and revealed a distinct theoretical stance. Researchers in education and educational psychology contributed criticism from another perspective. Moreover, Coleman's practice of strict limitation of inferences to the research data at hand had not been followed by sociologists working on educational policy prior to the first of these reports. Pettigrew and Back (1967) reviewed the use of sociology in the school desegregation process, giving numerous examples in which social and psychological theories were usefully applied, often without the collection of specific research data.

My purpose in this discussion will be not so much to evaluate what *has* been done, as to ask what is the best that we *might* accomplish within the next few decades. For this purpose we need to look beyond the disputes of the present to the part that general policy models and research might play in resolving them; and beyond present ambiguities in findings to the possible character of a justifiable future "moving consensus," including the procedures by which it may be attained.

The first Coleman report, *Equality of Educational Opportunity* (Coleman et al. 1966), was an extensive cross-sectional national study of pupils' achievement (measured by standard test items), school facilities, and teachers' characteristics in American public schools, conducted in accordance with a provision of the Civil Rights Act of 1964. Information about pupils' family background and attitudes was also gathered. The model used in the report was that of multiple linear regression, with an emphasis on the proportion of variance accounted for as one or another group of measures was introduced into the equation. Separate regression equations were estimated for various ethnic groups.

The report came to the surprising conclusion that the inputs into public schooling previously used in indicators of equity, such as per pupil expenditures, facilities, and curriculum, were more equally distributed by race than expected. Moreover, these inputs had little apparent effect on students' achievement after student background effects were controlled statistically. Because these findings ran counter to the beliefs of both educators and reformers (Moynihan 1968), the report was initially given little publicity. It also concluded, however, that the characteristics of the student body in a school (and presumably in the classroom)[10] were a major factor influencing learning. As a special instance of this finding, it led to the conclusion that minority students whose schoolmates had relatively favorable family backgrounds and high educational aspirations showed greater achievement than others. This conclusion was eventually widely publicized.

This report not only presented new information about models related to educational policy, but also contributed to the increasing use of educational achievement as an indicator. In the planning and analysis of the research, an earlier concern with the inputs into public education was largely supplanted by a concern for outcomes. Since the report, indicator statistics on educational achievement have been developed and publicized through the National Assessment of Educational Progress, through reports on trends in the Scholastic Aptitude Test scores, and through increased local reporting of the results of periodic achievement tests and competency tests in the public schools.

10. A substantial body of subsequent research continued to deal with aggregate characteristics of schools in relation to student achievement. But Bidwell and Kasarda (1980) stress the need for systematic theory and more detailed observation related to the processes of teaching and learning in individual classrooms.

The report was interpreted by many, including Coleman himself in separate statements, as supporting public action to desegregate the public schools. This support was strengthened by the finding that the mix of students in a school affected the learning of students within most minority groups more than that of majority students (Coleman et al. 1966: 302–04). Effects on the learning of majority students would not have been of concern if the only value considered had been equality, but these effects were valued significantly by many of the participants (T. Miller 1981: 330).[11] These findings were relevant to policy, if they were valid; if a suitable means could be found to improve the racial mix; and if that means did not have more important undesirable consequences. The means actually adopted was court ordered busing—chosen not simply for this reason, but for others as well; this means was not on the public agenda at the time of the report. As we shall see, Coleman later concluded that the finding seemed weaker in the light of subsequent evidence, and that the policy adopted had important undesirable effects.

The incomplete correspondence between research variables and policies is also illustrated by another variable studied in the report: teachers' characteristics, including their education and their scores on a vocabulary test, were judged to be second only to student body characteristics in accounting for variation in student achievement (Coleman et al. 1966: 316). Yet teacher characteristics are not themselves a direct policy variable any more than student body characteristics; the relevant policy variables are policies for educating, attracting, and retaining teachers who have the desired characteristics. The need to attract and retain able teachers often leads, in turn, to advocacy of higher expenditures per pupil, a variable that was not shown to be equally efficacious.

We can do policy-related research on variables such as student mix and teacher characteristics, but this is not the same as doing research on policies that might affect students or teachers. In the application of natural science, basic research is usually followed by the engineering development of products or devices that are then tested carefully in use; the eventual success and reliability of these products requires this further testing and is far from guaranteed by the soundness of the basic principles underlying their design. Coleman's 1966 report did not therefore deal directly with policy alternatives. It was a study of existing schools rather than of new arrangements such as would result from desegregation policies. It was cross-sectional, and did not involve experimental or quasi-experimental intervention, because of the time schedule imposed on the researchers by Congress. When the courts decided that desegregation was required (on grounds largely independent of the re-

11. T. Miller (1981: 321–22) suggests, however, that with an alternative statistical model stressing interracial rather than interindividual comparisons, the effects of school characteristics (primarily teacher quality but also student body composition) are greater for whites as a group than for blacks.

port), they had to implement that decision as a universal national right, governed by judicial precedents, rather than encouraging local experiments that might tell how well it would work.

Although policy-related work is not always submitted to the quality control provided by fellow researchers, the 1966 report received extensive scholarly review. It was favorably reviewed in the *American Sociological Review* (Sewell 1967). A Harvard faculty seminar in 1966–67 led to a publication that largely sustained its findings (Mosteller and Moynihan 1972). The regression model of school effects implicit in the report came under repeated criticism, however, from economists (Bowles and Levin 1968; Cain and Watts 1970; Hanushek and Kain 1972). These lines of criticism are important because we must ask whether it is the subject matter or the discipline that accounts for different modeling styles. They made the points that:

1. Because of nonresponse, the sample was not representative.

2. A clearly specified theoretical model is needed as a foundation for such analysis. One example of such an econometric approach, using such models in an effort to distinguish effects of ability and of years of schooling on income (though not addressed to this report), has been presented by Griliches (1977).

3. Regression coefficients, rather than proportions of variance explained, are needed for policy inferences. These would correspond more closely to parameters in an economic production function (Hanushek 1979), but the high correlation between family background and school inputs in such a national sample make it difficult to assess the magnitude of the effects of school inputs in a production function type of analysis.

4. The finding that teacher characteristics were associated with achievement deserved more emphasis and would have led to more encouraging policy conclusions than the negative ones flowing from the analysis of per pupil expenditures. A critique by Hanushek and Kain (1972) in fact concluded that the report had tried to accomplish too many and conflicting goals, and should probably not have tried to assess the causes of educational achievement.

Coleman (1972a), responding to this line of criticism, argued that the disciplines differed fundamentally. He observed that economists tended to seek "educational production functions" that would guide policy toward an optimum allocation of resources. The approach of the 1966 report, however, was a more exploratory one. "In economics," he contended, "there is frequently an assumption that the form or structure of the causal system is well known and that the empirical question at hand is to estimate values of coefficients for that structure." In contrast, "sociology contains less theoretical basis for deriving specification of models, but . . . there is often willingness to specify such a model to the extent required for path analysis. . . . However, it appears that the statisticians, who are less model-oriented, . . . shaped the report's policy in this respect" (Coleman 1972a: 168–69).

A further study of Atlanta school children by Heyns (1978) showed that a cross-sectional study such as the 1966 report might well have been biased in its estimates of the contribution of schools to inequality among racial and income groups. Using series of test scores obtained in both spring and fall, Heyns found that summer learning was associated with a greater increase in inequality among these groups than was learning during the school year. Thus the 1966 report may have been excessively pessimistic in this respect concerning the schools' capacity to promote equality of opportunity.

Additional research and criticism in the area of school effects since the 1966 report has drawn on methodological developments in disciplines other than economics. Factor analytic methods have been used for constructing indexes from the data of the report in reanalysis (Mayeske et al. 1973). Causal or path models have also been used extensively in estimation of the effects of various variables on learning (e.g., Bidwell and Kasarda 1975).

Coleman's next conspicuous venture into educational policy-relevant research involved the question whether school desegregation caused "white flight" (Coleman, Kelly, and Moore 1975). Indicator statistics had shown declines in the proportion of whites in urban public schools, but the causes of these declines were unclear. Coleman's study, sponsored by the Urban Institute rather than by a governmental agency, concluded that school desegregation in large cities had made significant contributions to these declines. This finding was also hotly debated, as the political forces supporting equity for blacks were aligned in favor of busing. The report provoked a storm of protest, accentuated by Coleman's willingness to go outside the academic role and testify to public bodies about the implications of the findings. The controversy was carried on in the publications and meetings of the American Sociological Association, but eventually the findings—limited to a particular category of large cities—were sustained by review. One sharp criticism was provided by Pettigrew and Green (1976), but their research nevertheless narrowed the range of disagreement. A subsequent study by Armor (1980) reviewed much of the intervening literature, concluding that court ordered mandatory desegregation produced white flight and detailing the course of this effect over time. Pearce (1980) also showed that school desegregation at the metropolitan level can produce beneficial side effects by reducing tendencies toward racial segregation in housing.

In initiating this line of research, Coleman wished to open debate among social scientists to more than one side of the issue. He also judged that the predominant findings in relation to school desegregation had changed since the first report: "The achievement benefits of integrated schools appeared substantial when I studied them in the middle 1960s. But subsequent studies of achievement in actual systems that have desegregated, some with a more rigorous methodology than we were able to use in 1966, have found smaller effects, and in some cases none at all" (Coleman 1975: 77).

Contrasting this report with the first, we see a discontinuity in focus that might have occurred for two reasons: (1) different aspects of the same problem were chosen for emphasis in the two researches; or (2) the system itself, which was under study, changed. In spite of the fact that major changes resulted from desegregation, greater continuity in research might have been possible if researchers had concentrated on problems of policy choice and on the models appropriate for this choice; we shall illustrate such models in the next section.

In this second report, an aspect of Coleman's research style becomes more evident. He was not studying the overall merits of competing educational policies, but presenting expert advice based on specific researches on particular aspects of those policies. By choosing to present evidence on white flight only, he was deliberately running the risk of seeming to oppose a major policy aimed at racial equality.

A key difference between "applied social research" and "policy analysis" is illustrated by this choice. Two possible social indicators were involved here: students' achievement (a policy indicator) and the racial composition of the student bodies (a contributory variable, less clearly a policy indicator). But racial composition presumably affected student achievement as well as other values. Rather than making a summary or synthesis of these effects and weighing them against one another—which would have demanded more time, required more value judgments about various consequences, and made the report either less scholarly or slower to appear—Coleman and his coworkers chose to concentrate on one aspect of the problem. This selective approach was intended to remedy a selectivity in the first report and in the debate about it.

Could the importance of white flight have been anticipated? It was not conspicuous in the critiques of the 1966 report, but Cain and Watts (1970: 240) did mention selective migration as a process by which school quality might have affected the average family background of pupils at a given school. A general model should allow us to see aspects of a problem that might otherwise be ignored; but in this case the problem seems to have been brought to our attention by developments in the cities after implementation of busing, more than by prior theoretical or policy models.[12]

This research seems part of a policy cycle in which an initial policy leads to unanticipated consequences that must then be dealt with (Wildavsky 1979:

12. The omission of this variable is not simply a case of misspecification of the model in the usual sense. If court ordered desegregation has special effects, over and above effects of desegregation that arises from changing neighborhood characteristics, then model development must be explicitly concerned with the effects of *policies* for desegregation. A deliberate and central focus in model construction on effects of policies can lead to proper specification of policy models in a way that basic research may not. Thus Dentler (1980: 209), frustrated by the lack of progress in research on school desegregation, argues that these issues should be relocated "in the mainstream of educational policy analysis."

Chap. 3). The findings of the 1966 report did not say how the desired mix of students in the school was to be accomplished. The policy chosen by the courts to do this, led to a situation where, in a free society, some of those families with the resources to move away or send their children to private schools took these means to escape, while others lacking these resources could not. Conceivably a more complete model, or trials and experiments, could have shown these results. But at the same time, there were few other ways available to declare symbolically that black children, like other black citizens, were entitled to "the equal protection of the laws." The courts had moved into an area where legislatures had failed to act; but their constricted role in policy formulation, which prevented them from usurping the functions of representative legislatures, also made it difficult for them to engage in policy analysis, to trade off various consequences against one another, or to propose experiments.

We are dealing here with a problem of societal "steering" or guidance (Etzioni 1968) in which various institutions play different parts. Legislatures had failed to protect a minority against oppression by the majority. The courts had taken a step to remedy this neglect, after a long delay, but in a rigid fashion that took into account only a few of the values involved in policies for education and equality. The question is whether thoughtful policy analysts could contribute to the improvement of this process. If they could see policy problems as a whole, transcending both ideological and disciplinary barriers, they might develop a continuity of analysis (Zapf 1977: 210–11) that would anticipate problems that the political institutions would later face. The steering function of society might be more effective if there were not such delays in sensing these new problems, and such great swings from one side to another in researchers' concerns. This continuity may, however, be difficult to realize, since researchers as citizens align themselves enthusiastically on sides of public questions; since they are drawn along (or held back) in public controversies by the changing hopes and fears of their fellow citizens; and since the public and private sources of their research support are also affected by the dominant policy perspectives of the moment.

The third Coleman report (Coleman, Hoffer, and Kilgore 1981; 1982)[13] was based on the first wave of a national longitudinal survey, "High School and Beyond." The third Coleman report dealt with the relative effectiveness of public schools and of various types of private schools in promoting learning and intergroup equity. It concluded that after a number of student background variables were controlled statistically, private schools produced better educational results than public. These differences were attributed in part to the

13. We cite both the prepublication (1981) and published (1982) versions because the former made a considerable contribution to both public and expert discussion, while the latter incorporates some revisions in response to that discussion.

greater discipline and amount of homework in private schools. A particular uniformity of treatment of students across racial lines was observed in the parochial schools, leading to the conclusion that they approximated the "common school" ideal more than did the public schools.

The authors contended that the constraints placed on the public schools, associated with the legal requirements to educate all students at least to the age of sixteen and the stress on students' rights, had undermined discipline and impeded learning. They seemed to support tuition tax credits as a policy measure, presenting data indicating that black families would be somewhat more likely than white families to respond to an income supplement by sending their children to private schools.

This report resembled a policy analysis more than did either of the previous two reports. It focused on a policy-related choice—whether students should be placed in public or private schools. It considered two values that might be affected by this choice—learning and racial desegregation—as well as other conditions contributing to them. In addition, it dealt with some possible effects of a policy instrument that might be used to effect such transfers—income supplements. Even though the report did not meet all the requirements of a policy analysis (as we shall see), it included some of them.

In the published version of the report, the authors make a useful distinction between research results and policy consequences (1982: 7–10). They note that if a causal model could be inferred validly from their data, it might yield accurate predictions of marginal effects of policies transferring a few students from public to private schools, but that a policy transferring substantial numbers might change the nature of the school systems in question. If a legislature were deciding whether to provide educational vouchers, it would have to estimate possible effects of changes in student body composition, in the quality of teachers at private schools owing to limited supply, or in federal regulation of private schools.[14] In a more complete policy analysis, these considerations must enter even if they are not illuminated by the research at hand.

The scholarly review of this report has included special issues of the *Harvard Educational Review* (1981) and of *Sociology of Education* (1982), and has involved not only sociologists and educators but also economists and psychologists. A major issue here, as in the preceding reports, is the distinction between effects of policy variables, acting through the schools, and nonpolicy variables. Especially at the start of this longitudinal study, no "before" data

14. Coleman et al. (1982: 8) also note that the commitment on the part of student or parent may be greater when the family pays for the school, and may be reduced under a voucher system. This point relates to valid causal inference rather than to the scale of the intervention. It is unfortunate that these considerations were not included in the version of the report released in 1981. Such marketlike reactions were considered in the white flight study (Coleman et al. 1975), and should be considered routinely by public policy analysts. A general analytic approach like that of microeconomics might help to avoid such omissions in the future.

were available on student achievement; thus the special difficulty of attributing causation from a cross-sectional study without control for prior ability (Alexander et al. 1981), which drew controversy after the 1966 report, was still present. Analysis of later waves in this longitudinal study should improve these inferences. One critique of the report by Page and Keith (1981) uses the literature of educational psychology to distinguish, among the measures of student accomplishment collected in the research, some that were more related to school instruction and others that were less so. Using the latter as controls, they reanalyze the data and conclude that the effects of private schools are less than the report claimed.

In the third report we again find a focus on certain selected aspects of possible policy choices—the effectiveness and equity of public and private schools rather than the overall effects of policies such as tuition tax credits or vouchers. The policies under consideration in 1981 were no longer the same as those of 1966; thus although school effectiveness and equity are common themes of the first and third reports, they are viewed with respect to different policy alternatives. The universe of schools studied was enlarged to include private schools, but the study providing the data was not planned with specific reference to these current policy choices. The economic effects of tuition tax credits on the federal budget and on income distribution are not part of the study.

We may now summarize the conclusions to be drawn from this series of reports, returning to the features of applied social research listed at the start of this section.

1. These studies are specific empirical researches dealing with particular aspects of a policy problem. They reflect the capacities of teams of experts to deal with parts of policy problems and to make contributions that are set off from commonsense discourse by systematic analysis of data. They are also sometimes set off from a more general expert review of the field by concentration on the data at hand. Researchers who approach problems in this way are often doing what is expected of them by clients or decision makers. However, research dealing with one aspect of a policy problem can bias discussion toward that aspect at the expense of others. The study of topics on which the researcher is expert or has recently analyzed data does not guarantee a balanced policy analysis and invites the use of the findings for one sided policy advocacy (much as the use of a single policy indicator, by itself, can define a public problem in a particular way).

In order to guide intelligent public discussion, communities of policy analysts might center their work about the choices available for policy and a broader range of criteria that need to be brought to bear on them (Chap. 10). On this approach the relevant expertise would not be that of individual disciplines, and the overall model under discussion would not be limited to components being estimated in a particular piece of research. This overall model would play a part in the policy literature analogous to the tested and established theory of particular disciplines, but would supplant that theory in guiding the design of research and the definition of significant arguments in research reports.

2. The empirical and atheoretical nature of these studies needs to be supplemented not only by the framework of a general policy-related model, but also by the theoretical contributions of the various disciplines relevant to that model. In the case of educational policy, sociology needs to be supplemented by economics and educational psychology, at the least. This is an extensive task of synthesis and delays the attainment of consensus, as the reviews of the first and third reports show. A major obstacle to the attainment of this consensus is the prestige accorded to the separate academic disciplines, instead of the field of educational policy itself, in defining the terms of the discussion. Between the general field of policy analysis and the particular expertise that should be found in schools of education, this sort of "technical community" (MacRae 1976b) needs to be developed.

3. These researches deal with the past; in a sense, all research must deal with data from the past, but some of it makes claims to generalize for the future as well. If we can speak of the future, do we need to wait nine years between the first two reports to study "white flight"? Do we need to wait fifteen years to compare the public and private schools? I do not mean to say that these three conspicuous reports exhaust the literature; but in scholarly communication as well as in the public mind we need to stress more continuity and more anticipation in our analyses. Here a concern with policy models can be of aid. The experiences of other countries can sometimes tell us about the effects of comparable policies there. The setting out of general policy-related models can perhaps provide greater incentives for diverse researchers and research centers to fill in the necessary details as well as to revise the structure of the models. And insofar as some of this work may already have been done, major reports that grasp the public's attention, such as background articles in the press or journals of opinion, should try to stress the cumulative judgment of the research community rather than placing exclusive emphasis on the findings from one particular study. If clients seem to be asking for specific studies, it is our responsibility to remind them that these studies should be the tip of an iceberg, the base of which is the continuing work of communities of scholars.[15]

15. The community of scholars in a field such as educational policy analysis must, however, function as a community. Ideological battles among its members can weaken outsiders' respect for everyone in the field. Excessive emphasis on research results to the neglect of practitioners' knowledge can accentuate implications, such as those sometimes drawn from the 1966 report by laymen, that schools are largely ineffectual in producing learning. Cohen and Garet (1975) remind us that policy need not consist of discrete decisions, but can involve general trends in public knowledge and beliefs as well as in public acts. Those who claim expertise in public matters have

4. The focus of these reports changes over time. This is due in part to the fact that the system under study, or its parameters, may have changed. "White flight" may have seemed only an empty threat, or a problem of unknown magnitude, when busing was first contemplated. To analyze it might even have increased the legitimacy of opposition to the implementation of busing. Similarly, the transfer of minority students to private and parochial schools may not have seemed an important issue at the start of the period of busing. In retrospect, however, we must wonder whether we could have seen the problem more nearly whole at the start.[16] We could not easily have done so through a report commissioned by Congress, or through the eyes of the courts, where the only value emphasized was equality. But the lack of continuity in this series of reports is due only in part to these restrictions and to the fact that the system under study has changed. If we can anticipate future developments with the aid of general models, theories, and value systems, and if individual scholars and analysts can coordinate their work through reference to these general models, we may be able to design and use research more effectively as well as to combine it with other sorts of knowledge.

One response to these problems has been that of incrementalism (Chap. 3): if we cannot encompass a problem in its totality, then perhaps we should be content to deal with it one part at a time. We can do this either in sequence, by a remedial approach to problems (Braybrooke and Lindblom 1963: 102–04), or by a division of labor that allots parts of the problem to different specialties or to organizations with different functions and goals. This response, however, can magnify a partial truth (that one person cannot always see the whole problem) into an excuse for doing less than the best that we can. We must indeed expect to make errors, either because we have adopted partial approaches or because the world is changing, and to correct them repeatedly. At the same time we can try to reduce the costs of making and correcting these errors, by suitable design of the system that corrects them.

We need therefore to devise a system of generating information that will equip us best, over the long as well as the short run, to cope with problems such as those that arise in educational policy analysis. Such a system need not be exclusively national in emphasis, as these three reports were. It can combine state and local studies and policies with national ones, making use of experiments and quasi-experiments in which one locality generates information that will be of use to another. It can also combine long- and short-run studies. An exclusive devotion to basic research can risk irrelevance to policy

---

a responsibility not only to speak out but sometimes to be cautious. A note of caution is also warranted by the fact that a number of relevant review articles have been published (e.g., Hanushek 1981) without yielding clear models or estimates of coefficients in them.

16. Coleman et al. (1982: 224) argue that "policy research should be directed to areas of policy conflict, not to more diffuse goals." This approach increases relevance, if the research can be done quickly enough, but decreases continuity.

unless this research is constantly used to deal with current practical problems. A preoccupation with short-term "firefighting," however, can prevent us from developing models that help us to look ahead; it can produce particular studies that are relevant to current problems, but without the collective thought of an expert community behind them. Models of pupils' transfers between schools in relation to school characteristics, for example, might conceivably have alerted us sooner to problems of "white flight" and allowed localities to monitor this effect. Studies of private schools, although not called for by Congress or immediately used by the courts, might have been undertaken earlier if we had focused attention on the general values and disvalues resulting from educational policies.

If the relevant literature can be focused on educational *policy* and its effects, we may conceivably draw together research on production functions, classroom processes, the psychology of learning, migration, and other topics that are relevant to decisions in this area. The resulting models will not have the elegance of basic natural science; but they may retain enough continuity over a decade or more for earlier findings to serve as foundations for later ones.

The drawing together of this literature will not be easy. Even if we were to contrive for all of it to be shared by a single technical community interested primarily in educational policy analysis, review and consensus would be at least as difficult as they are in the individual social science disciplines. Review articles, syntheses of previous findings, replications, and the interrelation of research designs require the orientation of a group of scholars to common goals and concepts. The use of the best possible methods for estimating parameters in models and for separation of effects of policy and nonpolicy variables (to which we return in the next section) also requires rigorous standards of relevance and inference in a domain where practical choices can create pressures to publicize findings quickly. We shall enlarge on these suggestions below after discussing models for crime reduction.

An Inclusive Framework for Educational Policy Models

We have suggested that in a social goal area such as educational policy, research could be better organized and made more continuous in its concerns by being focused on a single general policy model. An example of a framework for development of such a model is shown in Figure 4–1. We are imagining a community of practical researchers working on parts of such a model, suggesting changes in its structure, justifying their own contributions in relation to a literature on the general model or variants of it, and prepared to make use of the current state of knowledge about the model in discussing proposed policies.

Fig. 4-1. Schematic Policy Model for Education*

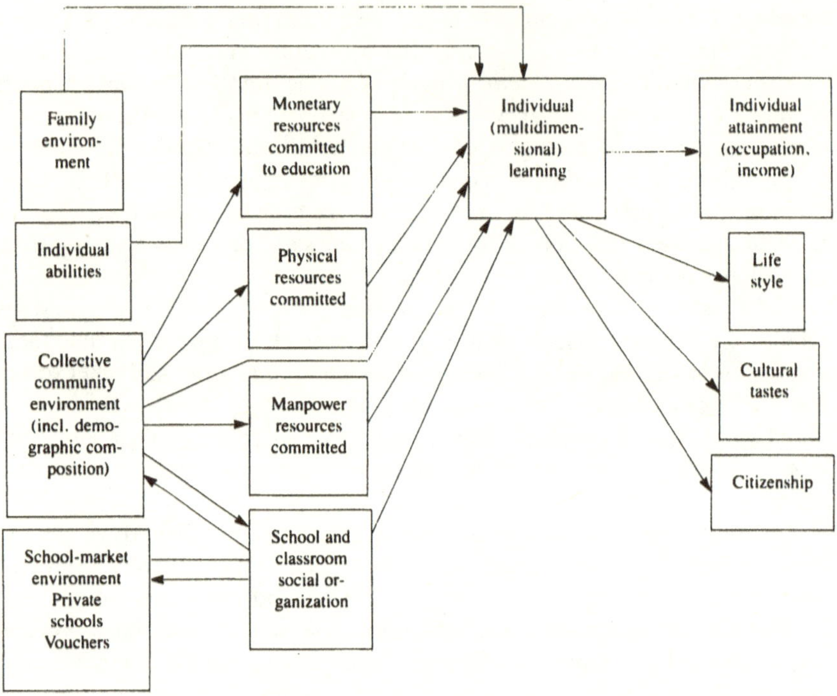

*Includes variables proposed by Land (1975a: 20).

We must recognize, however, that such a research community and the corresponding model cannot be isolated from several other types of research. First, of course, the community will use the findings of basic research when they are relevant. Second, a community studying one social goal area will have to be concerned with connections between that area and others; if a possible policy affects not only education but also income distribution, for example, they may have to compare it with other policies that affect income distribution. Third, they may wish to connect a model for education with models of the causes of more general end-values, to which education and other social goals may be alternative means.

The blocks in Figure 4–1 illustrate types of variables that should be considered; the arrows between them, possible causal paths.[17] On the left are placed

17. Arrows in causal models often indicate linear and instantaneous relations; but for optimization we often need to consider nonlinear ones (Chap. 5) and, for accuracy, lagged or other effects that are better represented by differential equations. The blocks in such models may also

four blocks representing types of exogenous variables: family environment, individual abilities, community environment,[18] and school-market environment (the set of competing opportunities for schooling that families face). Although for some purposes each of these may be treated as exogenous, each can also be imagined to be influenced by other variables, often including policy variables. The family environment of a student can be influenced by parents' education or by parental involvement in the education process; income can be changed by transfer policies. "Abilities" refer to capacities of students to learn, prior to a given educational experience; such abilities may, however, be changed by education, in which case the entire process shown in the model is cyclical with earlier learning creating a base for further learning. For the lower two of these variables we note two-way causal relationships reflecting the possible effects of school variables on the composition of the community (through migration) and on the decisions of families whether to send their children to private schools.

The second column from the left shows four blocks related to the operation of a school—three concerning resources (money, physical resources, and manpower), and the fourth relating to the school's internal organization. In the literature, these variables usually refer to characteristics of a school or school district, for which policies may be chosen or data may be available. They can refer, however, to characteristics of classrooms or to particular aspects of education having specific goals and outcomes. They are often the points of entry for policies, including choices made by school administrators, but we also indicate that they can be affected by the community and the school's environment. They have played a prominent part in the production function approach to educational policy, as they must if they measure policy interventions. It is an oversimplification, however, to link them directly to students' learning as Figure 4–1 does; they do not automatically cause learning, nor do they cause it uniformly even within a single classroom (Dreeben and Thomas 1980). They may affect the activities of teachers and pupils, which in turn work through pupils' experiences to affect learning (Harnischfeger and Wiley 1980); more complex models are necessary to describe these effects and the psychological mechanisms through which they work.

The third column contains only one block, the multidimensional learning that takes place in a given period of time; this can be caused by school factors as well as by social relations, personal communication, and the mass media outside the school. We speak of learning rather than achievement to empha-

---

refer to sets of variables. The variables in this model overlap with those proposed by Land (1975a: 20), as well as deriving from our discussion of the Coleman reports. Further proposals for development of related models have been made by Felson and Land (1980: 83–89).

18. More generally we might speak of the "extrafamilial" environment, including messages in the mass media that originate outside the local community.

size that we are studying effects produced by the school, not simply the characteristics of students after a process of education. This implies that in any actual research model, students' achievement and other characteristics prior to the process must be measured and included (Heyns 1980).

The right-hand column contains four blocks related to possible outcomes of schooling. With respect to attainment (a policy indicator concept), the model is thus related to studies of status attainment, the labor market, and the economy. The rewards to individual educated persons cannot simply be summed; however, to yield the social benefit from education, which can be greater or less because of externalities (Walters and Rubinson 1983). We might also wish to look at other side effects such as those of baby-sitting, the reduction of delinquency, the socialization of the labor force, or the propagation of peer culture.

Since we are concerned with questions such as the effects of educational policies on various ethnic and racial groups, and the effects of different types of schools, the diagram of Fig. 4–1 may stand for distinct (but interrelated) models that might hold for various groups and school types. The extent of equality or inequality in achievement, or in the school's contribution to it, will then be predicted by differences among outcomes in these separate models.[19]

Presentation of the model in a single diagram precludes showing a number of features that should be included in the concerns of the relevant technical community; we shall discuss some of them.

Our characterization of the sources of individual abilities should include heredity. This variable is not usually discussed by social scientists; use of it in the past has been discredited on the ground that unwarranted inferences were made about its importance, especially in relation to racial differences. Nevertheless, it has been shown to be significant in a study comparing children of adoptive and biologically related families (Scarr and Weinberg 1978), and we mention it to illustrate that the construction of general models must involve an effort to transcend the prevalent concerns of a particular time.

The uppermost arrow in the diagram, connecting family environment to individual learning, includes effects such as those of learning resources in the home (encyclopedias, conversation, etc.) and the presence of parents in the home, as well as possible parental involvement due to paying tuition for private school (Coleman et al. 1982: xxviii, 139). Some of these effects may be distinguished from school effects by the study of summer learning (Heyns 1978).

A causal connection from "collective community environment" to "individual learning" includes effects such as those of the media and the peer

19. Goldberger and Cain (1982: 107–08) suggest, however, that some of the coefficients in the separate models may be assumed to be the same for the sake of parsimony, at least at an initial stage of analysis.

group outside school. It also calls our attention to effects such as that claimed for the neighborhood school as a source of increased parental involvement in school affairs and of possible learning.

From the blocks in the second column we might have included additional arrows toward the left, by which resources committed to schools would affect parents' desires to remain in the community and to send their children to public rather than to private schools.

Also in the second column, the block "manpower resources committed" refers to teacher training and experience, to various aspects of teacher quality, and by implication, to average class size. Manpower can be committed to various functions that affect learning differently (Bidwell and Kasarda 1975). The bottom block in the second column, "school and classroom social organization," is intended to stand for the host of within-school variables that affect education, such as administrative leadership, distribution of resources by types of students and subjects studied, and the general school climate as it affects student motivation, discipline, and homework (Coleman et al. 1982: Chap. 5). It also includes the composition of school and classroom in terms of student background and ability.

Individual learning, the only block in the third column, refers to changes in knowledge or achievement over a given period of time. Such changes will be due in part to the school and in part to other influences. The most common indicators of learning are based on standardized tests, and measures of change must take into account random error in these indicators. There may be declines due to forgetting—a decay of human capital—and decreases in the value of learning as technology or other conditions change. Yet the student who has learned, even though his test scores later decline over time, may be capable of relearning material more quickly than someone who has never acquired it; thus we must interpret test results carefully when we use them over a period of time after an initial learning process.

Individual attainment (the upper right block) is affected by causes other than learning, not shown in the figure. There is first a direct effect of family resources on job placement and on opportunities for further education. The possibility that mere credentials, rather than learning, are responsible for attainment must also be considered; indeed, some of the protests against minimum competency tests seem to be claims that even the lowest on the scale of achievement, if they remain in school, are entitled to the credential of a high school diploma. General economic conditions and the state of the labor market also affect attainment, illustrating the fact that general values can be affected by causes that lie in multiple social goal areas.

The center blocks in the right-hand column, "life style" and "cultural tastes," are also related to household production and the educated person's increased capacity for it that may result from education. Finally, "citizenship" in the lower right block represents a presumed advantage of education, a collective good, whereby more reasoned decisions are made and the collective community environment (at the left) is improved. Our evidence as to the value of education in this respect, however, is limited; surveys show differences in the levels of participation and information among educational groups, but we cannot be sure as to their causes or results.

The development and strengthening of a technical community working on models such as this would increase the continuity and the eventual relevance of research on educational policy. Their work would involve improving the structure of such models, estimating their coefficients, and drawing policy inferences from them. They might not only supply research reports and criticisms, but also join in the design of longitudinal studies (Eckland 1980) and experiments.

Models for Crime Reduction

A second major social goal area illustrating this specific and less theoretical style of model construction (as well as an economic approach) concerns policies for crime reduction,[20] relating to the prevention of crimes that might be committed by persons not in prison, and the incarceration and rehabilitation of criminals. Other related values that may be sought through these policies include retribution, equity, citizen satisfaction and security (from the authorities as well as from criminals), cost reduction, and the welfare and rights of accused persons and prisoners. Such values can sharply limit or modify our choice of policies. For simplicity of exposition, however, I shall deal only with policies and models in which the dependent variable is a crime rate.

The area of crime reduction is subjectively social because both crime and punishment are perceived in terms of social definitions that can vary over time and among groups. As the social definition of a type of crime changes, some people's inclination to commit it and others' to report it to the authorities will change, and the effectiveness of policies for crime reduction will change accordingly. As prospective criminals learn of policies that are intended to deter crime or rehabilitate criminals, they can form notions of the expected implementation and effects of these policies by talking with other people, or by reading, or by viewing television. As these perceptions change, the validity of our models is likely to change as well.

20. The value concept here again requires careful definition. A phrase often used to describe this variable is "public safety." We can be unsafe, however, because of natural events such as storms and earthquakes. If we limit our attention to risks caused by persons, we still find there are many that are not categorized as crime, including the risk of unemployment.

This area is largely nonexperimental, in the sense of unavailability of random controls, because (at least for serious crimes) the imposition of major penalties by random processes seems morally inappropriate. Even for minor crimes, the random assignment of types of penalties in various jurisdictions would probably be resisted. Quasi-experiments have, however, been performed: these include interrupted time series based on increased policing in particular areas, with and without the use of other areas for comparison. Interventions with comparison groups facilitate causal inference more than do purely observational studies, but the inferences are still not as clear as for randomized experiments. A small number of randomized experiments have been performed to compare rehabilitative treatments or types of police action.[21]

We immediately encounter the problem that there is less consensus on the definition and measurement of crime than on that of the outcomes of education. This lack of consensus is due both to problems of measurement and to disagreement on concepts and values. Measurements are difficult because victims are not conveniently gathered together in one place (like a schoolroom) where they can be tested or questioned. Most crime indicator statistics and related data are gathered as incidental results of the activities of police, judges, prisons, and parole boards, and based only on the crimes that come to their attention. These activities ignore the crimes that are not reported; but independent measurements of crimes that harm specific, identifiable other persons have also been developed from victimization surveys, providing estimates that differ somewhat from police statistics.

The concept of crime itself is also a source of contention. The lists of acts classed as crime vary among societies and historical periods, even though they share important common elements. Equally important are differences in judgment as to the social cost or undesirability of particular crimes; more radical criminologists, focusing on social inequities as possible sources of crime, have tended in recent years to deemphasize the social costs of crime to middle class property owners and to stress costs to accused persons who are harmed by the police. Black (1983) points out that numerous acts classed as crime are efforts by the actors to remedy other wrongs. Wilson (1975: xix) circumvents part of this problem by using the word "crime" to refer to "predatory crime for gain, the most common forms of which are robbery, burglary, larceny, and auto theft."

The relevant policy models may differ among different types of crime. Models of causes and control of drug abuse are different from those for theft or homicide. Changing technology introduces the possibility of computer em-

21. Rewards, such as subsidies at parole, are more easily randomized than punishments (Rossi et al. 1980). A randomized study of treatments of domestic violence by the police has also been conducted (Sherman and Berk 1983).

bezzlement, or of selling or using newly available illegal drugs. The models with which we are concerned must thus be subdivided according to types of crime, or in other ways depending on the causal processes at work.

In addition, different types of crime are not all of equal cost or disvalue to society. These costs are difficult to estimate, not merely because of lack of information but more importantly because of lack of a well-developed and applicable valuative system. Many thefts might appear to be transfers from rich to poor, such as we may encourage in other realms of policy. We cannot then count the cost to victims of robbery and burglary as a net social cost. Perhaps, then, we must also try to estimate the cost of the fear, the unpredictability, the psychic trauma of having to rearrange one's life, or the feeling of injustice when one has been victimized. Studies in which respondents judge the "seriousness" of crime perform part of this weighting, but have not as yet provided values that can be compared with the monetary costs of crime reduction policies.

My purpose here, however, is not to assert a single "correct" view of indicators and models of crime reduction, but to show that many of these models have developed along different lines from the policy-related models of natural science and economics. Some of the differences among models of crime result from the fact that they are based on different ethical systems. Those writers who stress the process by which societies define or label crime seem more concerned with explaining these social definitions than with reducing crime as conventionally defined. But if we accept the prevailing legal and social categories of crime as undesirable behavior, at least for purposes of discussion, we can see that many models intended to deal with it have been relatively nontheoretical and have changed over recent decades. In these respects they resemble the models we have discussed in connection with educational policy. At the same time we shall see important contributions to this literature from economics.

I shall limit the discussion to models that deal relatively directly with crime reduction, deemphasizing less immediate and less manipulable causes (Wilson 1975: 52) such as unemployment, income inequality, discrimination, and family instability. Among the approaches (or parts of models) that remain, four seem to have been dominant:

1. Incapacitation: removing persons who have committed crimes from the free status in society that would allow them to commit further crimes. The policies involved include imprisonment and capital punishment. The literature on effects of incapacitation has been reviewed by Cohen (1978), showing a diversity of estimates. Ehrlich (1981: 316), also estimates these effects and judges them to be small in comparison with other effects of imprisonment. In more general terms this approach relates to limiting the composition of the free citizen population; it could therefore apply to a category of persons who

are considered likely to commit crimes. Setting aside earlier theories of physical types predisposed to crime, we can still note the recent controversies about alleged criminal tendencies of persons with XYY chromosomes. Moreover, although we do not usually "preventively incarcerate" present citizens, we often consider possible criminal tendencies in selecting immigrants.

2. Deterrence: In its major form this is the effect of apprehending and punishing criminals on the likelihood that other persons, not now imprisoned, will commit crimes.[22] This presumably introduces a special counteracting motive by increasing the expected cost of crime to the criminal. The likelihood that crimes will be committed can also be influenced by removing the means for crime (e.g., guns), or increasing the effective cost (in money, time, or inconvenience) of acquiring them.

3. Rehabilitation: the return of persons convicted of crime to noncriminal ways of life so as to decrease the likelihood that they will commit further crimes, as by providing motives and skills for legitimate employment after their release.

4. Specific prevention: creation of incentives toward legitimate alternatives to crime for segments of the population not now accused of crime. Models of this sort center about crime prevention that reduces particular and local causes of crime; they have been used to justify interventions ranging from settlement houses, to work with gangs, to encouragement of job placement and political participation.

These partial models are not necessarily mutually incompatible. Public resources aimed at reducing crime may be directed at persons inside or outside prisons, and at appeals to various types of motivations. Conceivably there may eventually be tradeoffs or syntheses among these alternative approaches. But in recent decades we have seen in this area a variety of models, relatively simple in nature, entertained either by researchers and the public at different times or by researchers in different disciplines.

We shall trace models for crime reduction through a similar period to that of the Coleman reports, but in this case with a greater variety of researchers. During the 1960s and earlier a set of less punitive models, consonant with a reformist orientation in sociology, received more serious consideration. Crime came to be considered more as a construct created by society in its definition and labeling of criminal acts. Insofar as imprisonment was considered a means for reduction of crime, its rehabilitative function was stressed. Much effort was put into discovering social sources of crime, which could be modified so as to reduce the incidence of crime; into work with urban gangs; and into creation of alternative social structures in which urban minority

22. Those who have been in prison may also be deterred from future crimes even if not rehabilitated.

youth could participate.[23] Law enforcement was not emphasized, however. Bordua and Reiss (1967: 295) then observed, concerning sociological criminology, that "sociologists . . . will have to face up to the hard fact that offenders must be caught before they can be humanely rehabilitated if [sociological] research is to be relevant to police organization and policy." Later, looking back on this period, Cook (1980: 212) wrote that "a decade ago, criminologists tended to view deterrence as an archaic theoretical construct associated with Bentham, Beccaria, and other somewhat naive scholars from the distant past."

A reorientation came in the 1970s. Prior to this time hundreds of evaluation studies of rehabilitation programs had accumulated in the literature, but had not been drawn together. But "in 1966, the New York State Governor's Special Committee on Criminal Offenders recognized their need for . . . an answer" to the question, "What works?" in the field of rehabilitation (Martinson 1974: 23). Robert Martinson was given the task of reviewing the existing evidence. The conclusions of this review were negative, and as a result publication was considerably delayed. Eventually, however, the results of the review were published (Martinson 1974; Lipton, Martinson, and Wilks 1975) and played an important part in persuading the relevant expert community that efforts at rehabilitation had generally been unsuccessful. The methods followed in the study were largely those that we would recommend: tests of actual interventions, and a thorough synthesis of the literature.

As a result of declining confidence in rehabilitation, interest returned to the deterrent functions of punishment. A more theoretical economic type of model, including calculation of expected benefits and costs by the prospective criminal, dominated this approach. A first line of theoretical investigation was begun by Becker in 1968 (1976: Chap. 4), with an abstract optimization model in which resources in the criminal justice system were to be allocated among various types of crime, and between producing certainty and severity of punishment, so as to minimize a social loss function. Ehrlich (1981) later proposed modifications of this model.

A second line of investigation centered on that part of the model dealing with the deterrent effect of capital punishment. An empirical study by Ehrlich applied an economic model to aggregate United States crime statistics for 1933–69, and estimated a deterrent effect of capital punishment corresponding to a reduction of seven or eight murders for one additional execution per year (1975: 414). The study was cited before the Supreme Court and generated extensive controversy. Unlike the 1966 Coleman report, it was first published in a scholarly journal and was sought out for policy purposes shortly

23. An example was Mobilization for Youth, an organization created on the Lower East Side of Manhattan because of a local concern with delinquency. Initially aiming at militant community action, it evolved toward manpower training (Helfgot 1974).

thereafter. In its model of the potential criminal's decisions it included the extent of criminal opportunity, measured by the percent of offenses followed by arrest—a practice criticized later by Cook (1980).

Two especially interesting lines of criticism developed. In a review of the literature under the auspices of the National Academy of Sciences, Klein et al. (1978) concluded that the use of Ehrlich's findings for policy purposes was premature. These critics suggested that *a priori* models depending on rational utility analysis should be supplemented by exploratory research such as had been used in the investigation of the relations between cancer and smoking. Criminal behavior, they argued, is one of several fields involving an interface between sociological and economic relationships, in which the development of models is in its infancy (Klein et al. 1978: 357). This observation paralleled one by Coleman (1972a: 173) concerning the exploratory character of sociological as contrasted with economic research, as well as an analysis by Gibbs (1975) of deterrence; it suggests a long period of research work before results can be justifiably used in public policy debates. Klein et al. (1978: 351) continued in this vein to point out that models in this area are necessarily complex and require extensive development before they can be applied practically: "We might well classify this research effort as being at the stage that econometric modeling reached in the 1940s and 1950s. A great deal of further painstaking developmental work was necessary to build up econometrics to the point at which the numerical findings ceased to be just interesting from an academic viewpoint and became applicable to public policy analysis" (358). This criticism might appear to support the emphasis by Sheldon and Land (1972) on the need for basic research on social indicators; but it differs in that the proposed research concerns models directly aimed at policy questions rather than general models of social change.

A second line of criticism has been developed by Cook (1980: 240–43), who suggests that Ehrlich's measure of criminal opportunities—clearance rates, or the proportion of crimes followed by arrest—was not a valid measure of the concept in question. He contends that this measure is affected not only by the activities of the justice system, but also by the degrees of care exercised by criminals in response to those activities.[24] He goes on to propose that more research in this area make use of interrupted time series studies of actual policy interventions.

Interrupted time series have been used by Phillips (1980) to examine the effects of individual instances of capital punishment. British data on homicides, available in weekly series over the period 1858–1921, allowed him to

24. The presence of simultaneous causal processes is a pervasive problem in analyses of crime and justice policy. Fisher and Nagin (1978) emphasize this problem; and Balkin (1979) points out that for similar reasons, victimization rates reflect not only the risk of crime but also the protective and defensive actions of prospective victims.

examine short-run changes in the homicide rate following a series of twenty-three heavily publicized executions. The result was that on the average homicides did decline after an execution; but they did so for a period of only two weeks, followed by a rise in the homicide rate in the following three weeks—as though the effect was simply a postponement. These findings could not have been revealed by Ehrlich's data or his theory.

Figure 4–2 presents an overall schematic model such as might be used by a technical community in the area of crime reduction policy. Because the processes leading to particular types of crime and to their treatment differ, such a model should be imagined to be specific to a particular type of crime. Moreover, criminals may also be distinguished as to the variables that will affect their behavior, and to the magnitudes of these effects; thus P. Greenwood (1983) stresses the importance of dealing selectively with those who are expected to be high-rate offenders. The variables indicated by asterisks are applicable to the individual potential criminal.

The left-hand column deals with variables that often seem to be non-manipulable: the socialization of potential criminals, through family and other sources; the effect of economic conditions on the presence of legitimate opportunities as alternatives to crime; the demographic composition of the population, for example, the proportion of youth; and social integration, especially in community or neighborhood. Legitimate opportunities vary with general economic conditions, but also vary among individuals and can be altered by policies that provide support to ex-prisoners (Rossi et al. 1980). The composition of the population in an area is not simply exogenous because, as for education, the prevalence of crime in a given area may lead to migration, especially by those who have the resources to migrate and to support themselves elsewhere (dashed reverse arrow). This analogy between education and crime points to two sorts of connections between the models in different social goal areas: they share both analogies and common variables. An effect in one model may also be a cause in another: migration, affected by fear of crime, may affect schooling.

The second column includes several types of variables that are considered antecedents to the commission of a crime of a given type. These include access to relevant tools (such as handguns) and skills; the propensity to commit the crime, which may be affected by all the variables in the left-hand column; and the presence of opportunities, such as unguarded homes with property that can easily be sold. Contributing to opportunities (negatively) is community defense, which in turn derives from community social integration, from specific policies for community defense (not shown), and from concern due to the perceived level of crime.

The central variable of the model, alone in the third column, is commission of crime of the type in question. This has three general types of consequences (outgoing arrows to right, from top to botton): (1) It (or a suspicion or belief

Fig. 4-2. Schematic Policy Model for Reduction of a Given Type of Crime

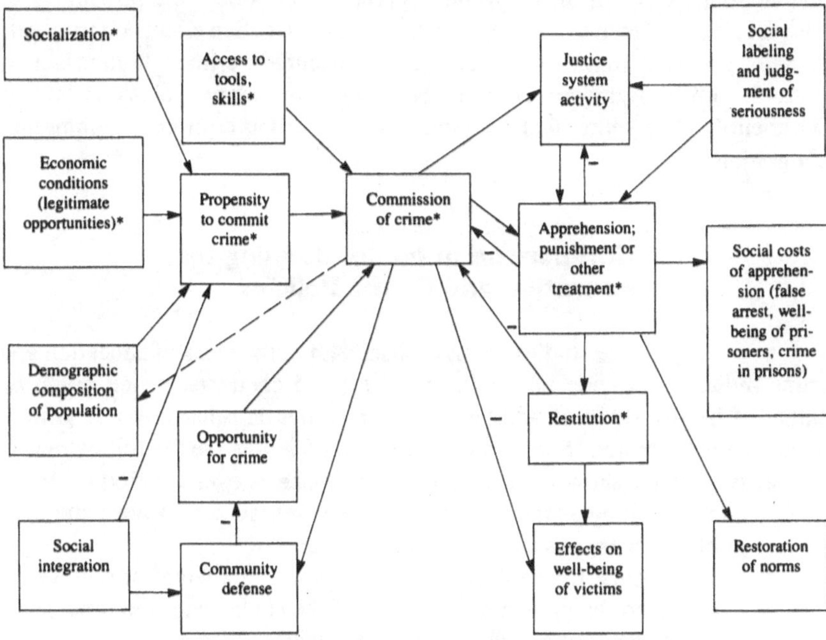

*Variables applicable to the individual potential criminal.

that it could occur) can produce activity from the justice system directed toward that type, or other types, of crime. (2) An individual crime can also provoke specific action by the justice system in apprehension and punishment of the criminal. (3) Regardless of apprehension, a crime can affect the well-being of victims, either as specific victims or as anonymous recipients of detrimental effects due to embezzlement, computer crime, and the like. Well-being can be affected directly, through anxiety about future crime, or through the personal costs required for defensive activities. The prevalence of crime, especially if unpunished, can also reduce social integration.

Several other variables also are expected to restrain the commission of crime, even when the propensity and opportunities are given: the deterrent, incapacitation, and possible rehabilitative effects of the punishment or other treatment given; and as a particular example, the effect of a policy requiring restitution by criminals, which should also increase the well-being of the victims.

Finally, the right-hand column shows a causal factor, the society's labeling of crime and judgment of its seriousness, which influences both the priorities of the justice system and the effectiveness of apprehension (including activity by citizens to aid the authorities). The center block in the right column shows several social costs of an intensified apprehension policy: the possibility of false arrest or excessive action by the police, and negative effects on the well-being of prisoners (removal of freedom and resources; crime within prisons).

At the lower right corner is a "restoration of norms" block reflecting Durkheim's observation that the punishment of crime reinforces community integration.

Comparison of Model Building for Education and Crime Policies

The studies we have discussed in the areas of education and crime reduction provide interesting parallels and contrasts, which might be extended by the study of additional case histories in other areas of policy-related social research. Some of their common features are the following:

a. Both fields show marked changes in emphasis over a period of decades, resulting as much from changes in researchers' values and in public concerns as from research findings;
b. In both fields, the contact of the studies with the world of politics encouraged the publicizing and use of the studies at some times, and their suppression and de-emphasis at others;
c. In the field of educational policy especially, the studies cited dealt with particular consequences of policies rather than presenting full policy analyses.

Let us review these features more fully.

a. Changes in emphasis. In the educational research we have reviewed, the initial focus of attention in the 1966 report was equality of achievement. The 1975 white flight report dealt with possible unintended consequences that might undermine the effects of desegregation, and might affect other aspects of urban life as well. The 1981–82 report dealt with achievement and other effects of school type, but also with possible effects that would flow from policies such as tuition tax credits. The authors of this report favored encouraging student transfers from public to private schools primarily for their effects on aggregate learning, and secondarily on the ground that it would not increase inequality but might decrease it.

Some of these changes in emphasis may have resulted from research findings. When the 1975 report emphasized the negative effects of white flight, Coleman (1975) also cited research showing that desegregation policies were less effective than he had anticipated. We may still ask, however, whether greater continuity could have been achieved in the study of policies and in development of an overall model. If time is required to learn the actual effects of desegregation policies, time and experimentation may also be needed to learn the effects of vouchers or tax credits.

In our review of crime related research we noted the emphasis during the 1960s on rehabilitation and on creation of legitimate opportunities, together with neglect of police action and deterrence as policy possibilities. In retrospect, and in the commentaries of some researchers at the time, this neglect seemed unjustified. After Martinson's negative review of rehabilitation studies, and with a change in the temper of the times, deterrence and incapacitation received more emphasis. Rehabilitation and the creation of legitimate alternatives then drew less attention. This change may have been due in part to research findings, but it was also in keeping with a change in the public mood.

b. Interaction between research and politics. In both fields the pressures of political advocacy, both inside and outside the expert communities involved, modified the customary scientific standards for making findings public and assessing their quality. Findings sometimes proved embarrassing to sponsors (in the case of the 1966 Coleman report and Martinson's study of rehabilitation) and were delayed in publication or not publicized. At other times the findings (and the researchers) were sought out to bolster policy positions, to an extent that may have gone beyond the findings and their scientific weight. These political controversies were paralleled by divisions or changes of emphasis within the expert communities.

The style of publication and review has differed somewhat, however, among the studies we have discussed in these two areas of policy research. The 1966 Coleman report was based on a study required by Congress, leading to some expectations of immediate policy interpretations. This report contained no policy recommendations, but it was written for public rather than scholarly readership and it provoked policy discussions. Martinson's report was prepared under similar auspices and was similarly delayed in publication because it ran counter to expectations. Ehrlich's controversial article was published in the *American Economic Review,* facilitating scholarly review; its later review by a National Academy of Sciences panel also focused on its scientific quality rather than the valuative controversy about capital punishment. Indeed, the predominant conclusions of the review panel were very cautious as regards policy recommendations.

We are left with the question whether the initial involvements of research in policy formation were beneficial for further development of policy models in these fields. The 1966 Coleman report left us with a concern for educational outcomes, and thus contributed to our further quest for educational policy models. Criminologists' earlier advocacy of rehabilitation led to the evaluative studies that Martinson reviewed, but this review was stimulated by state government rather than by the normal functioning of a scholarly community. The publicity received by Ehrlich's 1975 article stimulated critical analysis of models of deterrence, but may not have been necessary to generate this analysis. As regards the effect on policy, the courts might have reached the same conclusions with respect to desegregation or capital punishment without involving these researchers so deeply in political controversy.

If research or model building is to affect policy, it must be expected to arouse controversy. The expert community involved faces the problem of maintaining the relevance of its research to policy choice, while avoiding premature involvement that divides the community or aligns it too soon on one side of an issue. If the integrity and continuity of the expert group is a major concern, we might ask whether researches on education or crime could be given adequate scholarly review prior to their use in policy recommendations.[25]

c. Particular aspects of policy models. The 1966 Coleman report was commissioned by Congress to deal with equality of educational opportunity. It took the important step of enlarging the scope of debate from inputs to outcomes. Nevertheless, after that report the focus of attention continued to remain on effects in the public schools only. Could the research and policy analysis community have then placed more emphasis on demographic trends, so as to anticipate the white flight issue? Could they have dealt with private schools as a possible option? These might have seemed politically conservative options, and concern with them might have appeared to signal an unwillingness to see desegregation succeed; but in retrospect we can see that such researches would have been useful if undertaken sooner and in a broad policy perspective.

In 1975 the white flight report showed the opposite selectivity in its emphasis. It dealt only with demographic effects and not with educational. It conveyed the impression that, whatever the educational benefits of busing might have been, they were insufficient to outweigh the costs of white flight. Because the research did not analyze data on educational effects, these were given little attention in the discussion that ensued.

The 1981 report more nearly resembled a policy analysis. It compared students' achievement in public and private schools, but also explicitly considered possible effects of a policy such as tuition tax credits on numbers of students, by racial and income groups, who might be expected to transfer to private schools if such a policy were adopted. It dealt with concerns that had been expressed about segregation and divisiveness in private schools, but again did not deal explicitly with possible educational effects that might result from changes in the composition of their student bodies (Murnane 1983). It also neglected the purely economic effects of tuition tax credits as a "tax expenditure" affecting the federal budget or as an income transfer to the well-to-

25. A more rapid utilization of research can be obtained through the expert's assumption of the role of confidential adviser to a policymaker or leader; proposals are then evaluated publicly in political or commonsense terms without necessarily involving the prestige of scientific communities.

do. In the strict sense of policy analysis, it "analyzed no variables that are school policies," sets of instructions that tell responsible personnel exactly what they should do (Murnane 1984: 268).

The studies we have mentioned in the area of crime policy, although they showed a shift in concern from rehabilitation to deterrence, did not change in the emphasis of their dependent variables to the same degree; crime rates were a continuing focus. Those earlier studies that had dealt with rehabilitation and the creation of legitimate opportunities did give a particular stress to equity for disadvantaged groups; studies of deterrence, however, gained tacit support from a different notion of equity—that the criminal rather than society should pay for the reduction of crime.

Directions for Policy Model Development: Technical Communities and Multiple Methods

We have implicitly criticized the ways in which the expert communities entered into the political process in the above two areas. Experts must have some relation to politics if their knowledge is to be expected to aid policy formation. Some domains of policy-relevant science can be kept relatively free from politics if they contribute only general findings, unrelated to specific policies but usable as components of policy models; but there are other domains in which more practical recommendations are made. If these latter domains are placed entirely in the "world of action" (Coleman 1972b), then the opportunity for peer quality control over them will be greatly reduced.[26] Our problem is whether the institutions that generate, criticize and disseminate policy models can be improved.

What seems to be lacking in the case histories we have presented is a continuity of focus by the expert community on certain problems and types of policy options; internal critical discourse that can maintain shared standards of quality in the face of strong disagreements on values and political alignments; and collective encouragement by that community of experts' contribution to the policy process in such a way as to present a broad rather than a segmental view of the policy problem at hand. These shortcomings seem to result not only from insufficient concern with these issues, but also from inadequate organization of these expert communities, insufficient prestige for this type of work, and limited prospects for careers devoted to it (Chap. 10).

To remedy these shortcomings we need expert communities of a special kind—required for all types of policy models but especially for nonexperi-

26. Coleman et al. (1982: 225–26) nevertheless emphasize the risk of "suppression of results by extensive social science filtering," contending that this can arise from the policy-related values of social scientists as well as their standards of empirical inference. They therefore seem to approve of scientists' releasing findings to the media before the usual processes of scholarly review.

mental social policy research. These communities must differ from basic disciplines in seeking general policy-relevant knowledge, not knowledge for its own sake (Scott and Shore 1979: Chap. 6). They must differ from the relatively less organized communities of applied researchers by developing clear standards of relevance and quality and strong motives for highly qualified persons to pursue careers within them. I have referred to such communities, maintaining standards of quality as well as contact with the world of practical choice, as technical communities (MacRae 1976b).

The development of expert communities of this sort will not be easy. Experts are human beings, moved by the controversies of the moment. They have political allegiances, and may be reluctant to study problems or to present findings that run counter to them. A broad view of a policy problem may transcend their expertise and increase the risks that their public reasoning will express their personal values and conjectures rather than their expertise. The policy areas examined in this chapter involve controversies about deeply held values and about the distribution of resources—unlike the relatively consensual policies we discussed for national strategic policy or the rehabilitation of mental patients. Research in subjectively social areas, in the absence of experimentation, does not provide the strongest evidence for policy choice. Even natural scientists have been divided in their policy stands on issues concerning nuclear strategy and the environment. The adversary legal tradition, embedded in American politics, also encourages partisanship rather than a quest for more consensual means to the general welfare. Nevertheless, institutions for the maintenance of expertise may ameliorate these problems.

As we seek improved institutions for the development of policy models, we face additional problems because of the diversity of subject matter and of styles of models involved. The models discussed in this chapter have run the gamut from mathematical economic models to highly specific program evaluations. The theories underlying these models are equally diverse. Our task in summarizing the inferences from our two case histories of their use—and conjecturing about the universe from which these cases are drawn—is to suggest institutional conditions under which the most useful models will be developed. These conditions can best be seen in terms of three requirements that a model, useful for public policy and based on research, must satisfy:

1. It must work. Regardless of the prestige enjoyed by various sorts of disciplinary theories, or the pressures exercised by political interests and ideologies, the model must be tested repeatedly in practice if possible.[27] "Working" in practice also resembles "success" (Chap. 3) in satisfying public judgments

27. I am assuming here, in spite of the important symbolic component involved in the assessment of public policies, that a program's success can be defined objectively and intelligibly to laymen. If this cannot be done in the short run, then condition (2), the consensus of a technical community, becomes more important.

on various value criteria and thus dealing explicitly with side effects (Cook and Campbell 1979: 64).

2. The analysis of the model must be based on the shared expertise of a special kind of technical community, concerned with policy models (Chap. 10). Such a community, qualified in the area in question, should be able to appraise the models and proposals put forward, and be heard by the public before extensive resources are invested in corresponding programs or in extensive experimentation.

3. The first two conditions must be consistent with the working of the political process. The public must have some opportunity to see whether the program works, through translation of scientific judgment into intelligible terms. Political controversy should not disrupt the necessary consensus in the technical community, whose technical standards must be maintained and distinguished from *ad hominem* standards of political acceptability. At the same time, the technical community must be prepared to give timely policy advice, comparing policy alternatives as means to clearly stated goals. Even if its members disagree on the goals, their shared expertise should allow them to assess one another's policy models.

We shall now discuss these three conditions, and particularly the second, in greater detail.

The relative importance of a model's working in practice and of technical consensus depends in part on the costs of practical tests relative to theoretical analysis or smaller-scale scientific trials (Chap. 3). A program for deterrence of nuclear war must rest largely on theory, although local wars provide some relevant evidence. Policy changes affecting capital punishment cannot be undertaken lightly. Proposals for management of the national economy, some of which have been carried out with only limited expert consensus in their support, also involve considerable costs and should therefore benefit from an extensive debate in which economists explain their positions to the public. Fairweather's procedure for developing communities of ex-mental patients, in contrast, can be tried repeatedly (not without some possible human cost) and thus depends less, in the last analysis, on theory. We should be able to experiment with schools and prisons, though we are restrained by laws and values that require uniformity of treatment and guarantee certain rights to students, teachers, prisoners, and the public.

Our expert knowledge of whether models work depends on the power of the methods used to test them. In Chapter 3 we discussed the use of successive trials in field experiments as a means of increasing the chance that our policy models can be translated into programs that work. Experimental and quasi-experimental evidence, insofar as it is available, will be of great value in our development of policy models in education and crime reduction. Opportunities for intervention are limited, however, and much of our model development in these areas may have to be based on observational studies.

In the face of these problems—the risk of politicization, the multiplicity of methods, and the weakness of evidence—we require special safeguards to maintain common concerns and common standards in the development of policy models within technical communities. A major focus of these concerns and standards, and of the development of shared expertise, must be the development of general policy models in particular social goal areas.[28] The contributions of individual researchers in technical communities must thus be judged not by their importance for disciplinary theory, but by their importance for development of general models capable of translation into practice. The models that we have illustrated for education and crime reduction might provide some chance for anticipation of future policy concerns. This chance is greater if the corresponding policy problems are of long-run interest (Weiss 1978: 44–45).

The specific studies characteristic of subjectively social policy research can benefit from organization within such a framework of policy models for particular social goal areas. Replication, review, and synthesis of various studies are needed even for randomized experiments (Gilbert et al. 1975: 171–72), but are needed still more when experimentation is not used. To develop and improve these models we must direct our research to questions that the models pose, often by crossing disciplinary boundaries or studying questions ignored by basic disciplines. Professional schools, dealing with particular social goal areas, or the field of policy analysis, may provide a better home than the academic disciplines for many of these models.[29] Research for improving these models can be either comprehensive, aimed at improving the overall structure of the models; or partial, aimed at improving the form and accuracy of their component parts. In the latter case, the basic disciplines may have more to contribute.

A researcher who obtains results related to one of the component parts of such a model may be better able to recognize the partial nature of his findings, and the standards of the corresponding technical community will better support this recognition. Standards may develop as to the type of argument that is desirable to justify support of a policy. Distinctions between those areas in which the empirical support is strong and those in which it is weak may result from reviews conducted in that community. If end-values or other value concepts are defined with sufficient clarity, the community may also be able to

28. Policy models can also make use of general features shared among various social goal areas, as in the case of various types of urban services (Angrist et al. 1976).

29. Professional schools, although usually drawing on contributions of various disciplines, also have a characteristic bias in policy analysis. Because their central task is the training of professional practitioners, they tend to favor policies that work through the training and practice of their graduates rather than through means that bypass professional practice (e.g., training physicians rather than promoting exercise or improving the environment).

make judgments as to whether particular sets of values, combined with the existing evidence, will justify support for particular policies.

The consensus of a technical community, like that of a scientific community, is manifested by its members' assessment of published work and of one another's contributions. Unlike a scientific community, however, a technical community must assess its members' practical contributions; it may be necessary to supply a record of their practical advice for eventual scrutiny by their expert colleagues. Judgment of this record can be supplemented, in the area of public policy, by a general system for publication and review of policy analyses.

The work of these technical communities, though centered about general policy models, must also include the more specific models that correspond to particular programs and the variations needed to improve them. Experimental program evaluation can be conducted in many policy areas somewhat independently of the development of general policy models; Fairweather's approach is an example. In the detailed design of particular programs, in nonlinear and delayed effects, and in the effects of particular managers and circumstances, programs differ from one another and from the more general representations that our techniques of simulation and model construction usually permit. In spite of the attractiveness of uniform models that are intended to be relevant in many places, programs will always be carried out differently in different places: local conditions often require this and programs may benefit from diversity. Much specific knowledge is therefore relevant to model assessment, and general or theoretical knowledge must be combined with it.

Social intervention in the form of experiments or quasi-experiments can sometimes be aimed at general model development. If an experiment can be done on one component of a larger model, as in natural science, its results can be used to estimate parameters in the model. An experiment can also be designed to permit analysis in terms of a theoretical model of component effects, as was done by Rossi et al. (1980) in studying a program that provided unemployment insurance to released convicts.

More often, however, experiments and quasi-experiments are designed so as to tell us whether an intervention worked as a whole but to do so in less theoretical terms. A number of examples of this sort can be found in the field of crime reduction. Ross's (1973) study of the British introduction of a Breathalyser test for drunken driving yielded an interrupted time series showing a large initial effect, declining over time. Apparently the campaign publicizing the policy produced a reduction in drunk driving greater than did the eventual enforcement of the law; the later decrease in this reduction suggested a subjective reinterpretation of the law by drivers. A carefully designed experiment was used to test the effects of police patrolling in Kansas City (Kelling et al. 1974); although it showed no significant effect, it has generated considerable critical discussion. A number of other experiments in police assignments and patrolling have been summarized by Wilson (1975: Chap. 5), but their clearest common result is that "a massive increase in police presence on foot in densely settled areas will probably lead to a reduction in those crimes . . . that require their perpetrators to use the city streets" (1975: 96).

The technical communities we propose must therefore be concerned with specific as well as general models. In their general concerns they will be nearer to the basic sciences, but organized distinctly from them. Their specific work will be of greater relevance to particular decisions in the short run; but their recommendations of program designs for these specific tests will be guided by their general models.

The consensus within a technical community as to what is good work must also be maintained in the face of political pressures. Some criticisms motivated by political interests nevertheless go to the merits of the issues and require serious scientific answers; major consequences can be at stake (Rossi 1980: 896). Other criticisms, however, are directed at the values, political affiliations, and other presumably irrelevant characteristics of the researcher, and must be resisted.

The likelihood of external pressures is increased, however, by work that stresses particular aspects of a policy problem without reference to a more general policy analysis or model. The three Coleman reports exemplify this danger. The 1966 report was a path-breaking work, done in record time under a congressional mandate, and should perhaps be exempted from this criticism. After the publication of that report, however, Coleman (or others in the educational policy community) might well have published a series of interpretative works in the form of analyses of policies to produce desegregation, eventually including busing. They might also have looked forward to other sorts of policy interventions and suggested how they might be approached in future research and analysis.[30]

Coleman's 1975 and 1981 reports were also centered on particular types of information that dealt only with parts of the policy problems to which they were relevant. In each case, these researches led him to be associated with a particular policy position. It would have been more responsible to the political community to state a full justification of that policy position, or some other position, or to tell what further information would be needed in order to reach such a position. Such a broader policy-analytic perspective, even though involving considerations other than research, might have permitted a more general consensus among scholars and informed citizens as to the questions being addressed and the merits of the arguments.

30. There remains a question as to how the standards of publication in empirical science can be applied to speculation about policies that have not yet been tried. An analogy, however, lies in the present custom of suggestion of speculative interpretations of one's data or of new lines of research.

In the field of crime reduction policy, the fluctuation of research perspectives was even greater. The lack of concern with police action and deterrence during the 1960s led to a substantial irrelevance of research to policy. The testing of rehabilitation programs was an advance in our knowledge of policy outcomes, but academic communication was not structured so as to encourage continuing review of the results of these tests. The careful scientific review, and the development of models, proposed by the National Academy of Sciences panel suggests one useful general line of research related to these indicator variables.

Technical Community Versus Client Support

The specificity of some practical research can result not only from the nature of the subject matter, but also from the need for the support of clients or sponsors with immediate practical interests. These interests can be limited in scope and fluctuate over time. The needed continuity in development of policy models may thus require a continuity of support for model development and for the careers of those who pursue it, and not merely an appropriate form of scientific communication. Practical research is in part a collective good: though often done to benefit a particular program or locality, it can benefit others as well. But for this latter benefit to be realized, the research must be designed and supported so as to serve not only immediate needs but also the development of general policy models.

Of all the types of models we have considered, program evaluations give rise to the most specific models. In this respect they resemble the practical trials of prototypes of products—except that programs are not as well standardized as products. Specificity is both a necessity and a virtue in the assessment of possible interventions; but our findings must also be made relevant to later interventions and not simply to the one at hand. The guidance of the next intervention may have to take a more general form.

We usually narrow the focus of our model estimation in evaluative studies in several ways (MacRae and Wilde 1979: 270–74): our definition of the problem, our criteria, and our models are centered about the program in question. There can be a gain, however, from a broader perspective if the source of support allows or encourages it. The possible "external" benefits, in the form of information, produced by one evaluation for the choice of similar programs elsewhere should lead us to use such "laboratory" experience somewhat more generally. A greater use of theory in postulating models and identifying possible outcomes to measure has been suggested (Chen and Rossi 1980), and can lead to measurement of a wider range of outcomes than the official program goals. Policy models can have continuing relevance even though specific demands for them may vary over time, and a case can be made for "broad-focused applied studies" (Rossi 1980: 891–93). Larkey and Sproull (1981: 240) recommend research on fundamental questions related to practice, including the gathering of longitudinal data when a question is enduring. Similarly, the valuative perspective we have presented suggests that we have other sources for the choice of value variables than the goals specified by organizational managers. We may also design the relevant research carefully so as to be able to test the accuracy of the component parts and the effects of intervening variables in the model as well as testing overall effects.[31]

It is also possible, however, to transfer the results of one evaluation to another setting, if the program being evaluated is self contained, by treating the program as a product and evaluating it as a whole. This was done for the Fairweather Lodge, and shows that the transferability of experiments and evaluations depends in part on the role of the researcher. Fairweather and his colleagues were not working for particular clients who had established a program to be evaluated; they created a succession of programs themselves. This approach is possible only if the researchers have the resources and skills to carry out such a program. If, however, they are working with a type of program that they cannot design, or if they do not have two decades to pursue a continuous line of work, they must rely more on the organization of a scholarly or technical community to preserve the necessary continuity and insure that the results are usable by a wider audience than the initial client.

Suppose, for example, that we are designing a local program evaluation and wish to generate information that will be useful for general policy models. We need first to include variables in the research that are of general interest and comparability; prime candidates for these variables are policy indicator concepts. Conversely, the choice of new indicators can be guided by the need to be relevant to local evaluations. Unfortunately, the limited generalizability of specific program evaluations often results from the particular needs of sponsors *and* from the lack of a technical community that would encourage researchers to stress variables of more general interest. We need, also, to try to maintain standards of method that will meet the scrutiny of a technical community, rather than to rush to produce evaluations that satisfy a justificatory need of an involved party. These requirements may run counter to the needs of clients, but they will be supported more firmly if they are built into an expert or professional tradition affecting the education and careers of evaluators who are also policy analysts. Such a tradition requires support, not merely from professional norms, but from sources of finance that can free practical research somewhat from the demands of particular clients.

31. Standards of this kind, supporting the development of practical knowledge beyond the task at hand, exist in the field of biostatistics. They have contributed to caution in making practical judgments that are not grounded in theory. This caution was shown by leading statisticians in the controversy as to whether smoking causes cancer, when the evidence was nonexperimental and the physiological processes of causation had not been clarified (Brown 1972: 47).

Not only the outcomes and intervening variables in the evaluation need to be related to the wider concerns of the policy analytic community, but more important, the treatments themselves need to be described in reliable and standardized terms. Evaluation researchers are increasingly recognizing that if the treatment is characterized simply as "the X program," any findings about it will be of very limited generalizability. Were the effects due to the personality of the program director? To the organizational structure and staffing? To the timing of the intervention and the stage at which the evaluation was conducted? Or to the relations with the local environment? Answers to questions of this kind may not seem to be a local sponsor's highest priorities, but they are of great importance for any efforts to develop programs elsewhere on the basis of the findings. This emphasis leads to the design of comparative evaluations, in which not only treatment and control, but also various treatments are systematically differentiated in ways that can be repeated elsewhere. It leads also to the preparation of review articles, drawing disparate findings together and criticizing them.

These broader types of work, generating information for policy models that have benefits beyond a client's immediate needs,[32] need to be encouraged by the provision of resources with a longer and broader perspective than those of particular short-term contracts. Some clients can provide such a perspective: the Air Force, working through the Rand Corporation, allowed Wohlstetter to restate the basing problem. Universities, foundations, and government can do so as well. This broader perspective is not the same, however, as allowing research to be guided by the criteria of basic science; rather, a research community is needed that will seek long-term practical goals through its own commitment to them. We need to create another type of institution between Coleman's (1972b) "world of the discipline" and "world of action." The policy models that such a community develops must be justified in terms of their prospects for aiding practical choices in foreseeable ways. If the universities can provide continuing support for basic scientific research by persuading donors and taxpayers of its utility, they may be able to provide similar support for the development of policy models, providing regular careers and higher prestige for able people who develop them.

32. This justification extends to evaluation studies that may not be welcomed by the organization under study—a desirable but difficult type of work (Wildavsky 1979: 212).

Part III.
General Indicator Concepts in Policy Choice and Research

5 Net Economic Benefit

In addition to the practical policy models considered in Part II, centered about social goals, we must also develop causal models aimed at the more general end-values introduced in Part I—economic benefit,[1] subjective well-being, and equity. These causal models can interrelate the more specific models for social goal areas. By reminding us to justify those goals in more general terms, they can provide a safeguard against treating particular social goals as ultimate ends. In Part III we shall devote a chapter to each of three principal end-values: net economic benefit (Chap. 5), subjective well-being (Chap. 6), and equity (Chap. 7). In addition we devote a chapter (Chap. 8) to a special class of contributory variables, those dealing with social integration, because the concern with social systems as whole entities is both important and neglected in the literature on policy indicators. The four concepts treated in these chapters deserve special consideration for inclusion in an organized system of policy indicators.

The two end-values discussed in this chapter and the next—net economic benefit and subjective well-being—are the most relevant to the functions of tradeoff or value reconciliation. Net economic benefit is directly useful for tradeoffs, because its use involves assigning comparable monetary values to the dependent variables of policy models centered about particular social goal areas (e.g., health, education, safety, environmental quality). These monetary values can be assigned, in principle, by comparative judgments on the part of the persons affected. Subjective well-being is a more difficult common measure to use in such comparisons, because its use requires the development of empirical models showing how much well-being would be produced by an increment of each social goal variable. Equity, though an end-value, must be defined in terms of the distribution of other particular values; distributional equity is thus most useful as a tradeoff variable when it is defined with respect to other general end-values. Equity with respect to health and to safety, for example, cannot be compared unless these two types of equity are themselves expressed in common valuative terms. We are therefore especially concerned

1. We referred in Chapter 1 to economic "efficiency"; we note below that net economic benefit expresses the same value.

with net economic benefit and subjective well-being as indicators that can be used for tradeoffs.

Although general end-values have an important part to play in organizing policy indicator systems, they will not necessarily have the highest priority as policy indicators in their own right. General end-values can help us to allocate resources among competing uses; without them our choices among social goal areas might lack reasoned guidance. Yet the public statistics that serve for problem definition and monitoring of conditions may well be based on particular social goals. Statistics on general values can focus public attention on general problems; but unless we can act on these general problems, we may be driven back to the level of social goals as a more practical focus for problem definition. Economic indicators can suggest macroeconomic policies, but a decline in national happiness may not so readily suggest corresponding policies. We may thus debate questions of education or crime rather than of general well-being. Our inclusion of general end-values—especially outside the economic realm—thus rests on the organization that these variables provide for other indicators and not merely on their immediate use by the public for problem definition.

Economic Statistics as Examples for Imitation

The social indicator movement arose in reaction against the "economic philistinism" of the statistics of the time (Gross 1966a: ix). It proposed that a wider range of substantive concerns be reflected in the public statistics that are used to define "how well off we really are." Yet in form, if not in substance, it sought to imitate economic indicator systems. It dealt with national statistics, seldom connecting these to specific local models of causation. It sought to develop macro models, analogous to those of macroeconomics (Land and Felson 1976) as well as to construct national accounts. In these tendencies, it seemed to seek a calculating procedure similar to that of economics but directed at different values.

Our approach to policy indicators, however, leads us in a different direction. We include one major economic indicator, net economic benefit (corresponding to national product or income), as a central variable in a proposed set of general end-values, looking to continued comparison in public debate between this indicator and subjective well-being. As a major similarity between net economic benefit and other indicators, we stress the possibility of reconciliations or tradeoffs among more specific values. We place low priority on national models and national accounting for end-values other than economic benefit—desirable though they may be in the longer run. Finally, we note that macroeconomics has only limited value as an example for emulation because its causal models have no immediate parallel in other fields, and because its dependent variables are not organized into a general ethical system like those of microeconomics.

This chapter will therefore deal with the similarities and differences between economic and other indicators. We consider first the formal properties of optimization of net economic benefit, a means of value reconciliation that might be copied in our use of other general end-value indicators; the connection between particular measures of economic benefit and the national product statistics; and the extension of this reconciliation to nonmarket areas. This extension of the estimation procedure, involving questioning persons rather than simply observing the market, leads us to explore the general definitions of objective and subjective indicators. Finally we show that some macroeconomic indicators other than the national product require the same sort of valuative scrutiny that we proposed for social goal areas in Chapter 1.

Our inclusion of economic benefit as an end-value departs from the practice of the social indicator movement. This inclusion is based partly on the logical structure of applied welfare economics, which can serve as an example for value maximization in other fields, and partly on the connection of benefit-cost analysis of particular policy choices with the national income and product statistics. It also rests, however, on the status of preference satisfaction as an end-value.

To the critics of economic values, these values may seem to symbolize material things at the expense of human relationships, wasteful production in a time of increasing scarcity, calculation in terms that count dollars rather than persons equally, and the elevation of money to the pinnacle of our value system. Although we may not be able to persuade these critics that economic values are intrinsic values, it will be useful to state our reasons for classifying them in this way. At the heart of economic reasoning is the assumption of given preferences by individuals among states of affairs. People are assumed to act rationally so as to satisfy these preferences. If they satisfy them, they are assumed to be better off.[2] In this perspective, they are attaining an intrinsic value: preference satisfaction is not, in this view, a means to anything else. Conceptually, as we shall see, it is also closely related to subjective well-being.

How then (the critic may ask), did we move from this nonmaterial, qualitative notion of the good life to the monetary, material measures against which the social indicator movement has protested? The customary Pareto criterion for preference satisfaction is applicable in principle to sources of preference satisfaction outside the market as well as to marketed goods. The monetary aspect arises, however, from the connection of markets with this criterion. In a perfect market (the source of elegant theorems though not always approxi-

2. In more detailed analyses it is recognized that this relation holds only for adequately informed preferences.

mated in reality), a system of prices transmits information between the buyers and sellers of various commodities so as to promote the satisfaction of their preferences by Pareto improvements. Another step towards marketed (and largely material) commodities takes place when we compute the national product, or the net benefit of a proposed project. In principle many sources of well-being can be valued in monetary terms, but in actuality this valuation is easiest to carry out for things bought and sold in the market.

The Optimization of Economic Value

The economic approach to both policy analysis and national product indicators involves the comparison of competing uses of resources in terms of a general ethical criterion. This criterion is net economic benefit—the criterion used in benefit-cost analysis, which also corresponds to the national product or income. In maximizing net economic benefit, economists have raised the following issues, which are relevant to the maximization of any general end-value and especially to variants of utilitarianism: we must recognize the contribution of ethics, over and above that of science, to the formulation of welfare criteria; we must ask how effective nongovernmental processes can be in furthering the value in question; we must allocate resources so as to produce the value in question efficiently, considering not only benefits but also costs in similar valuative terms if possible; we must recognize the requirements that are imposed on our causal models by the need to maximize—they cannot be simply linear.

We shall therefore present an elementary account of the basic assumptions involved in this process of maximization. To show how it is arrived at and used, we shall proceed through several preliminary steps. We shall show how economists analyze the performance of a hypothetical perfect market in terms of the Pareto criterion; then introduce the use of market failure as a condition for government intervention; and then present the criterion of net economic benefit, used in benefit-cost analysis to estimate whether a particular government intervention will increase the national product or income. Throughout all these steps we shall stress the generality of the procedure, which allows us to compare commodities or policies in terms of a common monetary measure.

Welfare economics—the branch of economics concerned with well-being and with recommending policy choices—is based on a monistic ethic of preference satisfaction (MacRae 1976a: Chap. 5). It uses this ethic to specify tradeoff procedures among more specific competing values produced by alternative policies. Individuals' preferences are considered as given, and market or other arrangements that satisfy them most fully are considered most desirable. One expression of the ethic of preference satisfaction, due to Vilfredo Pareto, considers each individual's set of preferences as a rank ordering of states of the world, or of states in which that individual may be.[3] A clear improvement in the collective preference satisfaction for a number of persons can be seen when, as we move from one policy (and the state of affairs it produces) to another, no one arrives at a less preferred situation (becomes "worse off") and at least one person attains a more preferred one (becomes "better off").[4] A move or change that has this property is said to satisfy the Pareto criterion, and a situation from which no further such move is possible is said to be Pareto optimal. This approach seems to make few assumptions, as it takes individuals' existing preferences as they are and does not favor policies from which anyone would suffer in his scale of preferences; but precisely because of the assumptions it does and does not make it is a particular ethic, distinct from other ethics, leading to decisions and nondecisions about desirable states of affairs.

The Pareto criterion can be applied not only to comparing particular sets of policies, but also to evaluating the functioning of markets in general. Economists have studied the theoretical properties of perfect markets and have shown that such a market, transmitting information through a system of prices, would lead to Pareto optimal results (Bator 1958: 351). The conditions for a perfect market are not always met, however. Thus a classification of types of *market failure* (departure from these conditions) can be made and can be taken to justify possible intervention in the market by government or other collective actors (MacRae and Wilde 1979: Chap. 5). Market failures include inadequate competition; effects external to the market, which people receive without compensation or exchange; and collective goods, which when furnished to one consumer are automatically furnished to many. This line of reasoning provides an extremely general model that sets necessary conditions, given the Pareto criterion, for instituting particular public policies.[5]

The market failure criterion, setting necessary conditions for government action, needs to be supplemented by further conditions; when it suggests government intervention, we still need to decide whether particular interventions are desirable. These further conditions will lead us eventually to benefit-cost

3. This assumption appears quite general; but in application it is associated with "comparative statics," by which we compare various possible equilibrium states of the economy and neglect the processes of transition between them. The orderings in question are not comparable among persons; thus without further valuative assumptions one person's gain cannot be balanced against another's loss.

4. The latter person's preference is thus satisfied; but this is not the same as subjective satisfaction (Chap. 6).

5. The market failure approach, though widely used in policy analysis, can be criticized on two grounds: (1) that other criteria than the failure of the market, such as equity, merit goods, or dynamic problems, should also be considered (MacRae and Wilde 1979: 171); and (2) that government may not in actuality perform in the ideal way necessary to correct market failure (Wolf 1979). This second criticism is mitigated, however, if realistic estimates of the net benefit of programs can be made.

analysis. We must first show, however, how the foundations of benefit-cost analysis are related to the optimization performed by a perfect market. To show this relation, I shall sketch a theoretical approach to the optimization problem, used in discussions of public finance, for a single good (Musgrave and Musgrave 1976: 53–54); in the next section I shall consider a more practical approach to this problem of policy choice, involving the estimation procedures of benefit-cost analysis.

We pose the problem of optimization first for a single good sold in a perfect market—a special case of the more general optimality theorem mentioned above. For this case, the quantity of the good produced at the normal equilibrium of the market, without government intervention except as is necessary to maintain the market conditions, can be shown to be optimal. This judgment of optimality rests on a comparison of static equilibrium situations (*comparative statics*) that is characteristic of microeconomics but not of macroeconomics or of subjective indicators, as we shall see later in this chapter and in the next.

Our criterion for analyzing this problem will now be that of benefit-cost analysis—the maximization of net economic benefit—rather than the Pareto criterion. The monetary calculation of benefit-cost analysis was introduced to replace the Pareto criterion because that criterion often fails to prescribe a choice between alternative policies. When a new policy, relative to an old one, makes some people better off and others worse off on their preference scales, the Pareto criterion cannot recommend the change to the new policy. But if we started from the new policy and were considering the old, it could not recommend that we change to the old either; again someone would be worse off. Because of this indeterminacy, Kaldor (1939) proposed that such a new policy might be chosen, even if some persons became worse off on their preference scales as a result of it, if these losers might (hypothetically) be compensated by the gainers. This principle of hypothetical or potential compensation—a strong valuative assumption—has served as a foundation of benefit-cost analysis, according to which only an increase in total net benefits is required to justify a new policy, and it is no longer necessary that no one lose from it.

A special case of the equilibrium of a hypothetical perfect market—the partial equilibrium of that part of it that concerns a single good—may be described by the conventional diagram of demand and supply curves shown in Figure 5–1.[6] The horizontal axis Q measures the quantity of the good that would be purchased in a unit time, and the vertical axis P the price at which it is sold. The demand curve D shows the quantity that would be demanded per

6. The partial optimum that we shall discuss cannot be assured to be desirable unless the larger market is perfect.

Fig. 5-1. Demand and Supply Curves

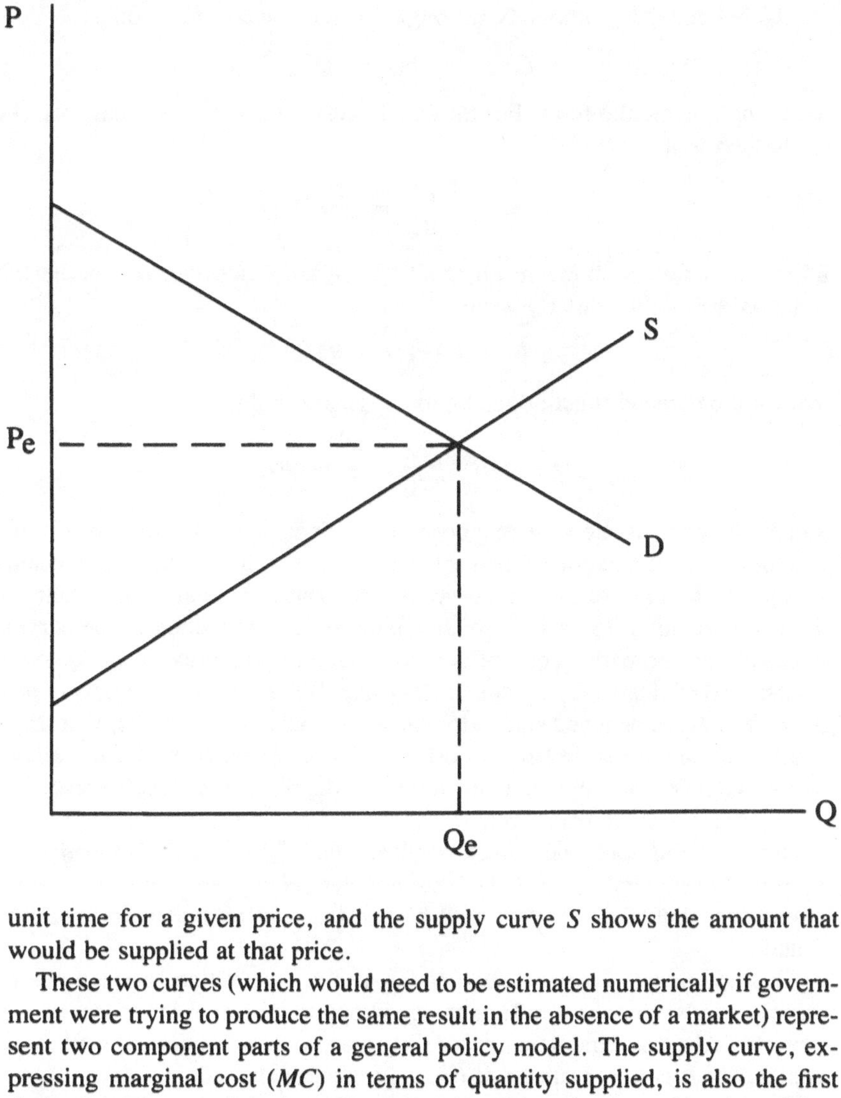

unit time for a given price, and the supply curve *S* shows the amount that would be supplied at that price.

These two curves (which would need to be estimated numerically if government were trying to produce the same result in the absence of a market) represent two component parts of a general policy model. The supply curve, expressing marginal cost (*MC*) in terms of quantity supplied, is also the first derivative of a cost function (*C*) embodying the optimal combination of factors of production for each quantity supplied. It therefore represents one component of a policy model, connecting factors of production and their cost with the output of goods or services—a model resembling the production models we have discussed for social goal areas. The demand curve, expressing marginal benefit (*MB*) in monetary terms as a function of quantity purchased, is also the first derivative of a benefit function (B) connecting quantity consumed with value realized.

Algebraically, for simplicity we might express the cost function as

$$C = a + bQ + cQ^2 ,$$

assuming a parabolic form. For the supply curve, equal to marginal cost, we could then write

$$S = MC = \frac{dC}{dQ} = b + 2cQ ,$$

a linear function as shown in Fig. 5–1 with b and c positive. For the benefit function we might similarly write

$$B = d + eQ + fQ^2 ,$$

and for the demand function, equal to marginal benefit,

$$D = MB = \frac{dB}{dQ} = e + 2fQ ,$$

a linear function as shown with e positive and f negative. The optimal solution is found by maximization of the net benefit (benefit minus cost), which occurs when the difference of the derivatives is zero. This is equivalent to equality of the two derivatives ($S = D$), which obtains at the intersection of the curves, the condition for market equilibrium. At the equilibrium price P_e and quantity Q_e the market clears (supply equals demand). However, even if Q_e were produced by government and sold to the same consumers at a price less than P_e, it would still be optimal in benefit-cost terms. The benefit-cost criterion relates to optimal levels of production and not to the distribution of benefits and costs among consumers and taxpayers.

Note that the nonlinear form of at least one of the B and C functions is essential for maximization. If both were linear (as is often assumed in causal modeling), their derivatives would be constant and no maximum could be found.

This optimization for any one good is also an implicit comparison among various goods, since cost reflects foregone opportunities for other uses of factors of production, and demand reflects foregone opportunities for other expenditures. It is thus a means of reconciling particular values.

This optimal equilibrium is arrived at spontaneously in a perfect market for one good; but the same criterion for optimality also applies to an analogous condition that we may wish to reach by calculation and government policy. We illustrate this condition by a pure collective good, which when one consumer receives it is automatically available in equal quantity to others. This is an instance of market failure, in which the optimum point will not be reached spontaneously. Individual consumers will attempt to be "free riders" and allow others to pay for production of the good, which they will then receive automatically. The result will be less than optimal production of the collective good.

To introduce these distinguishing features of a pure collective good we shall contrast it with the private type of good discussed in connection with Figure 5–1.

When the demand curve for a private good results from the combined demand of a number of buyers—each buying his own portion of the good at the equilibrium price P_e—that curve can be understood as having resulted from a horizontal summation of individual demand curves. At a given price each purchaser buys his or her own quantity, and these quantities must be added to give total demand at a given price. Figure 5–2 shows this relationship, with D_1 and D_2 being the demand curves of two purchasers and the heavy curve, $D_1 + D_2$, being their horizontal sum. This summed demand curve again intersects the supply curve at equilibrium, giving rise to the equilibrium price P_e and quantity Q_e. Now, however, we see that the equilibrium quantity Q_e is divided between the two buyers in parts equal to Q_1 and Q_2 respectively.

An analogous diagram (Figure 5-3) can be drawn for a pure collective good such as national defense or clean air, which might be supplied through public policy. Consumption of such a good is "nonrival" in that one person's consumption does not detract from the quantity available to others. For such a good, individual demand curves D_1 and D_2 must be summed *vertically* to yield $D_1 + D_2$; once a quantity Q is produced, that quantity automatically becomes available to all consumers. The supply curve, here as in the case of an actual market, reflects the costs of furnishing various quantities of defense, or of reduction in the concentration of a pollutant,[7] in a unit time interval. If production is carried out by government, these are the opportunity costs of withdrawing resources from the private sector.

An optimal solution (no longer a spontaneous equilibrium) can again be expressed (Lindahl 1967 [1919]; Bowen 1943; Samuelson 1954) in terms of the intersection of the supply and demand curves, but there is now no single price; for individual preference satisfaction, consumers 1 and 2 would ideally pay different prices P_1 and P_2 for the same quantity Q_o that each receives. The two individual prices now sum to the total optimal social price P_o. This type of solution, like the market optimum for a private good, represents a reconciliation among various valuations for this collective good and for foregone opportunities.

7. Such an analysis tacitly assumes that the bad resulting from pollution is subject to compensation based on rational judgment by those affected. Thus, if smoke increases the laundry and cleaning bills of nearby residents, they might be compensated for this cost and for the esthetic cost of less attractive clothes and home furnishings. If, however, pollution causes health hazards about which the recipients are uninformed, or about which we think they ought not to bargain, then analysis in terms of prices is open to criticism.

Fig. 5-2. Private Good: Horizontal Summation of Demand Curves and Equilibrium

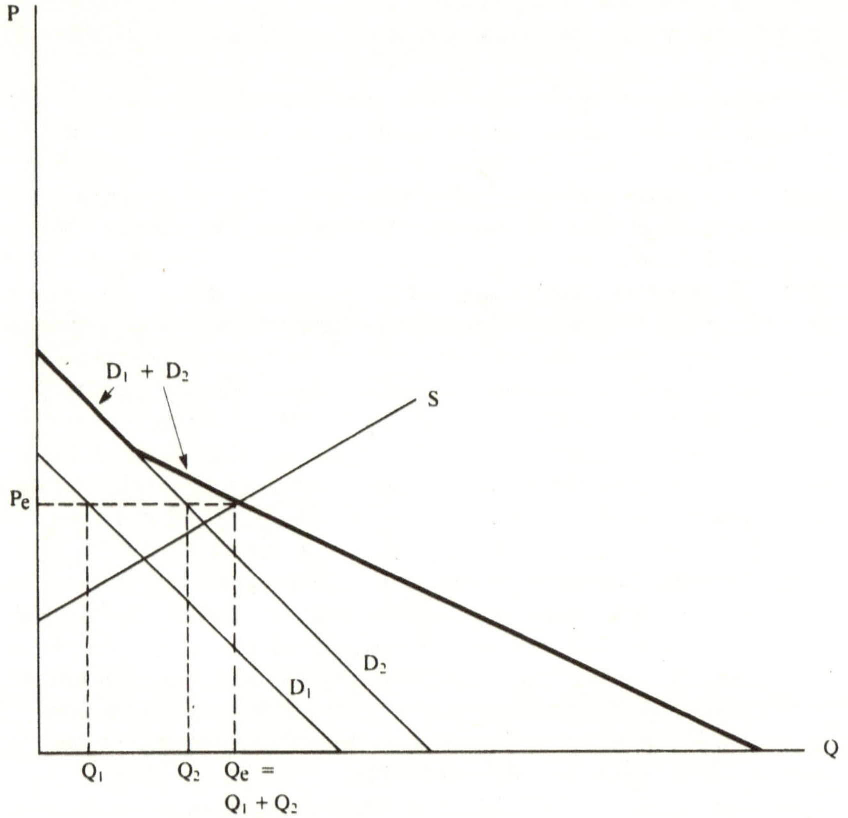

For a perfect collective good of the sort shown in Figure 5–3, the estimation of Q_o (if it is possible) would indicate the amount of the good to be produced. That quantity would then be automatically available to consumers 1 and 2; no problem of distribution would arise, but there might be problems in obtaining the appropriate payments P_l and P_2. When benefit-cost analysis is used to judge whether a project is desirable, it typically focuses on the quantity to be produced efficiently rather than on the distribution of prices (taxes) paid. For the sake of equity the taxes that are used to pay for a project should ideally be collected from persons likely to benefit from it (e.g., those who reside in the area affected; or gasoline users for road construction), but we cannot usually seek the exact correspondence between benefit and payment that Fig. 5–3 suggests.

This general approach—maximization or marginal analysis—can be used even when markets are imperfect and shadow prices must be estimated (Dasgupta et al. 1972: 244–45). Conceivably it may also be applicable to the choice of levels of production of goods or services in terms of nonmonetary measures of recipients' needs or demands. We have noted in Chapter 2 that the claims for satisfying various sorts of needs (as for health and housing) are often made in noncomparable terms. Conceivably we might characterize needs as deficiencies in welfare. We might then try to describe the needs of a population in terms analogous to consumer demand curves, with the quantity of each needed good related to the derivative of a social welfare function; in this way we might try to distinguish more urgent from less urgent needs. In conjunction with cost curves using this common measure, this information could be used in principle to set optimum levels of production. This sort of calculation, if based on monetary values, is understandably resisted by those who are concerned for the needy, as poor people cannot express intense

Fig. 5-3. Collective Good: Vertical Summation of Demand Curves and Optimum

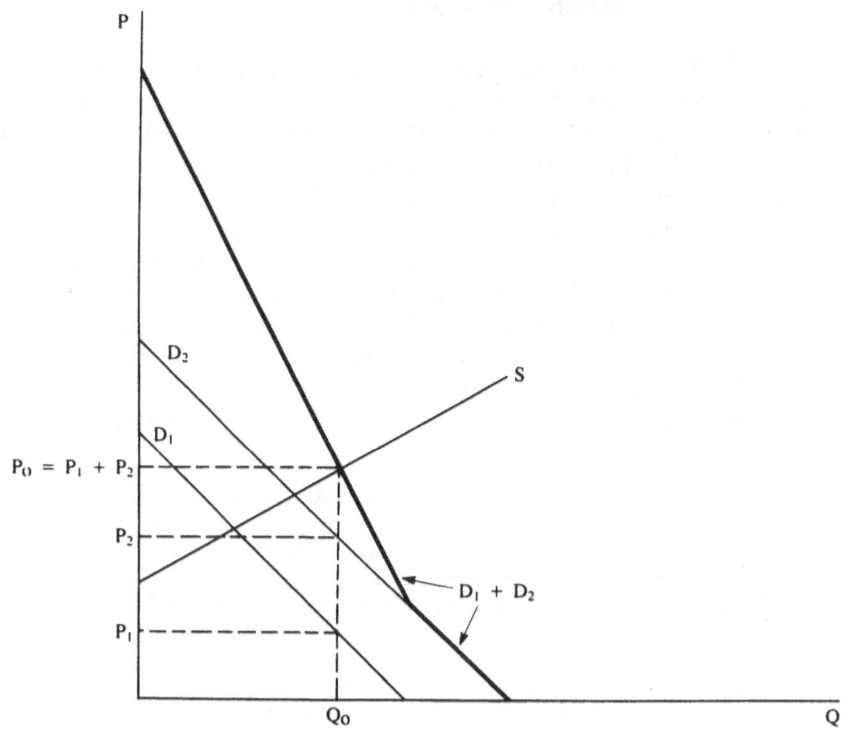

desires by willingness to pay large amounts of money. If, however, there were a more equitable basis of weighting individual demands—say, in terms of the number of hours one was willing to pay—then the reasoning might be more acceptable.[8] We shall return to this possibility in Chapter 7.

So far we have described an abstract procedure for estimating an optimal level of production of a good when we know the benefits and costs of that good as functions of quantity produced and consumed. We have noted, however, that it may not always be easy to estimate these functions, especially in regard to demand or benefit. If we ask people about their valuation of the good and they expect to have to pay in terms of their reply, they may not tell us the truth.[9] It is nevertheless possible to estimate demand or benefits from data on market behavior, or to use the responses of one population (not required to pay in relation to their responses) as a guide for policies that affect another. We shall now examine the procedures that have been used to estimate benefits and costs, starting with those related to market information (also related to the national product) and then considering those for nonmarket goods, services, and conditions.

Estimating Net Economic Benefit and the National Product

Benefit-cost analysis involves the estimation for proposed public programs, of benefits, similar to the satisfaction of consumer demand, and costs, similar to the costs reflected in a supply curve. As Haveman and Weisbrod (1975: 171) point out, benefit-cost analysis is "an attempt to replicate for the public sector the decisions that would be made if private markets worked satisfactorily." Thus the criterion of net economic benefit is equivalent to that of market efficiency. In Figure 5–1, if a quantity less than Q_e were being produced because of some imperfection in the market, the corresponding price that would be offered by buyers (on the demand curve) would exceed the price asked by sellers (on the supply curve), and the difference between the two would correspond to a net social benefit of producing another unit of the good. If, through a government program, we produced further

8. If an equitable basis of interpersonal comparison could be chosen and measured, we might still draw on economic measures for *intra*personal comparisons of values. This is the approach used by benefit-cost analysts when they calculate monetary benefits or costs to individuals and then weight them by some function representing the marginal utility of income. To carry out such procedures in more general terms, however, it is important to compare economic and noneconomic measures of well-being.

9. This difficulty has led to the proposal of ingenious incentive schemes to induce "sincere" expression of consumer or voter preferences for collective goods in quantitative terms in order to avoid this sort of bias (Tideman and Tullock 1976; Hylland and Zeckhauser 1979).

units and increased Q gradually toward Q_e, we would generate additional net benefits of this sort until, finally reaching Q_e, we could no longer do so. An excess of benefits over costs is required for the program to be preferable to no program; if there is a budget constraint, however, only those programs that together lead to the greatest net benefit should be funded.

Measurement of the benefits of alternative policies requires that we assess the preferences, in monetary terms, of potential consumers and others who would be affected by those policies. Measurement of costs requires knowledge of production possibilities and preferences of owners of resources. Preferences are subjective variables, but the economic approach estimates them wherever possible from market behavior. If a water project will permit the growing of new crops on irrigated land, the prices that consumers would pay for those crops are used to estimate their value. Similarly, cost estimates are based on free-market prices where possible. Even for goods and conditions not sold in the market, economists seeking to estimate their value tend to prefer indirect inferences based on choice behavior like that occurring in the market.[10]

Thus the prices that are a central feature of the perfect market have been taken as measures of value for benefit-cost analysis (Harberger 1971: 785). Insofar as actual markets resemble perfect ones, there is one type of objective variable—price—that not only reflects widely held values but is also embedded in an organized ethical system.

This use of price as a basis of estimation of value, however, is an approximation. The marginal benefit that a consumer receives from a purchase is not, in general, what was actually paid, but the most he or she would have been *willing* to pay. Similarly, the producer entering into exchange has paid out for each unit produced (apart from fixed costs) not its full market price, but only its marginal cost of production. The corresponding vertical differences between the consumer's marginal benefit and price ($D - P$ in the figures above) and between price and the producer's marginal cost ($P - S$), which represent marginal net values to consumer and producer from a transaction, are known as the *consumer surplus* and *producer surplus* respectively. The sum of these two surpluses is the difference between the consumer's marginal benefit (measured by the demand curve) and the producer's marginal cost (measured by the supply curve).[11]

10. Some economists have estimated demand for publicly produced goods from voting statistics on referenda, in relation to community income (Deacon and Shapiro 1975; Noam 1979). We discuss the estimation of nonmarket values further in the next section. For both costs and benefits, the existence of the opportunity to invest money at interest leads us to value a given sum less in the future than in the present, and thus to discount costs and benefits expected in the future.

11. Estimation of these surpluses, when they correspond to conditions far from those of actual markets, may require the use of questioning rather than market data. Producer surplus (economic rent) involves more complex problems of estimation than consumer surplus (Mishan 1982: Chap. 10).

The adequacy of price as an approximation of the value defined by demand depends on whether the goods and services resulting from a proposed policy are a minor contribution to an existing market and do not change prices in that market appreciably. Thus, if a dam were built and provided irrigation for the growing of new crops, and if these crops were sold as a small part of the supply offered in a free market, the additional willingness of consumers to pay for them would be measured by the preexisting price times the quantity produced. Those intramarginal consumers who had a substantial consumer surplus, that is, who valued the crop more than the price, would have already bought it, and the new amount of the crop would go only to marginal consumers who valued it at approximately its market price. On the other hand, if a new bridge were built where there was not a preexisting market for crossings, the benefits produced would be indivisible "on the grounds that [a] large increment is made available as a unit" (Merewitz and Sosnick 1971: 135). In this latter case, consumer surpluses would have to be estimated, since some of the new users of the bridge would not have been previously buying crossings in the market at all.[12]

Benefit-cost analysis includes procedures for monetary measurement of values in social goal areas, such as health, safety, and education. Insofar as these conditions are purchased in a perfect market, expenditures to obtain them can be used as approximate estimates of their monetary value. Alternative estimates of their monetary value may also be based on their expected effects on earning power (human capital), reflecting effects on production and earnings through changes in the discounted earnings streams of persons who were ill, victimized, or educated. The effects of health, safety, and education are not limited to market production, however; the ill health of retired persons affects their well-being even though they are not in the labor force; the pain of injury affects the victims of crime or accidents directly, over and above its effect on their production; fear of crime affects well-being; and the benefit from education through appreciation of science or literature, or through household production, is a social benefit even if it does not pass through the market. These latter effects, which work directly on positive or negative "consumption," are treated further in the next section.

Benefit-cost analysis is not only a technique for policy choice but also an expression, with respect to particular policies, of a value concept that is the basis of a national economic indicator—national product or income. As Haveman and Weisbrod (1975: 176) express this relation: "When a project is evaluated, the analyst should bear in mind that his immediate target is to de-

12. The estimation of consumer surplus from a single demand curve involves an approximation; but Willig (1976) and Randall and Stoll (1980) have estimated the extent of this approximation to be small when consumer surplus is small relative to the consumer's income.

termine if the value of the output in the economy (the national income) *with* the proposed project is greater than the value of the economy's output *without* it."

The analysis of the relative merits of particular projects is thus conducted so as to show whether they make a net contribution to national output or income.[13] These terms refer, more specifically, to the national product, a major conceptual basis for economic indicators. The *net national product* is a measure of "the money value of the flow of goods and services" in the economy and includes net investment, that is, investment minus depreciation. The *gross national product* (GNP) does not deduct depreciation because of the difficulty of estimating it, and is thus more widely used (Samuelson 1970: 169–77). Economists can correct the GNP by subtracting flows that do not contribute to welfare and by adding sources of welfare that are not reflected in money flows (Nordhaus and Tobin 1972); these problems of estimation closely parallel problems of estimation in benefit-cost analysis.[14] Among the problems that are shared by both types of estimation are the avoidance of double counting, and decisions as to whether and how to count benefits and costs that are not transmitted through the market.

The national product can be measured in either of two ways: as a flow of final goods produced, by summing the "value added" in all the processes of production in the economy, or as a flow of earnings, by summing the payments for these contributions in the form of wages, interest, rents, and profit. The latter approach is better suited to the estimation of well-being if we are concerned with who receives benefits (e.g., questions of equity among income groups), rather than simply with a monetary sum. Similarly, a particular project that is subject to benefit-cost analysis may generate tangible outputs to which we can assign monetary values; but when we ask who received the corresponding benefits, we must trace the ramifications through the economy to individual recipients. These ramifications can extend through the market for the outputs, including not only direct benefits to consumers but also effects on complementary or substitute goods. They also extend back through the productive process to suppliers at more than one remove. For many projects, the

13. On the equivalence of these terms see Samuelson (1970: 169).

14. An alternative view, however, is that such economic indicators do not exhaust the measurement of well-being, but measure only *economic* well-being by methods in which economists have special competence (Harberger 1971: 785–86; Okun 1971). Distributional equity, for example, is recognized by these writers as valuable but excluded from this procedure of measurement. From this perspective, net economic benefit is still an end-value but it coexists with other values in a system of multiple values. As Nordhaus and Tobin (1972: 4) express it: "Economists all know [that maximization of the GNP is not a proper objective for policy], and yet their everyday use of GNP as the standard measure of economic performance apparently conveys the impression that they are evangelistic worshipers of GNP."

estimation of the distribution of benefits across income groups would be extremely difficult—at least commensurate with tracing the ultimate incidence of a proposed tax.

The relevant policy indicator statistics, if they are to be disaggregated, must thus be gathered at the level of individual income payments or individual consumption. Our models of causation may not be able to predict the distribution of income changes as easily as they predict the total increment to product or income made by a project, but we need to know individual levels of end-value measures in order to compare the individual incidence of alternative value concepts, and also to examine distributional equity. Juster et al. (1981: 42–43) have suggested that social accounts that consider behavioral and psychological variables together with economic variables should be constructed "from records obtained at the micro level."

The advantages of directing both indicator measurement and policy analysis at the same variables in this way are multiple, and may be sought for other policy indicators as well. First, this sharing of variables reflects a consensus among experts as to the values that are being sought; in spite of the criticisms that have been directed at economic values, the consensus of that discipline on values as well as on theory has been a source of its strength. Second, this sharing of variables can sometimes permit the cross-checking of results of particular programs (if they are large enough in aggregate) with changes in general indicator statistics for a nation, state, or locality. The contribution of education to individual income, for example, may be compared with estimates of the overall contribution of education to the economy (Becker 1975: 196). Third, as a result of this cross-checking, possible positive or negative external effects of particular types of programs may become visible when overall statistics are studied. If particular programs appear to further social advancement for individuals or groups, then the lack of a comparable advance in the well-being of such groups over time can sensitize us to possible external costs or positional goods (Hirsch 1976: 4–7).

We need not, however, compute the entire national output in order to evaluate a particular project; all we need to know is what that project will add or subtract, relative to its alternatives. Braybrooke and Lindblom (1963: 86) distinguish between this sort of marginal evaluation and the large task of ranking social states in terms of a social welfare function. The marginal type of reasoning embodied in benefit-cost analysis is applicable in principle, however, to any ethic that seeks to maximize a social welfare function.

The national product (or product per capita) as an aggregate indicator statistic is obviously a significant source of problem definitions. Suitably interpreted, it can help to define problems not only for macroeconomic policy but also in broad areas such as the productivity of the labor force. This type of problem definition does not, however, tend to identify the sort of projects that are usually assessed by benefit-cost analysis. One source of this discrepancy is that benefit-cost analysis is considered appropriate regardless of whether the public considers a problem to exist; in a period of prosperity the possibility of increasing production further by a new project could be worth analyzing, even if the public had to be persuaded that the project was desirable. As we noted in Chapter 1, the definition of problems by comparison of indicator statistics with expectations, identifies felt deficiencies but need not focus our attention on those areas where we can act most effectively.

A second source of the discrepancy between the use of national product statistics and benefit-cost analysis lies in the fact that the national product is subject to modification by macroeconomic policies. It is thus seen as part of a complex of indicators, linked together by macroeconomic models, including measures of inflation, unemployment, interest rates, and money supply. The goals of macroeconomic policy are not, however, limited to efficiency in the perspective of comparative statics, as were the goals of microeconomics or benefit-cost analysis. Rather, they deal with stabilization and growth in the economy—questions of change and not simply of optimum level. In evaluating change we may conceivably be optimizing a summation of well-being over time, seeking only to maximize production or income as before; but the public definitions of the problem, as affected by these indicators, may equally well center about economic change itself. We shall return to the question of how economic change is to be valued in the final section of this chapter.

Estimating Nonmarket Benefits and Costs

The value concept of net economic benefit involves preferences translated into monetary valuations, usually in terms of prices. This translation is made most directly when the resources required for a project, or the products from it, are bought and sold in the market; even then, however, imperfect market prices need to be replaced by shadow prices (Margolis 1977). There remain, in addition, a wide variety of benefits and costs of projects—and of corresponding elements of our aggregate well-being—that derive from goods and bads that are not bought and sold. These include the externalities from production and consumption, the general qualities of the environment as we experience them, the use of time in many ways, and the values resulting from social relations. Many economists would forswear efforts to estimate monetary values for all these things, but some have proposed ingenious measures for conceptualizing (Becker 1976) and measuring their economic value. One rationale for social indicators and social accounts, in fact, has been to place monetary values on those aspects of well-being that are not captured by the present economic accounts (K. Fox 1974: Chap. 3).

The key to the measurement of benefits and costs in this larger sense is therefore the measurement of the preferences of those affected—consumers, producers, or recipients of externalities—in monetary terms. If we could estimate what consumers would be willing to pay to receive a nonmarket good or avoid a nonmarket bad—another aspect of shadow pricing—we could use that estimate in assessing the value of public policies that provided more or less of that good. The resulting estimates, like actual prices, are likely to reflect the existing income distribution.

We may start with a broad classification of types of benefits and costs that people may receive without monetary transaction at the time:

1. The use of durable consumer goods. Home owners receive the same sorts of benefits from their dwellings as do renters, but they do not pay rent; this equivalent additional income is included in the national income accounts (Nordhaus and Tobin 1972: 9). Other consumer durables similarly produce flows of well-being from their use, which are represented in the market only at the time of purchase.

2. Leisure time. Presumably time that a person may allocate freely is a greater source of well-being than time obligated to work; we thus often consider leisure time as of special benefit. Similarly, a parent obliged to care for young children may be sacrificing well-being. Neither work nor family duties are always unpleasant, however; the satisfaction from work and household production should also be examined. The well-being resulting from leisure time also depends on the other resources available to the person—money, goods, social relations, availability of facilities for recreation, or "consumption capital" (Stigler and Becker 1977: 78). Time may thus be regarded as a factor of production, which in conjunction with other resources can be used to produce well-being (Morgan and Smith 1969).

3. Household production. The "new home economics" has called attention to various basic commodities produced in the household (Becker 1976: 207). Not only unpaid household work by homemakers (Sirageldin 1969), but also cooperative activities by household members, contribute to a family's well-being by adding to the value of goods purchased in the market. This added value is a form of in-kind income but is less likely to be counted in official statistics than in-kind transfers from government.

4. Collective goods and externalities. As we have noted, this large class of nonmarket goods and bads poses important questions of estimation for policy choices. Particular progress has been made in measuring the values of collective goods and externalities. An important feature of these measurements, however, is the identification of the precise goods or bads—or features of them—that the consumer values or disvalues. It may not be a home as such that the consumer buys, but a combination of space, convenience of location, and environmental quality. If we could estimate what part of the price of the home was being paid for location, we might infer the value to the buyer of living near a park or of enjoying cleaner air or freedom from airport noise. This decomposition of the good into more fundamental aspects, the prices of

which are in turn estimated, has been proposed by Lancaster (1966) in terms of "characteristics" of goods, and by Michael and Becker (1973) in terms of more basic "commodities."

The various principles of measurement that may be used for assessing the benefits and costs of environmental conditions from market data have been summarized by Freeman (1980): (a) Payment for complementary goods: when a marketed good such as recreation is consumed together with environmental quality, the price of that marketed good may provide information about consumers' valuation of the environment. (b) When a marketed good or service is used to alleviate an undesired environmental condition (for example, cleaning of homes or removal of litter from roadsides), willingness to pay for this "defensive expenditure" may be used to estimate the cost of an undesired environmental condition. (c) When environmental conditions vary among localities, prices of homes may reflect those conditions.

These methods may be extended to publicly produced goods and bads that have specific locations and thus constitute part of the environment at those locations, such as recreation facilities, smoke from electrical generating plants, schools of various qualities, or local taxation. Valid estimates require careful statistical control of other relevant variables and well-specified models (Harris 1980). When market data are not available for estimation of prices, interview or bidding techniques may be used for similar estimates; these methods are especially appropriate when data for individuals are needed.

The specific methods used in estimating monetary values for indicator variables may be illustrated by citing representative studies (for a detailed review of environmental studies see Fisher and Peterson [1976]). We shall illustrate first environmental variables, then the direct evaluation of time and life. Not all the indicator domains or social goal areas listed in Table 1–3 have been evaluated equally successfully in monetary terms, but these examples will illustrate the methods used.

One method that is frequently used is the estimation of contributions to housing value by various conditions associated with location. Regression equations predicting housing value are thus formulated in terms of size and features of houses, but also of environmental conditions. The effect of air quality has been assessed in this way by Brookshire et al. (1980), who compared this estimate with one from a survey bidding game and found the latter to be lower by a factor of about two. Decreases in valuation of homes due to aircraft noise have been estimated by McMillan et al. (1980) and by Nelson (1980), who found regression coefficients varying by a factor of about two, over six cities. Increases in property value related to proximity to urban water parks were estimated by Darling (1973), who found that these estimates varied considerably in comparison with survey estimates.

A somewhat different approach concerns the valuation of recreation facilities in terms of the expenditures that people make for transportation to them;

by this means, demand curves can be estimated. The pioneering article of Clawson (1959), based on a suggestion by Hotelling, initiated much of this research. Other studies using this approach are discussed by Merewitz and Sosnick (1971: 148–53). The assumptions of the initial study have been refined, and subsequent studies have shown greater methodological sophistication (Cicchetti et al. 1976; Krutilla and Fisher 1975).

The estimation of the costs of pollution through measurement of "defensive expenditures" is more difficult because many of these expenditures are undertaken by government rather than purchased in the market. What is required is not simply an estimate of the costs of cleaning or removing pollution or waste, but the price that those affected would be willing to pay.

In addition to these inferences based on economic behavior, there have also been efforts to measure willingness to pay by means of direct questioning. Interview studies of this sort have to be carefully designed, as these questions are more difficult to answer than the customary opinion questions. One approach is that of the bidding game in which the respondent is asked whether he would be willing to pay each of a succession of prices suggested by the interviewer. This method has been applied by Knetsch and Davis (1966) to recreation and by Gramlich (1977) to the reduction of water pollution. Conceivably it could be used to study the relations between preference satisfaction and subjective well-being.

Closer to the measures of subjective well-being which we shall discuss in the next chapter are measurements of valuations of time and life. The valuation of time is an important ingredient in transportation planning, since one consequence of the development of new transportation systems is the saving of time (Quandt 1970; Hensher 1976). The value placed on time, in this sense, is the difference between the value of having free or leisure time, on the one hand, and that of spending time in some less desired way such as driving a car, riding in public transportation, or waiting.

In the field of health policy there have been efforts to measure the values that persons place on their own life expectancy; in this case the alternative is not some other use of time but a shorter life span. Because of the extremeness of the alternatives, direct questions about the monetary valuation of an additional year of life are typically not asked. More frequent are questions about the values of various probabilities of living additional years (Acton 1973). Estimates of this sort differ in their presuppositions from the conventional estimates based on discounted expected earnings: they can include in the valuation of life aspects such as consumption, enjoyment, or process benefits (Juster et al. 1981: 33), and not simply production.

This summary of approaches to measurement of nonmarket benefits and costs in monetary terms has been intended to illustrate the extensiveness of the literature, and the ingenuity devoted to the problems of measurement, in this area. Many nonmaterial and nonmarket conditions are thus susceptible, in principle, to monetary valuation. This valuation depends, however, on preferences that may be considered fixed relative to our policy alternatives. To assess the value of policies that alter preferences, we require other criteria.

Net Economic Benefit: Objective or Subjective?

Benefit-cost analysis rests on the assessment of preferences in monetary terms. Ordinarily these assessments are made from market data; sometimes, however we go beyond them to ask questions of the persons affected. The combination of these two methods—objective market data and interviews or questionnaires about subjective matters—suggests that benefit-cost analysis partakes of two opposite qualities. "Objective" and "subjective" indicators have been distinguished throughout the development of the social indicators movement (Chap. 6 deals specifically with subjective well-being). These terms have represented observable things and personal feelings respectively. This distinction seems to break down, however, when we try to use it to classify net economic benefit; indeed, this problem of classification requires us to reconsider the objective-subjective dichotomy and reformulate it. As we do so, we shall connect some concepts and themes of this chapter with those of the next.

The objective-subjective classification of social indicators has been questioned by Andrews and Withey (1976: 5), who propose several dimensions of classification in place of that single dichotomy. A two-dimensional alternative, proposed earlier by Fienberg and Goodman (1974: 74), is to classify both the phenomenon under study and the measure used as possibly being objective or subjective, generating a fourfold table. An elaboration of this two-dimensional approach in the form of an eight-cell table has also been proposed by Andrews (1981: 392–93).

We can draw on these efforts at improvement to propose the modified array shown in Table 5–1. This six-cell table again cross-tabulates characteristics of the phenomenon being measured (rows) and of the measure (columns).[15] Each cell represents a combination of a particular type of phenomenon and a particular type of measure.

We begin with approximate classifications of the rows and of the columns as objective and subjective (in parentheses). For the rows, we call a phenomenon objective (top row) if the corresponding sensory inputs are shared or available to coobservers and subjective (bottom row) if such an input is avail-

15. The table corresponds roughly to one presented by Allardt (1976: 228), in which the mode of observation (objective/subjective) defines the columns (welfare/happiness) and the nature of the things being observed defines the rows. Allardt's rows, however, distinguish between material resources and human relations as sources of value.

Table 5–1. Objective and Subjective Phenomena and Measures Related to Well-Being

		Measure (Objective)	(Subjective)
		Direct Observation of Phenomenon Consensual	Less Consensual
	External; shared observations (Objective)	a. Conventional objective measures (housing density, food intake)	b. Judgment or perception of shared phenomenon
Phenomenon		Manifestation of Phenomenon Behavioral	Verbal Self-report
	Relational; between person and environment (Subjective)	c. Purchases, decisions to accept jobs, spending time with others, striving	d. Verbal expression of preferences, attitudes, satisfactions
	Mental (Subjective)	e. Suicide, physiological measures, bodily movements, facial expressions	f. Verbal expression of global well-being, happiness

able only to one person (Andrews and Withey 1976: 6). We shall refer to the respective phenomena as "external" and "mental."[16]

It is useful, however, to insert a middle row in the table. Some important phenomena involve relations between the mind and the external world; these

16. Thus Fienberg and Goodman (1974: 74) refer to "phenomena that are subjective in the sense of being inside people's heads," in contrast to those that are "directly observable." This is a restriction of our definition of subjective phenomena above, which would include a physical event seen by only one observer. Our classification is not exhaustive; there remains a nonexternal and nonmental category, including physiological states and personality characteristics of the person in question.

include most preferences, attitudes, and satisfactions.[17] All these types of phenomena are mental, but they can be directed toward external goods, objects, or conditions, and are then somewhat constrained by the external world. The difficulty of classifying preferences arises in part from the need for this intermediate category; but we shall also see by examination of this category that the measurement of preferences resembles the measurement of certain indicators of subjective well-being.

Classifying the columns as objective and subjective gives rise to additional problems. We begin with a second dimension proposed by Andrews and Withey (1976: 5), *consensus*: "the extent to which people agree on how to characterize a given phenomenon."[18] An example would be the difference between a measure of neighborhood housing density (objective) and a rating of the crowdedness of a neighborhood (relatively subjective) (Andrews 1981: 393). This dimension is applicable, however, only to the top row in Table 5–1, since this is the only row for which shared observations of the phenomenon are available.

The measures corresponding to the second and third rows have also been divided by our classification of the columns into objective and subjective, but on a different basis from that of inter-observer consensus. Since the phenomenon in question provides sensory inputs to only one person, consensus on these inputs is logically impossible. We therefore classify measures of these phenomena according to whether they are based on self-reporting by the one recipient of the sensory inputs, or on consensual reports of some behavioral manifestation (an "indicator" in the methodological sense of the term) of these sensory inputs. Examples of these manifestations are suicide as a measure of low subjective well-being, and physiological variables such as pulse, breathing, perspiration, or brain waves as measures of emotional or mental states. Bodily movements, facial expressions, and even verbal behavior might also be observed and reported consensually, however. Thus we cannot distinguish objective and subjective measures in terms of the degree of consensus on the manifestations; observers can agree just as well on what a respondent said in response to a question as they can on whether a person committed suicide or smiled. Rather, we distinguish the columns for the second and third rows according to the type of manifestation studied; a physiological measure or a behavioral response made in the normal course of life (which may itself include verbal behavior), has been distinguished from a verbal self-report in a research situation. We thus insert special headings above the top row (cells

17. Andrews (1981: 393) makes use of a similar category of "subjective, concern-level phenomena"; we discuss "concerns," which relate the mental to the external, in Chap. 6.

18. "Subjectively social" characterizations (Chaps. 3–4) show consensus at a given time, but change over time or between cultures.

a-b) and the last two rows (cells *c-f*) indicating that the meanings of the columns are not the same for all rows.

The second row, for relational phenomena, deserves special attention. Both columns are well represented by economic measures: purchases and job decisions as behavioral measures, and indifference judgments and willingness to pay as verbal self-reports. The left-hand column includes spending time with others, and the right, attitudes; both can be measures of "attachment" to a social system (Chap. 8). In addition, the left-hand cell (*c*) includes "striving," a measure that Allardt (1976: 230) considers an objective indication of people's needs; while the right-hand cell (*d*) includes satisfactions. Each of these might have a counterpart in the other column.[19]

We thus see that economic well-being, or net economic benefit, is a subjective (relational) phenomenon relating persons to things preferred, measured usually through aggregate behavioral manifestations but sometimes through verbal self-reports. In its position in the table, it parallels several other significant indicator variables, suggesting its similarities with them as regards both measurement and theory. However, we reserve the term "subjective well-being" for a variable measurable with direct reference to individual feelings such as happiness and satisfaction (bottom row) unmediated by the market.

Macroeconomic Value Concepts and Well-Being

A special strength of the microeconomic ethic is its capacity to reconcile diverse particular valuations in terms of an overall value—preference satisfaction or net economic benefit. Yet some effects of the working of the economy are not easily incorporated into this system of valuation. Macroeconomics deals with variables such as employment, inflation, stabilization, and growth, which correspond to several distinct values. Even aside from questions of equity, many of the effects of unemployment and of economic changes such as inflation are not revealed by national product statistics but must be assessed independently. For these effects I shall suggest that indicators of subjective well-being may provide a more inclusive framework for tradeoffs than does the benefit-cost ethic.

The decrease in a person's welfare resulting from unemployment can be described only crudely by the dichotomous variable of the official unemploy-

19. Andrews (1981: 393) gives as an example of a subjective measure of a subjective phenomenon at the concern level "self-assessment of own housing," and as its objectively measured counterpart the question, "Did the individual move to a different dwelling in the same geographical area?" The latter is more nearly a measure of a preference, but it suggests that domain satisfactions might be cross-checked against behavioral measures.

ment statistics[20] or even by the decrease in earned income; it also depends on the person's obligations (number of dependents and their requirements) and resources (including possible aid from family members, other private sources, and government). But over and above economic aspects of welfare are the decline in self-respect, and of position in family and neighborhood, that may result, as well as effects on one's health[21] and on other family members. Policies designed to increase employment thus have multiple effects. Problems of employment and unemployment thus lead us to seek general valuative criteria by which the benefits and costs of such policies can be compared with others.

The classification of a person as employed or unemployed, as we have noted, is a statistical simplification. Some critics of the validity of unemployment rates have claimed that they are due in part to the unemployment of new groups seeking to enter the labor force (such as second earners in families); in part to the unemployment of teenage youth with more resources and fewer responsibilities than family breadwinners; and in other instances, that the employment involved is only part-time. It has been argued that the "new unemployment" is less serious than the old because "it is of short duration, voluntary, and because many of the unemployed are secondary earners or are protected by public or private compensation programs." A survey study of the unemployed has shown, however, that the differentials in dissatisfaction between various types of unemployed persons are less than might have been expected (Schlozman and Verba 1978: 333 and *passim*).

At the same time, claims are made that various groups ought to have the opportunity to work: the elderly, as part of a healthful existence; the young, because it prepares them for later work or removes them from harmful activities; and women, because it frees them from economic bondage to men. What is needed, ideally, is not merely a count of jobs and persons seeking them, but a consideration of the effect of each particular gain or loss of a job on societal welfare and equity. If employment provides self-respect, then we may ask what wages are necessary to provide self-respect, and whether it can be provided more cheaply in other ways if self-respect is indeed our goal. If employment provides resources to a lasting group such as a family, then we must ask how the earnings will affect various members of that group. Children may wish to get jobs in order to be independent of their parents, but they may gain

20. It is possible, however, to analyze the net benefit from an individual's employment in terms of preferences for income, work, and leisure, as well as from its effect on the economy. Distributional effects of unemployment on the incomes of various types of individuals and families are analyzed by Palmer and Barth (1977: 228–38). A fuller treatment of the shadow value of labor is given by Dasgupta et al. (1972: Chap. 15).

21. Additional possible effects of unemployment on morbidity and mortality rates have been suggested by Brenner (1976); but Cohen and Felson (1979) argue that Brenner's findings need to be examined further for possible spurious relations.

jobs at the expense of others. The same problem may arise from the employment of a second adult member of a well-to-do family, though it may seem archaic to ask whether a job should go to a breadwinner instead of a second earner. An economically efficient allocation of jobs does not automatically maximize well-being. The problem is concisely summarized in Marx's slogan: "From each according to his ability, to each according to his needs!" (Feuer 1959: 119).[22] We recognize this principle in special cases such as sheltered workshops for the handicapped, while trying to insure that for the vast majority, the labor market will both provide efficiency and assure standards of living above a basic minimum. But in developing policy indicators of unemployment, we need to specify more clearly what values we are pursuing.

By examining the welfare loss from unemployment in this way we can see, first, that national statistics are often approximations of the variables we wish to represent by them; and second, that survey studies of the subjective well-being associated with economic conditions are not only a useful supplement to economic data, but can sometimes provide a yardstick for comparison of otherwise separate value concepts.

Another major source of valued effects not encompassed by the ethic of net economic benefit is economic change. Indeed, one aspect of the decrease in well-being resulting from unemployment is the change in the individual's social status; thus in Britain, pension policy has changed to reflect "continuity of life-style rather than minimum needs" (Rein 1976: 21). In our discussion of net economic benefit we have omitted the effects of change, because welfare economics tends to compare one static state of affairs with another, ignoring the path between them. Benefit and cost are typically defined by comparing individuals' preference satisfaction between two well-defined and stable states.

The change-related or dynamic functions of economic policy that are most commonly separated from static welfare economics are those concerned with stabilization and growth. Stabilization of the economy is a major concern of public finance (Musgrave and Musgrave 1976: 13–16) and is associated with the goals of full employment and price stability. Our discussion of the costs of unemployment dealt in part with the effects of change; price instability also imposes social costs that are not easily estimated by benefit-cost procedures.

Stability of prices (or of their rates of change) may be associated with welfare through the improved environment it provides for calculation and planning, and through setting the expectations that contribute to people's subjective definitions of their well-being. We might, therefore, wish to assign a "cost" to inflation so as to compare it with other costs or benefits. The assignment of such a cost is complicated, however, by the fact that inflation has

22. Two approximate synonyms for need, in terms of our discussion, are (1) vertical equity, according to which inequalities are to be reduced; and (2) efficiency of production of well-being, according to which a greater amount of well-being can often be produced by allocating more resources to the needy than to others. See also Chapter 7.

other consequences that affect various groups differently: it impinges inequitably on debtors and creditors, and on the holders of money as against goods.[23] Inflation encourages consumption rather than investment (Flemming 1976: 111). But, most important for our present purpose, it fosters uncertainty and thus, in addition to impeding rational economic decisions, gives rise to anxiety about the future. Just as for unemployment, we need to know whether the effect we are evaluating is persistent or temporary. In the case of inflation, a predictable rate might be easier to adjust to than a fluctuating one; but governments do not have sufficient control over the process to assure this (Hagger 1977: 19).

Studies of subjective well-being in relation to inflation might help us to evaluate the effects of such changes in terms comparable with other costs. Similar studies might also be useful for assessing the costs of change that result from the normal workings of a competitive system. Social mobility, the decline of some industries and the rise of others, and other responses of the economy to exogenous variables, also have their social costs. Thus stability needs to be considered in a broader sense than the conventional macroeconomic one.[24] We must also consider the temporary costs associated with the transition from one static equilibrium to another: the readjustments involved in job search, retraining, and geographical mobility. It is often these transitional costs that generate the most intense resistance to change, but that are not counted in customary estimation of costs. An important aspect of these transitional costs is their relation to expectations. The psychic costs of changing job and place are often greater for mature persons who have built a network of friendships, habits and expectations. For young persons setting out on their careers and for others without close social ties, migration may bring with it the promise of new possibilities or the chance of escape from constraining obligations, and thus lead to increases rather than decreases in welfare.[25]

Growth is an additional economic condition that is widely valued without the nature of its value being well defined; we do not seem to have clearly answered the question, "Growth for what?" In a static sense, an increase in any aggregate welfare function is an improvement in terms of that concept of welfare. But we are also often concerned with the *rate* of growth as well as the

23. The distributional effects of inflation are analyzed more fully by Palmer and Barth (1977: 202–18). Dornbusch and Fischer (1978: 499) contend that aside from the distribution of wealth, "the costs of unanticipated inflation would be negligible."

24. The emphasis on economic indicators alone, apart from measures of other values, has been criticized by Lindberg (1982) on the ground that it gives power to economists at the expense of representation of the wide range of interests affected; and by Biderman (1979: 333) in that excessive use of the Consumer Price Index undesirably increases the power of the central bank.

25. Situations of this sort may also be studied with the aid of subjective social indicators. McKennell (1978: 400) shows that the group of respondents who are very happy but not completely satisfied are disproportionately young.

position to which it has brought us. A steady increase in production, if it did not involve excessive hopes, might well provide not only new levels of welfare, but also benefits associated with change itself; new resources, for example, may alleviate social conflict. The rapid growth of a boom town, on the other hand, can produce strains and dislocations. Advocates of a possible steady-state economy have raised some of these questions as well. Common to all these concerns is the need for more systematic valuative analysis of the concepts of macroeconomics.

Summary

The economic approach to measurement of well-being centers about the expression of various sorts of values in monetary terms. This commensurability of particular values allows us to weigh them on a common scale, applicable both to national indicators and accounts and to the assessment of particular policy alternatives. The economic approach to policy analysis—benefit-cost analysis—is, in fact, the estimation of the incremental contribution of particular projects to a social welfare function, the national income.

Although most of the measurements that contribute to the economic assessment of the national income (or product) are derived directly from records of production and payment, others require more ingenuity and go beyond the direct use of available economic data. Ingenuity is also required to estimate benefits and costs. For some that do not flow through the market, inferences can still be made through analysis of their contributions to the prices of goods that *are* sold. Other inferences, inaccessible to this sort of estimation, can be based on direct questioning or bidding techniques of measurement; these measurements resemble, at least in the method used, those used for subjective indicators of well-being (Chap. 6).

After discussing the variables that have been subjected to the common measurement standard of money, we considered several macroeconomic effects on welfare that are not fully included in the national accounts or in the procedures of benefit-cost analysis—those of unemployment, inflation, and economic change. Discussion of these contributions to welfare suggests that they may sometimes more easily be compared with others through the use of survey assessments of subjective well-being—the topic to which we now turn.

6 Subjective Well-Being

Economic well-being is usually measured in terms of incomes or production, whereas subjective well-being can be measured by asking people how happy they are. The two overlap, however; economic preferences are subjective and are sometimes measured by asking questions. They differ because, although both are rooted in utilitarianism, they have developed differently since Bentham's time.

The economic notion of well-being centers on preferences and things chosen, and among them it emphasizes things sold in the market. Preference satisfaction in these terms lies in consumers' getting what they prefer, especially if the thing preferred has a measurable price and demand function; or in producers' getting what they prefer, with measurable supply or cost functions. Subjective well-being, in contrast, is a state of feeling or emotion. At least since Bentham (1948 [1789]: i), some utilitarians have believed that mankind, being placed "under the governance of two sovereign masters, pain and pleasure," reveals its feelings through its choices. This is the viewpoint of economics. Other utilitarians deem this an empirical relation, not a tautology; they try to measure feelings directly, and then ask whether the satisfaction of preferences, or the receipt of goods or services, leads to desirable feelings. More important, they introduce contributory variables such as Allardt's (1976: 232) "loving" and "being," that are not things sold in the market, as well as simply "having." It is for these reasons, and because of the possibility of greater equity in calculation, that we seek to measure subjective well-being as distinct from economic benefit.

We deal in this chapter, then, with feelings rather than preferences.[1] The measures of feelings that are potentially most useful as policy indicators are based on end-values referred to as general or global well-being, happiness, satisfaction, or the subjective quality of life. Also useful are satisfactions with

1. In addition to feelings, other subjective variables include contributory variables such as expectations and perceptions of the world; data on these are sometimes published as social indicator statistics. Because these are less clearly value concepts, however, we do not give them high priority as policy indicators. We are thus mainly concerned with subjective well-being measured through the reports of the person in question (cells *d* and *f*, Table 5–1).

specific domains of life, which we treat as contributory variables. In aggregate time series used to define problems, satisfaction is more likely to fluctuate than happiness: examples are satisfactions with municipal services and with the President's performance. Although national statistics on happiness have not fluctuated so much, consistent differences among population groups have been found (A. Campbell 1981: 230–37), and these differences can be used to define problems.

Subjective well-being, like net economic benefit, is a potential basis for tradeoffs and reconciliation among more specific values. This possibility exists in analyses of particular policies and projects, even without reference to national aggregate statistics, just as benefit-cost analysis can be carried out without reference to the national product. In order for this potential to be realized, however, we must develop models that show the relative contributions of various specific policy outcomes, as in social goal areas, to general (global) subjective well-being. We shall discuss the problems of developing these policy models later in this chapter.

The practical use of indicators of subjective well-being also depends on accuracy of measurement; but at present, they lack the precision or the detailed statistical basis of economic indicators. Johnston (1978: 291) points out that "the bulk of the subjective measures now available represent 'poor measures of the right things.'" Measures of subjective well-being, however, if they could be made sufficiently reliable,[2] and if they measured a notion of well-being that we considered desirable, might remedy some of the ethical deficiencies of benefit-cost analysis.

One particular ethical advantage of using measures of subjective well-being is that, because they are based on sample surveys, they allow us to count each person rather than each dollar equally. Braybrooke and Lindblom (1963: 180), in proposing public statistics based on a censuslike classification of persons, note that such a classification leads "to counting one person's condition for no more and no less than any other's." This advantage holds only if we can avoid certain possible biases in the sample of persons who respond. If nonresponses come disproportionately from the poor and the deprived, our estimates of well-being may be biased against them. Especially if we are concerned with the persons who are least well off, we need to be sure that our samples (whom we reach and who respond) adequately represent groups such as migrant laborers, persons in institutions, and those with difficulties in communication because of handicaps or language. The emphasis that opinion surveys have

2. One possible reason for the lack of reliability of survey measures of subjective well-being is that we have not tried hard enough to interrelate them with more reliable measures. If physiological and behavioral measures (cells *c* and *e* in Table 5–1) reflect the same value concepts as do answers to questions, then we might bridge the gap between such measures and survey questions by questioning special samples of persons whose physiological states and behavior are also being observed.

placed on active citizens, by such practices as omitting institutional populations, may need to be replaced with a stress on persons who are less likely to participate.[3]

Types of Subjective Values and Measures: Happiness and Satisfaction

Subjective well-being has been measured by a wide variety of survey questions. Andrews and Withey, for example, in an effort to synthesize previous measures, included in their surveys 133 different questions aimed at measuring respondents' concerns with particular areas of their lives, and 68 different measures of general (global) well-being (1976: 32–43, 66–70). This diversity of measurement presents several problems. First, since these are presumably measures of an end-value, we must ask to what extent the various underlying concepts are concerned with one or more intrinsic values. Second, if these measures reflect a given underlying concept to different degrees, we may have to find which measures, or which combination of measures, best reflect this concept. A given measure can also reflect more than one underlying concept, leading to the same problems for each of the multiple concepts. Third, if we intend to use one or more measures as indicators of this end-value, we must also be concerned with the intelligibility of statistics based on them.

Three types of measurement of subjective well-being are of special importance because of their repeated use. One of the questions most frequently asked, and providing time series since 1957, is: "Taking all things together, how would you say things are these days—would you say you are very happy, pretty happy, or not too happy?" (A. Campbell et al. 1976: 24). A second major line of investigation involves asking respondents a battery of ten items referring to specific feelings during the past week, such as "pleased about having accomplished something" and "depressed or very unhappy" (Bradburn and Caplovitz 1965: 17; Bradburn 1969: 56). Cluster analysis of these items reveals two independent dimensions—positive and negative affect. A third major line of investigation involves questions about respondents' satisfactions with various domains of life experience (Campbell et al. 1976).

3. Hand in hand with these issues of sampling and response goes the problem of whether well-being can be measured validly through survey responses even when they are available. Interviewers require special skill and training. Distrust of a strange interviewer may lead people to say they are happy when in fact they are under stress. They may be reluctant to report on ecstatic or extremely painful experiences. Distrust of possible manipulation by the sponsor of the survey may lead politically oriented respondents to reserve their expressions of dissatisfaction for political action. Similar questions to those that economists have raised about biased expressions of monetary demand may arise for politically relevant expressions of dissatisfaction (Chap. 9).

The concepts measured in the first and third of these approaches are presumably happiness and satisfaction, because these concepts figure in the wording of the questions asked. In examining their validity as end-values, we might first ask whether they correspond to happiness as defined in utilitarian philosophy. Yet even within this philosophical school of thought there have been differences in judgment as to what sort of feelings are most desirable. Bentham simply identified happiness with pleasure and its opposite with pain, making no distinction among sources of these feelings; for example, he assigned no particular value to the happiness that was caused by poetry (Baumgardt 1952: 317, 481). Mill, on the other hand, insisted on distinguishing higher and lower kinds of pleasure, and contended that "it is better to be a human being dissatisfied than a pig satisfied; better to be Socrates dissatisfied than a fool satisfied" (1910 [1861]: 7–9).

Two alternative notions of good experience thus coexist in the utilitarian literature. One refers strictly to those experiences that people value, regardless of their sources; the other introduces a somewhat external consideration, implying that only those pleasures that are worthy of being enjoyed shall count as happiness.[4] This distinction between a person's simple valuations of experience, and a further criterion of what is worth valuing, is reflected in some criticisms of subjective valuation by proponents of objective indicators.

In addition to the differences among abstract definitions of happiness and related terms, we must also compare more specific operational differences. Should we rely more on questions that refer to the present or the very recent past (for example, a week), or to a longer and less definite period ("these days")? Are questions dealing with specific circumstances and feelings, such as Bradburn's, likely to produce more valid responses than abstract and general questions? Should we say that we are measuring subjective well-being, or "perceived" well-being (Andrews and Withey 1976: Part 1) or the "sense" of well-being (A. Campbell et al. 1976: Chap. 2)? These latter two usages suggest that well-being is an entity which its possessors may not perceive or sense with full validity. These operational problems also require us to assess the words and symbols used in questioning. Many words other than happy or satisfied appear in Andrews and Withey's list of sixty-eight measures of global life quality: best, good things, feeling, delighted, terrible, worry, progress, and others. Pictures of smiling faces, ladders, and thermometers are also used.

When we ask respondents about their satisfactions, in contrast to their happiness, we typically ask whether the respondent is satisfied *with* something—ranging from specific roles such as "your job," or items such as "your car," to

4. The diversity of meanings of happiness is even more evident when we consider a definition offered in another language over two millennia ago: "happiness is an activity of the soul in accordance with perfect virtue" (Aristotle 1942: 1102).

"your life as a whole" (Andrews and Withey 1976: 32–33, 67). Questions are often asked about satisfaction with particular *domains* of life (A. Campbell et al. 1976: Chap. 3) similar to social goal areas, such as marriage, health, friends, job, government, or neighborhood. We can also take as referents *criteria* or aspects of these domains, such as attractiveness, safety, freedom, and opportunity for success (Andrews and Withey 1976: 11–13). These domains and criteria are collectively known as *concerns*.

In developing policy models related to subjective well-being, we must show the effects of objective outcomes of programs (health, education, safety, and the like) on subjective well-being. We might try to make this connection by seeking the relations between policies in a given domain and satisfaction with that domain (e.g., housing policies with housing satisfaction). It seems convenient to assume that satisfaction in a given domain is caused only by policies in that domain, but this assumption requires checking. If a halo effect or a cross-domain effect led to a connection between satisfactions with family and with job, for example, we should be in error in judging job satisfaction to result only from experiences on the job.[5] If we can estimate the size of such effects, we will then be better able to make tradeoffs among domains by relating specific domain satisfactions to overall well-being (A. Campbell et al. 1976: Part 2, 61).

The notion of satisfaction is not only potentially more specific in its referents than happiness, but also implies a standard of reference—expectation or aspiration—that happiness does not imply to the same extent. We are satisfied with something when it measures up to our valued expectations. Thus the question can be raised, more for satisfaction than for happiness, whether a high score on such an indicator merely reflects low expectations due to "false consciousness," and conversely for a low score. We are led, then, to a second ethical critique of satisfaction—that it embodies cognition rather than simple emotion. A similar concern is expressed in Johansson's (1973: 213) concern that surveys of satisfaction might become mere "continuous pseudo-plebiscites": we want indicators to tell how well off people are, but not to express their judgments or votes that flow from this condition. Rather, the condition of the society as a whole is first to be measured and then to be considered in a deliberate formation of public judgments, opinions and votes—a two-stage process that is even more important for collective decisions than for individual decision.

5. Similarly, when a respondent is asked about satisfaction with urban services, his responses may reflect conditions other than the services themselves (Stipak 1979). We may then conclude either that the responses are poor guides for administering services or that they should lead us to seek other contributory variables, either as manipulable causes of well-being or as statistical controls.

Dimensional Structures of Subjective Value Concepts

We can interpret various measures of subjective well-being by studying their meanings together with their patterns of statistical association. One method often used for this purpose is factor analysis, which assumes that particular measures reflect various mixtures of underlying factors. It takes as input the pairwise associations between individual measures and constructs from them a space of limited dimensionality, in which each measure is represented by a point, and which will account substantially for the initial associations. The coordinate axes of this space are initially indeterminate with respect to rotation. Preferred axes in it can be located by seeking "simple structure" or axes such that as many points as possible lie on or near them. We can also examine whether these axes, or sets of measures whose points lie near one another, have recognizable conceptual content, and we can introduce hypotheses about these conceptual relations into confirmatory factor analyses. Other similar procedures used to reveal structures of association are cluster analysis and smallest space analysis.

These procedures may be carried out with questions concerning either global well-being (general happiness, satisfaction, and the like) or domain satisfactions. One of the first such analyses was that of Bradburn and Caplovitz (1965: 17). Using twelve questions dealing with relatively specific aspects of the respondent's feeling during the previous week, they found two distinct clusters of four and five questions, dealing respectively with positive and negative feelings. Items within either of these clusters had correlations of the order of +0.4 with one another, while across clusters the correlations between items ranged from +0.1 to -0.2. The cross-cluster associations, although mostly negative, were not nearly as negative as might have been expected if positive and negative affect had been opposite aspects of the same dimension.[6] Bradburn (1969: 58) repeated this analysis with three additional items, yielding two essentially uncorrelated clusters of five items each. He summed responses to form separate scores for positive and negative affect, and formed a third score, "affect balance," from their difference.

Andrews and Withey (1976: 88–89) conducted a factor analysis of twelve measures of global well-being, including Bradburn's three indexes. They identified a first rotated factor on which the highest loadings corresponded to the questions "How do you feel about your life as a whole?" (taking into ac-

6. The meaning of these clusters is based on joint occurrence of the reported feelings in the past week. This is not the same as the likeness of those feelings, however. Another sort of questioning, requiring respondents to place feelings on continua, might have led to the conclusion that positive and negative affect were opposites. Headey et al. (1984a: 122) present results of a factor analysis of measures of well-being and ill-being, showing oblique factors with a correlation of −0.53.

count the past year and the near future; a seven-point scale from "delighted" to "terrible"), and "How do you feel about how happy you are?" (same seven-point scale). Two other factors centered about positive and negative affect respectively. Their results might suggest that happiness and the delighted-terrible continuum identify a common dimension of general well-being, apparently resembling the utilitarian notion of happiness. Such a judgment requires further scrutiny, however, since Andrews and Withey (1976: 88–89) call their first factor "cognitive evaluation," reserving the term "affect" for the factors centering about Bradburn's measures; and McKennell (1978) also interprets happiness questions as having cognitive content. The loadings of positive and negative affect on the first factor were considerably lower than one might expect from assuming that affect balance, the difference between the two, was equivalent to the first factor. Possibly the distinctness between Bradburn's affect scales and the other measures of global well-being resulted from their different time referents—one week for the Bradburn measures, one year or more for some of the other questions. The Bradburn questions also represent only a sample of the domains of positive and negative affect.

A challenging analysis of the structure of subjective value variables has been made by McKennell (1978), contrasting cognitive and affective components in measures of happiness and satisfaction. He compared respondents' answers to questions on happiness and satisfaction, examining deviant-case groups that are dissatisfied and happy or that are satisfied and unhappy. This comparison showed that the former tend to be relatively younger, better educated, higher in income, and more participant. Examining the relations among questions on happiness, satisfaction, and positive and negative affect, he postulated separate factors, cognition and affect, that underlay these measures and were themselves revealed by the two deviant-case groups. Contrasting the deviant case groups on other variables, he concluded (1978: 407) that Bradburn's affect balance index is "probably the purest available [measure] . . . of the affective component," and that happiness questions, contrary to previous suppositions, involve a considerable admixture of a cognitive factor. He based this inference in part on the relative stability of the affect balance scale with respect to events outside the individual's immediate life space, and the positive partial association of affect with income when satisfaction (presumed to involve more of cognition) is controlled. Further evidence for this view has been presented by Andrews and McKennell (1980). Kammann et al. (1984), however, question the special status given to Bradburn's scales.

The question whether any particular measure of well-being—objective or subjective—coincides with our ethical notions of the good is of course central to the development and use of policy indicators. Among subjective measures, we may well judge measures of affect to be more relevant ethically than measures of cognition. McKennell (1978: 403) observed that: "With increasing income there is a clear tendency for affect to rise, but no simple relationship in the case of cognition. . . . It is intuitively appealing to interpret the first result as showing how the increased enjoyment of the material aspect of life which a larger income makes possible is directly reflected in the affective component. The cognitive component on the other hand is more complicated, reflecting the relativity inherent in judgments along this dimension." In these terms, a pure measure of affect would provide an absolute measure of well-being rather than one that depends on a person's level of aspiration. Such a finding might strengthen our judgment that affect is more validly related to well-being. Yet alternatively, we might take self-assessments of happiness as the more valid measure of this end-value and accept their cognitive component as part of this entity (Shin and Johnson 1978).

A similar type of spatial representation is possible for questions dealing with specific domain satisfactions or criteria. Correlations across individuals permit the mapping of a space in which specific concerns appear as points. Andrews and Withey (1976: Chap. 2) present several such maps based on smallest space analysis, in which general categories such as family, self, job, and larger society correspond to sets of concerns near one another with similar meanings. The configurations of these maps differ somewhat among population groups, as by age and race (1976: Appendix D). Such findings are of sociological interest and may also reflect the ramification of effects when a policy affects one domain of life directly and others indirectly.

Andrews and Withey (1976: 13, 134) also characterize each domain of satisfaction by possible scores on a set of criteria or more general values that a domain may afford to a person; examples are fun and enjoyment, financial security, success, freedom, and safety. These criteria, when combined with the domain satisfactions in a common smallest space analysis, seem to lie within the same dimensional structure. Conceivably in a factor analysis they might correspond to underlying dimensions.

This quest for underlying structure in satisfactions is analogous to a line of research in economics on structures of preferences. Demand curves and indifference curves represent interrelations among preferences for two or more things—money and commodities, or one commodity against another. A multidimensional surface of this kind would be estimated (in principle) not from correlations but from consumer behavior in which one commodity is given up when another is gained. Embodied in the particular goods and services, however, there may be more generalized basic commodities (like Andrews and Withey's criteria) that give value to each particular good. Becker (1976: 207) suggests that households produce numerous such commodities, making use of time and market goods to produce them, and that these include "the quality of meals, the quality and quantity of children, prestige, recreation, companionship, love, and health status." He sees these basic commodities as the underlying and relatively stable entities at which consumer purchases and household production are directed, and the referents of relatively stable preferences (Stigler and Becker 1977: 84). Lancaster (1966) also suggests that consumers seek similar underlying characteristics of goods rather than goods as such.

A parallel economic research topic, the estimation of "hedonic price indexes," involves the decomposition of the prices that consumers are willing to pay for a commodity into parts depending on various postulated general dimensions of that commodity. For example, the costs of automobile model changes have been examined by Fisher, Griliches, and Kaysen (1962) in terms of dimensions such as advertising, size and horsepower, transmission, power steering, and power brakes. Such components are not necessarily expressed in terms of more general commodities or types of benefit, such as convenience or safety, but they might be. Another parallel topic in the field of marketing is that of "multi-attribute models" of consumer decision, in which the attributes resemble the dimensions indicated above (Hughes and Ray 1974: Section 3). Multidimensional scaling has also been used in this approach to analyze data from interviews about products (Hughes 1973: 183–85).

There may be value in research drawing together the different disciplines' approaches to the underlying dimensions of choice and well-being. Even if we wish to depart from the economic assumption that preference satisfaction is synonymous with welfare, we may still try to make use of certain formal advantages of the economic approach for policy models. It recognizes a principle of diminishing marginal utility, such that a consumer's well-being will be best promoted by an optimal mix of commodities rather than a large provision of one alone. It recognizes the possibility of complementary or substitute goods, so that we must seek the effects of increasing the quantity of one commodity on the well-being contributed by others. And most important for policy purposes, it seeks to estimate the effects of interventions quantitatively; we must eventually proceed, even in the realm of subjective well-being, from the exploratory qualitative realm to that of asking "How much?" Measures of subjective well-being can be used for explicit tradeoffs only if we can say how much of one type of policy outcome is equivalent to how much of another in terms of this end-value.

Combining Time and Subjective Well-Being[7]

Policies for health and safety often require tradeoffs between the quality of life and its length; safety precautions, or medication, can increase the expected length of life but decrease its quality. A combination of quality of life and temporal duration has been proposed in the notion of

7. In Chapter 7 we discuss another use of time (conversion of monetary values to hypothetical time values) that may promote equity but that does not involve subjective well-being.

"quality-adjusted life years" (QALYs) by Weinstein and Stason (1976, 1977). This concept involves aggregating the expected years of life of persons who may be affected by health policies, but weighting them by the quality of those years. The assessment of quality is based on questions such as the following:

"Taking into account your age, pain and suffering, immobility, and lost earnings, what fraction, P, of a year of life would you be willing to give up to be completely healthy for the remaining fraction of a year instead of your present level of health status for a full year?"

Or, alternatively,

"Taking into account these same factors, what probability, P, of death would you be willing to accept so that, if you survived, you would have full health rather than your present health status for the rest of your life?" (Weinstein and Stason 1977: 719).

Responses to questions of this kind may be used to estimate the respondent's quality of life, relative to a standard of "full health."[8] The resulting estimates are then used as weighting factors for quality adjustment, to be applied to the numbers of years of life expected under various possible treatments or policies.

The criterion that the respondent is asked to use is actually a mixed one, combining expectations (age), capacity to function (immobility), earnings in the economy, and subjective well-being (pain and suffering). Presumably, however, we could devise alternative questions that would single out subjective well-being as a weighting factor if we wished.

This combination of quantity and quality of life in a single index suggests that we might use it to measure the effects not only of health and safety policies, but also of various other types of policies. If a policy increases well-being, its overall value would depend on the duration[9] as well as the amount of the increase. The use of time in this way as a basis for the measurement of value, like the use of nonmonetary survey questions, might reduce the inequities between rich and poor that arise through the use of a monetary standard in benefit-cost analysis. Such inequities result from attributing high monetary values to effects for which wealthy persons are willing to pay large amounts, whereas neither the duration of life nor its subjective quality has this close relation to income.

Juster et al. (1981) incorporate subjective well-being in a time-based proposal for social accounting, by recognizing "the intermediate nature of tangible goods and observable states of the world in generating psychological

8. I have argued elsewhere (MacRae 1980b), that years of full health need not be of the same value under different circumstances such as variations in income. A further problem is that when quality is measured by a probability in this way it cannot be negative; yet for some purposes it is desirable to define a zero point on this variable.

9. We assume that time is measured by clocks or calendars rather than subjectively.

well-being" (1981: 30). Psychological well-being, in this perspective, results from tangible and intangible stocks possessed by the individual; from the contexts (organizational, social, and environmental) in which the individual is located; and from "process benefits" associated with activities. Well-being is thus seen as basically subjective. This approach, like the one proposed here, requires us to develop policy models that show how objective circumstances affect well-being.

Suppose, then, that the concept of quality-adjusted life years were accepted as an end-value to be incorporated in indicator information; how would we generate indicator statistics for it? Let us acknowledge that there are problems of measurement, and concentrate on the problems that would arise if we had good measures of subjective well-being as well as of corresponding time intervals. The social indicator literature already contains information on quality of life as well as on death rates by age and on life expectancy; how would we combine them?

An annual measurement based on a single survey could not directly measure the sum of a respondent's QALYs over an interval such as a year unless we relied on retrospection. Apart from weekly or seasonal types of fluctuations, however, the distribution of satisfactions and dissatisfactions experienced by the respondents at one time might represent those experienced over a year. For a given population, therefore, the available data on subjective well-being would be a useful surrogate for statistics on QALYs.

We wish to know simultaneously, however, whether the quantity as well as the quality of life is changing. We must first set aside in a separate statistic the size of the population as a whole; by doing so, we can compare populations of different sizes, as well as consider separately the ethical question whether twice as many lives, other things the same, are worth twice as much. This might lead us to seek a measure, for a single "average life," of "quality-adjusted life expectancy." An indicator of this sort, based on capacity to function, has been proposed by M. Chen et al. (1975: 80–84). For practical purposes, however, it may be more meaningful to present estimates of aggregate expected QALYs only for periods no longer than a few years. Such calculations for subjective variables over long periods may depend on uncertain assumptions about the constancy of parameters over time.

A further problem is that our data on length of life, based on death records, do not usually pertain to the same individuals who report the quality of their lives. To combine the two types of information we may therefore have to deal with age categories, further subdivided by other social characteristics, and to present statistics for each such subgroup on quality of life and on probability of survival through an interval such as a year. Among income groups, these two features might be positively associated: the life of the very poor may be both nasty and short. For general reporting purposes, these group statistics might be the best that we could present initially. But for analysis of particular policies, we may be able to follow special populations over time and collect data on both survival and quality of life.

Such statistics, even in the crude form in which they would now have to be available, could be useful in providing an alternative measure of value in contrast to the economic standard. First, they would maintain advantages of existing measures of subjective well-being. Descriptively, they could show what segments of the population were experiencing high or low quality of life relative to what would be expected from their incomes; we know already that the association is far from perfect. They would treat equally persons who were paid unequally (by race or sex) or who were not producers for the market (homemakers or the elderly); we could see more clearly how possible policies might affect such groups in these terms. Second, they might illuminate possible tradeoffs between quality and quantity of life, if resources could be used to increase one at the expense of the other. Especially in studies of the elderly, they might throw light on the social benefits of our efforts to prolong life, as they affect not only the elderly themselves but also those around them. These sorts of studies reflect an important use of indicator variables for particular policy judgments.

Objective Causes of Subjective Well-Being: Individual Circumstances

The usefulness of indicator statistics on subjective well-being depends on our knowledge of its manipulable causes. In seeking these causes we shall start with basic models of causation that are not directly connected with policy; later we shall interpret these models in relation to policies, including some that have not been studied extensively and others that might be questioned on ethical grounds. We shall see that there are two quite disparate perspectives for building these models. One, analogous to the economic approach, assumes that subjective well-being is a mathematical function of current objective circumstances; such a relation may reflect instantaneous causation or a comparison of equilibrium states. The other grows out of an effort to test the first, as well as from psychological theory. It recognizes that subjective well-being cannot always be predicted from such a functional relation but depends also on internal processes and standards of comparison, within the organism or experiencing person, that make it dependent on temporal sequences of events and other subjective or psychic variables.

The method most often used to study causation in the first of these perspectives has been to examine individuals' differences in well-being in relation to their current circumstances. The background characteristics of individuals that are compared with their subjective states thus include variables such as sex, ethnicity, age, education, income, occupation, roles in family and community, and health. Some, but not all, of these variables are alterable by public policy. We shall refer to them collectively as *circumstances*, and note that they are typically known to us as objective phenomena. Numerous studies have examined the relations between these variables and subjective well-being.

To understand these relations, however, we must examine the meaning of these variables more closely—ultimately specifying models of causation in which these variables can be surrogates for others (as age is) or can act through multiple causal paths. We have listed them in approximate sequence from fixed to changeable—sex and ethnicity being essentially fixed; age changing slowly; formal education being largely fixed after the first few decades of life; income, occupation, family and community roles, and health changing more through the life course. The more changeable circumstances tend to be those that we can change through public policy, though often we can hope to do so only indirectly by altering the conditions under which individuals make their own choices.[10]

We can also interpret these variables as regards the *ways* in which they might affect subjective well-being. We shall first characterize these ways as though current well-being is a mathematical function of the current values of the independent variables—a persisting or steady-state effect if there is no change in these variables or circumstances. From this perspective we can summarize many of these effects in terms of three types of contributory variables:

1. *Resources* that can be used to satisfy demands and needs (A. Campbell et al. 1976: Chap. 11). Of these, money (income or savings) is the most typical and general resource. Akin to it, however, are health and the skills that result from education, facilitating action toward one's goals; because of this similarity, resources of this sort have been classified by Becker (1975) as "human capital." An individual also has resources as a result of roles in family and community, and depending on these roles, may draw on others to varying degrees for social support. One's general high or low social status, related to several of these variables, may affect available resources. The value of these resources can also depend on the time one has in which to enjoy their fruits (consume them or produce other things with them). Social resources and time have also been discussed by Becker (1976: Chap. 5, 12) in an economic perspective; he refers to "the sum of . . . money income and the value of [one's] social environment" as "social income" (1976: 257).

10. Other directions of causation are also possible: subjective well-being (or personality variables influencing it) can affect education, income, health, and the roles that a person occupies.

2. *Needs and obligations* are the opposite of resources, i.e., demands on resources and commitments of them; and subjective well-being might thus be expected to depend on the net of resources minus needs and obligations.[11] One may have money but also have numerous dependents to support, or have a chronic disease or handicap requiring treatment or equipment. One may have time, but be required to spend it taking care of young children or other dependent family members. One may have to spend effort and resources to attain a given goal, simply because of low social status, while a higher status person would not. Thus these two types of variables, resources and needs-obligations, constitute an extension of the economic perspective to include an aspect of social relations.[12]

3. *Social interaction.* At the same time, in a noneconomic perspective, we hypothesize that a person's well-being will increase in proportion to his or her participation or interaction with others. At least at the low extreme of this variable, we expect an isolated person to be less happy than someone otherwise similar who has more social interaction. W. Wilson (1967: 304), in an extensive survey of the literature on correlates of avowed happiness, concludes that "perhaps the most impressive single finding lies in the relation between happiness and successful involvement with people."[13]

Aside from these three types of variables, numerous empirical studies have examined the relations of people's subjective well-being to particular social and economic circumstances. Many of these are based on cross-sectional data. Andrews (1981: 395–96), reviewing the literature on social and demographic correlates of subjective well-being (not including health), notes that these associations are consistently weak; in a typical study, six such variables together "accounted for only 5% to 10% . . . of the variance in an index of feelings about life-as-a-whole." Fernandez and Kulik (1981: 841) also review numerous studies, noting the "strong, consistent, positive relationship" of health to subjective well-being.

Among the consistent correlates of subjective well-being in these studies are income and marital status (married persons as contrasted with single, divorced, or separated). Increasing age is associated with satisfaction but not happiness. Young persons who are married without children are happier than those with preschool children. Gender seems to make little direct contribution

11. This comparison is also sometimes expressed as a ratio, such as that of income to needs; a related aggregate measure is the "dependency ratio" of a population.

12. These additional resources and needs-obligations may contribute to present well-being, but cannot automatically be used for calculation of distributional equity (Chap. 7). For example, the obligation to repay a debt or right a wrong one has done ought not necessarily to be alleviated by public policy.

13. The benefits from social interaction are a special case of the "process" benefits cited by Juster et al. (1981); and types *1* and *2* overlap with these writers' notions of stocks and contexts. Interaction may also be associated with social integration (Chap. 7).

to life satisfaction, but whites are somewhat more satisfied than nonwhites (A. Campbell et al. 1976: 53).

Cross-sectional studies of this kind can be supplemented by longitudinal studies of subjective well-being over the life course, although these latter studies are far less frequent. Longitudinal studies may allow us to infer causation more accurately by comparing the individual with his or her own past. Panel studies over one or two years have shown plausible associations between life events and well-being (Campbell et al. 1976: 199–207; Headey et al. 1984b). To the extent that we can trace subjective well-being over longer time periods, we can also detect delayed causal relations and possibly effects of temporal change in one's circumstances; thus we might examine the effects of role transitions such as changes in marital status, parenthood, unemployment, occupational advancement, or retirement. The sort of information that we should ideally like to have is a periodic record, kept by or obtained from an individual over an extended period of years, including reports of subjective well-being and of related events. Even more useful would be time series data in which policy intervention occurred and a comparison group was used.

The methods used in actuality are only approximations to this ideal. One is the retrospective "life chart," in which respondents are asked "to draw a curve indicating their degree of happiness in earlier years of life, and then interviewed . . ." (Runyan 1980: 46–47). When this method is used, changes in interpersonal relations, success and frustration at school and job, and economic fluctuations such as depressions, are typical variables that appear to be associated with changes in the reported level of happiness.

Two sets of respondents, initially living in California, have been traced over several decades and asked questions about life satisfactions. One such group, composed of gifted children born in about 1910, was first studied by Terman in 1922 in connection with the long-run concomitants of intelligence. The subjects have been restudied at intervals over their life-course; in a follow-up study in 1972, they were asked to report on how satisfied they were with their experience in each of several respects (occupation, family life, friendship, cultural life, service to society, joy in living). Because the questions on life satisfactions were asked only late in the series, the chief value of the longitudinal study for our purposes is to trace effects of earlier events and personal characteristics on these variables. Reporting on males in the Terman sample, R. Sears (1977) found that among the six sources of satisfaction, respondents found family life and occupation to be the most important specific sources of satisfaction; joy in living was also ranked high. Occupational satisfaction, when treated as a dependent variable, was shown to be significantly associated with some variables measured decades earlier, including the feeling at age thirty of having chosen an occupation rather than having drifted into it. A study of the women in this gifted sample by P. Sears and Barbee (1977) also showed numerous significant associations between an index of present general satisfaction and earlier measures of variables such as self-confidence and perseverance (forty-four years earlier) and specific sources of satisfaction (twenty-two years earlier). These lagged relationships enlarge the range of our possible models, but do not deal explicitly with changes in circumstances.

A second group of respondents, the Oakland Growth Study sample, has been studied by researchers at the University of California, Berkeley. A longitudinal study of growth and development was begun by Jones and Stolz with a sample of eleven-year-old children in 1932. A major reanalysis of these data was conducted by Elder (1974) with the aim of assessing the long-term effects of the depression of the 1930s. Although no comparison group was available, experiences of children and families during and after the depression could be traced in detail. The retrospective life chart, used in a 1958 study of the sample, revealed recollections of painful experience in the depression years for many of them.

A further study by Runyan (1980) of this sample reported differences in life satisfaction in adolescence in relation to early or late maturation. Although the age of maturation does not seem alterable by policy, conceivably adolescent life satisfaction can be improved by policies that influence the composition of peer groups. Runyan also found that upward social mobility was associated with increased satisfaction (between ages six to ten and thirty-four to thirty-eight)—not an obvious conclusion in view of the possible strains and adjustment problems associated with mobility.

In summary, these longitudinal studies emphasize interpersonal and occupational relations as sources of satisfaction and show effects working over periods of decades—in addition to the pervasive effects of a major economic depression suggested by Elder's work.

The effects reported, or suggested, in these studies characterize well-being at one time as dependent on circumstances and attitudes at that time and previous times. They do not emphasize the possibility that change itself may affect well-being—the second of our major perspectives. But one review of such individual studies does suggest that change is relevant.

The social and demographic characteristics that we have considered in relation to subjective well-being differ from one another in their degree of fixity over the life course. Focusing attention on this difference, Inglehart (1977) has suggested that those characteristics of individuals that are most fixed (such as sex) have the lowest cross-sectional associations with subjective well-being because of a process of psychological adaptation. He postulates that "aspirations gradually adjust to changed life conditions" (431). Income, on the other hand, may change, and thus depart from the previous level of aspiration; for this reason, he suggests, one's income is more closely associated with overall life satisfaction than is one's sex. He goes on to show that respondents' reports of *changes* in well-being over the previous five years contribute importantly to explanation of variance in current well-being (445).[14]

Psychological Contributory Variables: Personality and Domain Satisfactions

When we seek to estimate the effects of objective circumstances, modifiable by policies, on people's general subjective well-being it is useful to include other contributory variables in our models as well. First, as A. Campbell et al. (1976: 16) point out, subjective well-being may be influenced by the individual's personal characteristics, apart from his or her circumstances. Certain social roles, conditions, or personality characteristics may lead people to interpret given situations differently: "Persons of different ages, races, income, or urban-rural residence may have characteristically different ways of evaluating the same objective situations, in no small measure because they bring different standards of comparison to bear on these evaluations. Similarly, people may be characterized by a general optimism or pessimism, and such inclinations . . . , along with other attributes of personality, may be expected to influence the way they perceive the world around them" (15).

It has been shown repeatedly that lasting personality tendencies are associated with subjective well-being. Wessman and Ricks (1966), for example, reported results of daily recording of moods by college undergraduates over a six-week period. Their main research goal was to find personality determinants of elation-depression, regarding both its average level and its variability. A general tendency toward a high hedonic level was associated with test scores two years earlier on a "depression scale" from the MMPI test battery, and thus seemed persistent. Costa and McCrae (1980) also contended on the basis of several studies that Bradburn's two components of well-being, positive and negative affect, were predicted by lasting personality tendencies of extraversion and neuroticism, respectively; they demonstrated associations between these two types of variables over an interval of ten years. Findings of this type show that personal characteristics are an important type of contributory variable that should be used as a statistical control in nonexperimental studies.

Second, once we have set aside personality variables or other general causes of well-being unrelated to policy, we can also usefully examine subjec-

14. The total variance explained rises to twenty-nine percent, nearly three times that explained by background variables alone. This general approach is also suggested by A. Campbell et al. (1976: 485). There is, however, a possibility of artifactual correlation between change up to the present in a variable and the present value of that variable.

tive variables that intervene between objective policy-manipulable circumstances and general well-being. A. Campbell et al. (1976: 16) suggest that these subjective variables include perceived attributes, evaluated attributes, and satisfaction with the domain in question. These writers have centered attention particularly on satisfactions in domains including housing, job, marriage and family, community, money, health, and government. Within particular domains, detailed questions about satisfactions with smaller subdomains (such as schools in the community or craftsmanship in housing) can show effects of possible specific policy emphases that might not be revealed by a study of more general domain satisfactions or global well-being (1976: 506). In view of the importance of people's standards of comparison and their changes, we might also measure these standards directly and trace them over time.

Our interest in domain satisfactions as intervening variables between policy and general well-being must be centered on those domains that correspond to objective policy outcomes.[15] Andrews and Withey (1976), for example, analyze the associations of feelings about oneself and one's personal life with general well-being. These feelings, however, are not affected directly by specific policies (except therapy) even though they may be affected indirectly by a wide variety of policies or conditions. We thus omit the domain of self; similarly, A. Campbell et al. (1976: 76n) exclude analysis of the self as a specific domain, in part because of its broad scope.[16]

The eventual model that we are seeking would include causal connections from objective (policy-manipulable) conditions to corresponding domain satisfactions, and from these to general or global well-being, with controls on variables such as personality that affect well-being.[17] Such studies have been infrequent (Andrews 1981: 396), but a number of studies have investigated parts of the overall model. The relations between objective conditions and domain satisfactions vary considerably among domains: A. Campbell et al. (1976: 382) note that the proportions of variance in domain satisfactions ex-

15. We set aside those symbolic policies that affect well-being directly.

16. Such a domain, including self-esteem and feelings of competence or efficacy, might play a part in a more complex model even if it were not affected directly by policy. School desegregation, for example, has been believed to affect the self-esteem of minority students and thus to increase their achievement and eventual well-being.

17. These relations to global well-being are usually treated as current functional relations, based on cross-sectional data. A number of studies have carried out multivariate analyses in which domain satisfactions were combined with objective variables (e.g, income) in a procedure such as multiple regression. We do not report these studies because the method risks giving the impression that the objective variables are of little importance because their direct causal contribution is small, and it does not allow us to assess the strength of the effects of the objective variables on the domain satisfactions. More systematic causal modeling would be appropriate with such data. If, however, we wish simply to control for variations in income, the method can be useful.

plained by corresponding objective conditions range from about six percent for financial or income satisfaction (in relation to income) to sixteen percent for satisfaction with friendship and forty-three percent for health.

The relation between domain satisfactions and general life satisfaction has been studied by A. Campbell et al. (1976: 85), among others, by means of multiple regression.[18] Most authors who use this method make it clear that it yields predictions, but not necessarily causal coefficients. The fact that the coefficients in regression equations that predict general well-being are not necessarily causal is reflected in a study by Burt et al. (1978: 369), who analyze similar data on the reverse assumption that satisfaction with consumption of any particular good is affected by the actor's well-being on more general dimensions. Yet because one may argue plausibly (even if not conclusively) that domain satisfactions cause general subjective well-being, we shall review some of the findings from regression studies that use domain satisfactions for prediction of well-being, seeking major contributory variables.

A. Campbell et al. (1976: 79–80, 85) find that a linear additive model adequately fits the relation between domain satisfactions and the overall sense of well-being, and that more complex models are unnecessary. They report raw partial regression coefficients for twelve domains for their national sample. Those above .3 are for family life (.41), marriage (.36), financial situation (.33), and housing (.30); between .2 and .3 are job (.27), friendships (.26), community (.25), health (.22), and nonwork activities (.21); below .2 are national government (.15), organizations (.12), and religious faith (.11). Findings of this sort led to their conclusion that intimate personal relationships were of major importance in accounting for subjective well-being. They also report modest variations in the values of regression coefficients among population subgroups; for example, health is more important for older respondents, marriage for housewives, and standard of living for lower income respondents (81).

A similar regression study is reported by Bharadwaj and Wilkening (1977) for a sample of respondents in northwestern Wisconsin. Dividing their sample into subgroups by age, sex, and income, they find that the highest standardized regression coefficient in each subgroup is for the domain of health or family. For those with low income, satisfaction with income enters as the sixth highest coefficient, but it ranks less important for respondents with higher incomes. Again, the personal situation seems more important than income, which may be causally more remote from subjective well-being.

Andrews and Withey (1976: 169) report a set of multiple classification analyses (analogous to stepwise regressions in that they successively reveal the most effective predictors) predicting global life satisfaction from feelings

18. A more complex model, including measurement effects as well as substantive causal effects, has been estimated by McKennell et al. (1980).

in twelve selected domains of life concern. The high predictors include feelings about "amount of fun" and "yourself," which we would not include as corresponding to specific policy domains; but together with them, feelings about "family life" and "money" are among the highest predictors.

A number of studies have reported raw correlations between general subjective well-being and domain satisfactions; we report only one, which makes interesting comparisons among subgroups. Freedman (1978: 39–40) examines associations between "how happy [people] were with many aspects of their lives" and "overall happiness," for men and women, single and married. For single men and women, the highest associations are found for the domain of "friends and social life," with the second and third highest being "job or primary activity," and "being in love." For married men and women, "job or primary activity" and "friends and social life" are much lower; "being in love" is among the top two for both sexes, "marriage" in the top three, and "personal growth" first for married men. "Finances" are no higher than seventh for any of the four subgroups. These associations again suggest the importance of interpersonal relations as a correlate of happiness, with job and finances on the list but less important.

A panel study over a period of two years by Headey et al. (1984) shows similar associations between changes in domain satisfactions and changes in well-being. The six domains with highest standardized partial regression coefficients (between .46 and .39) were friends, standard of living, sex life, marriage, leisure, and job.

A study predicting subjective well-being from more nearly policy-related "domain satisfaction" variables was conducted by Diemer and McKean (1978). In a municipal survey, they asked respondents to rate their general life satisfaction from zero to one hundred, and to rate their intensity of particular complaints about conditions that might be affected by urban policy or planning. Regression coefficients suggested that public transportation, growth planning, and cultural programs were among the specific domains of complaint that had the highest partial associations with general satisfaction.[19]

The Diemer and McKean regression study, like that of A. Campbell et al., used subjective variables as both independent and dependent. It dealt, however, with predictors that are related to policies (growth planning) rather than conditions (housing or community). It did not include domains such as marriage, family, and friendships, presumably because they are not related directly to municipal policy alternatives. Ultimately, we need to reconcile the results obtained from using these two types of predictors. In doing so, we shall face the questions as to what causal relations exist among the predictors. In particular, we need to learn whether variables accessible to policy (such as

19. Causal models predicting general well-being from both policy-relevant objective conditions and domain satisfactions have also been investigated by Hall (1975).

growth planning) have significant effects on variables closer to the person (such as family life).

From these studies it appears that major domain satisfactions affecting subjective well-being are those concerning one's close personal relationships, material and economic situation, and job or occupation. These satisfactions are not, of course, perfect reflections of the objective conditions; as we shall see, the perceptions and standards of comparison that a person brings to each domain affect the satisfaction or happiness reported for it.

Objective Predictors: Standards of Comparison, Adaptation, and Change

We have suggested that there are two main types of models for the causation of subjective well-being. In one type, used in economic utility functions, well-being is a function of current objective conditions. In another, however, well-being can depend on changes in those conditions or on comparisons of those conditions with expectations or other standards.[20] The possibility of this second type of causal relation is suggested by several researches dealing with the relation between economic and subjective well-being.

We move here beyond the economic notion of utility, which despite its original connection with happiness has come to be defined operationally by the satisfaction of preferences and its objective manifestations. We might nevertheless expect that direct measurements of subjective well-being would confirm its supposed connection with preference satisfaction. In many instances, a consumer who purchases a good and obtains a consumer surplus is presumably happier as a result, and one who has a higher income and spends it regularly on such goods is also happier. Even the mere opportunity for such spending may increase one's subjective feelings of well-being, as is illustrated by the role of "satisfaction with savings and investments" as a contributor to overall satisfaction with life (A. Campbell et al. 1976: 76).

It is important, however, for us to know the limits of validity of this presumed relationship between economic variables and happiness. The problem is not merely that some economic values are not ordinarily priced in the market, because "demand" for clean air and uncluttered streets can be measured with the aid of conventional economic concepts (Chap. 5). Rather, one source of the discrepancy between subjective and economic measures of well-being is the fact that human beings bring diverse standards of comparison to bear on their objective or material conditions. They perceive and evaluate these condi-

20. Following A. Campbell et al. (1976: 14), we use the phrase "standard of comparison" to refer to a variety of psychological standards against which a person judges his circumstances.

tions differently depending on their past experience and personality characteristics. Thus, if it is happiness that we seek to maximize, this maximand is not a simple function of a material or objective condition alone (Sen 1980: 362–63).

The distinction between the objective environment and the perceived environment is a persisting theme of psychology; but it has been especially brought to the attention of students of social indicators by work such as that of Easterlin (1974), Duncan (1975), and A. Campbell et al. (1976).

Easterlin has analyzed comparable surveys conducted in fourteen countries using Cantril's self-anchoring scale of well-being, and has shown that the association between per capita income and this measure of happiness is far less than we might expect on the basis of intranational differences. Comparisons between the highest and lowest status groups *within* thirteen nations, both developed and less developed, show an average difference in happiness on this scale of 1.8 points; but across nations, the national averages "for 10 of 14 countries lie within a range of 1.1 points" (1974: 102, 106), even though these ten countries cover a "tenfold range in per capita income." His findings suggest that persons in different nations take as standards of comparison particular levels of well-being characteristic of their own nations, and that improvements in national material conditions may not carry with them a proportionate increase in reports of subjective well-being. An analogous study of the subjective quality of life in fifteen United States cities also shows very low associations of this variable with aggregate objective indicators (M. Schneider 1976). Substantial correlations between subjective well-being and national per capita GNP are reported, however, by Inkeles and Diamond (1980: 93–95).[21]

A study by Duncan (1975) of the satisfaction of Detroit housewives with their standard of living in 1955 and 1971 also shows very similar distributions of subjective well-being in the two years even though the median real income had increased by forty-two percent. The same surveys nevertheless show "that individual satisfaction typically increases with increasing incomes in *cross-sectional data*" (1975: 270). Again the findings suggest that a new generation had set new standards against which to judge its satisfaction with its standard of living.

A comparison between economic and subjective measures of well-being is valuable because the two measures reflect somewhat different models of human activity and valuation. Whereas the consumer's economic welfare (utility) is often characterized in terms of rates of consumption of various commodities in relation to his preferences, he may also alter standards of judgment in relation to his life experiences. For example, in economic assessments of the returns to education it is customary to consider years of educa-

21. National income and GNP are objective measures of preference satisfaction; in comparing countries or historical periods we assume that they represent objective conditions.

tion as a continuous variable whose value increases as education increases. A consistent finding has nevertheless been reported by Bradburn (1969: 45) and by A. Campbell et al. (1976: 51–52, 136–38) that respondents with some college education, but without a college degree, are lower in global subjective well-being than are high-school graduates who have not attended college. With the exception of this group, both studies show a monotonic increase in well-being with education.[22] One explanation may be that those entering college have acquired a new reference group and raised their standards of comparison, and that their failure to graduate produces lasting dissatisfaction.

A similar disparity between economic (or ecological) and social-psychological models of human choice may perhaps be seen in the study of migration. In an economic model, one might characterize the welfare of persons at various locations in terms of their objective opportunities, as indicated by wage rates or unemployment rates, and use these variables to predict migration. It has nevertheless been shown by Lowry (1966) that in a regression of the logarithm of number migrating between urban areas in the United States on unemployment rate, wages, and other objective characteristics of origin and destination, raw regression coefficients for the characteristics of the destination are higher than those for the same characteristics of origin. Similar findings have resulted from other studies (Greenwood 1975: 400). It may be that people set their expectations by adapting to the standards of their place of current residence and that their motivations to migrate are less affected by the objective characteristics of their places of origin than might be expected from an economic model.

Whether this difference is due to adaptation or to habituation to the community of birth, it seems to reflect a tendency to use one's present location as a basis for comparison. A more explicit test of this type of comparison has been made by Fernandez and Kulik (1981: 849), who show "a small and perhaps unreliable effect of relative deprivation, whereby individuals with incomes below the neighborhood average appear less satisfied."

These studies suggest that the social groups in question—different societies, generations, and communities—have distinct standards of comparison that allow their members to judge their subjective well-being somewhat independently of their objective circumstances. If so, we face the problem of accounting for these subjective standards. Conceivably they may be functions of current and prior objective circumstances, together with the length of time that people have been exposed to them; effects of change and of habituation may exist, as we shall see in the next section.

22. The anomalous position of those with some college persists when income is controlled (A. Campbell et al. 1976: 52). Smith (1982) presents somewhat different findings, but reports a similar result for respondents who did not complete graduate school; he reports (in a letter to the author, 1983) that these effects do not appear when Bradburn's affect balance scale is the dependent variable. Perhaps a measure with cognitive content is necessary to reveal this effect.

It is also possible, however, that these collective differences in subjective well-being are not strictly functions of objective circumstances and time even in this more complex sense. If these phenomena are subjectively social, then exhortations, hopes, group cohesion, and other features of people's life situations that are shaped by communication may lead them to perceive their circumstances differently in ways that are very difficult to encompass in mathematical functions. Thus, the problems of the subjectively social sciences are relevant to our prediction of subjective well-being.

We are thus led to examine more closely the working of these standards of comparison as they affect well-being. Several psychological hypotheses about the nature of the reference points that people use as standards for comparison are reviewed by Andrews (1981: 403–09). We shall discuss some of these approaches below; but what is most important here is their common departure from the economic (objective) approach.

A similar argument, in even stronger terms, is made by Brickman and Campbell (1971: 287) on the ground of adaptation level theory: "The subjective experience of stimulus input is a function not of the absolute level of that input but of the discrepancy between the input and past levels. As the environment becomes more pleasurable, subjective standards for gauging pleasurableness will rise."[23] They are led from this theoretical postulate, and numerous substantiating instances, to conjecture that humans are condemned "to live on a hedonic treadmill, to seek new levels of stimulation merely to maintain old levels of subjective pleasure" (289).

A more complex argument for the time-dependence of affective responses to given circumstances (stimuli) has been made by Solomon and Corbit (1974). They propose that a wide variety of observed phenomena can be explained by postulating two concomitant but opposing psychic processes caused by the initiation and removal of reinforcing or aversive stimuli. The first is the production of an affective state corresponding roughly to the stimulus, as long as the stimulus continues; alone, this would resemble the economic assumption of utility corresponding to the presence or flow of goods. The second or opponent process sets in following the start of the first process, and reduces the initial affective state so as to produce a result similar to the habituation effect considered by Brickman and Campbell. But after the stimulus is removed, the opponent process remains and leads to an inverse affective state (after-reaction) which persists for some time after the removal. Extreme examples are the pulse of euphoria after an addict takes heroin, followed by intense craving when the drug is withdrawn; or the anxiety of a parachutist, followed by euphoria when he lands safely.

We see again in this theoretical model—regardless of the fact that its exact

23. This change in the adaptation level is seen as occurring independently of the level of aspiration.

parameters vary among types of stimuli—a time-dependence of affective responses to stimuli.[24] Solomon and Corbit also point out that the strength of the opponent process is increased by repeated stimulation; thus it depends on the individual's previous history as well as on events or stimuli within one immediate stimulus sequence. Both these features are in sharp contradiction to the dominant economic model in which utility is simply a function of the current flow of commodities (stimuli).

The possibility that subjective well-being is not simply a function of given objective circumstances, but can vary with time and the individual's history, has important policy implications; we shall elaborate them in the next section. But as long as this remains simply a possibility—a phenomenon that occurs frequently but not necessarily universally—we must be careful in proposing untried policies to increase subjective well-being. The relations reported between objective circumstances and subjective well-being actually vary somewhat among measuring instruments and among conditions. Easterlin's cross-national comparison used the Cantril self-anchoring scale, which by allowing respondents to set their own anchoring standards may have conduced to his finding of small cross-national differences; Gallup (1977) found greater differences with a different measure. In various studies of the relation between gender and mental illness, a recurrent finding has been that women are more liable to symptoms of mental illness than are men (Gove and Tudor 1973; J. Fox 1980); this seems inconsistent with Inglehart's hypothesis of adaptation to sex roles. And as a conjecture, we might ask whether at low levels of material satisfaction (e.g., hunger) it may be more difficult to adapt to a given level and to deem oneself satisfied with it. For all these reasons, our policy inferences from these findings will be tentative.

One final study suggests both the relevance of temporal change and the difficulty of predicting in advance what the effects of such change will be. We have seen that one of the most consistent correlates of subjective well-being, within one society and over short time periods, is income. It would then appear to be a trivial confirmatory test to examine the effect of changing someone's income. We might expect, for example, that a sudden increase in income would produce greater immediate gain in subjective well-being than would be predicted from the steady-state association between well-being and income. This conjecture would not, however, prepare us for the results of an experimental study recently analyzed.

Randomized experimental studies in which subjective well-being is a dependent variable are rare; but the experimental studies undertaken to examine the effects of income-maintenance programs provide one such instance. Thoits and Hannan (1979) made use of the Seattle and Denver income-maintenance

24. The stimulus is characterized here in objective terms, not mediated by perception; thus affect is represented as a determinate (but time-related) function of objective conditions.

experiments to see whether experimental increases in income altered the incidence of psychophysiological symptoms of distress over periods from four to twenty-four months after enrollment. They found only a limited number of significant effects, but all of these were negative, i.e., additional income brought *increased* distress, which showed no tendency to decrease during the period studied. They interpreted the effect by suggesting that the experimental treatment brought with it disruptive life events. Their findings, running counter to the presumed positive relation between income and subjective well-being, should remind us of the importance of time sequence in the relation of resources to subjective well-being, and of the difficulty of predicting specific real world effects from general theories.

Policy Inferences

It is harder to make policy recommendations for subjective well-being than for objective social goals or economic benefit. For subjective end-values we have not only a variety of concepts but also diverse measures of each particular concept, not always highly correlated. There also seems to be less agreement as to what is of value among these concepts than among objective measures, in spite of our criticisms of the economic and objective approaches. In addition to the question whether subjective or objective well-being is more valuable, we also face the choices between emphasizing "higher" pleasures as against pleasure generally, and between valuing pure affect, cognitive judgment, or some mixture of the two. Subjective measures used in combination with time do, however, provide a possibility of greater equity among persons than monetary measures afford, as do census-type measures of objective conditions (Chap. 7).

The findings we have presented on causes of subjective well-being suggest sorts of policy intervention that go beyond the provision of goods and services. Because research in this area has been based largely on cross-sectional data without interventions, time series, or policy trials, however, our range of possible interpretation is wide. Not only are our causal inferences uncertain, but policy interventions can encounter moral criticism. We may be concerned, on one hand, with manipulative or intrusive policies that run counter to our moral feelings; but even if we believe these policies to be desirable, it may be that government could have only limited influence over subjective well-being.

Reactions to public events. We may illustrate the possible limits of government influence on subjective well-being by examining the effects of major public events. Two significant events studied by Bradburn and his associates, the Cuban missile crisis and the death of John F. Kennedy, seem to have had remarkably little immediate effect on American respondents' subjective well-being—whatever other effects they may have had.

In the spring of 1962 the National Opinion Research Center (NORC) conducted a survey of four Illinois communities, studying relations between community characteristics and subjective well-being. After the Cuban missile crisis in October, respondents in two of the communities were reinterviewed. They were asked in both surveys whether they had worried about various topics during the week preceding the interview (Bradburn and Caplovitz 1965: 75–76). At the second interview wave, during a period when a feeling of crisis persisted, worries about a wide variety of personal matters had decreased; the proportion reporting worry "often" about the atomic bomb or fallout, however, had increased from six to twenty-four percent. "Most of [the] respondents . . . had developed a new concern during the period of the Cuban crisis, one that probably took their minds off their other problems" (76). Respondents' reports of psychosomatic symptoms changed relatively little, however.

Between the two waves of interviews, respondents' replies to four items in the "positive affect" scale showed modest decreases in the proportion who had "felt this way . . . at all" (ranging from nine to fifteen percent) (82); but negative feelings did not increase. The proportion who said they were "very happy" increased by five percent over the same interval. The authors concluded that "the trend data [on affect alone] do not support the idea that the international situation had a major impact on the feelings of our respondents" (93). Respondents did report worry about the Cuban crisis, but it seems not to have affected "their psychological state" (127), that is, their well-being.

In 1963 a similar panel study of respondents in several metropolitan areas was being conducted by NORC when the news of President Kennedy's assassination was heard. A subsample of respondents in two areas were reinterviewed starting on 26 November, the day after the funeral (Bradburn 1969: 213). Among the items in the positive and negative affect scales, two showed changes: there were increases of thirteen and twenty-one percent in the two sampled areas in the proportion saying they had been "depressed or very unhappy" in the past week, and increases of sixteen and twenty percent in the proportion saying they had been "excited or interested in something" (215). Responses to other items in the positive and negative affect scales changed in the opposite directions, however, leaving the averages of these two scale scores relatively unchanged. There was shown to have been a "strong and specific grief reaction" (221), but there was no corresponding anxiety reaction or net change in affect.

We conclude that the symbolic and emotional effects attributed to major public events need to be interpreted with care. Effects were shown by these two surveys; they involved fear and grief, but not changes in well-being. We should be careful, therefore, in concluding that a politically significant symbolic event causes changes in well-being.

Bearing in mind the caution that these studies suggest about governmental influence on subjective well-being, we may conjecture about larger and more controversial effects.

Types of policy intervention. From the studies and hypotheses we have discussed, the policy directions we may take to promote subjective well-being seem more diverse than those for objective or economic well-being. We may consider three types of processes for enhancing subjective well-being: (1) The provision of objective (material) means or facilities such as income, housing, hospitals, and schools; (2) The provision of social resources, or the reduction of obligations, that in effect increase an individual's social income; (3) The modification of mood or affect in the absence of corresponding change in amounts of material or social resources—through physiological effects, through the timing and variety of resources or stimuli, and through communication. For some means of this type, however, there are important social prohibitions as regards the permitted range of either government or individual action.

The first of these three types of policy intervention—providing objective things—overlaps considerably with economic analysis of policy choice. Such analysis needs, however, to be followed by a detailed study of the effects on subjective well-being of providing money or facilities. An expression of subjective well-being is much more difficult to connect clearly to a particular policy than is an expression of willingness to pay (Stipak 1983). The negative income tax experiments have furnished much information on the effects of providing money to the poor; but as we have seen, it does not immediately increase their subjective well-being. When facilities are provided, we need to see who takes advantage of them and what effects follow.

The second type of intervention—provision of social resources—may appear to relate only to the provision of services of various sorts, from medical services to home care visitors for the elderly. But public policy may also be able to aid members of groups in aiding one another as in the Fairweather Lodge—to increase the capacity of families and communities, for example, to cooperate and contribute to their members' well-being. We shall explore indicators related to this type of cooperation further in Chapter 8.

The third type of intervention, which does not proceed through the simple supply of resources, is of special interest. It is this type of intervention that we are most likely to forget when we characterize policy variables exclusively in objective or material terms; but it also raises ethical problems as to whether it should be exercised by government.

We set aside those means of changing moods or subjective well-being that depend on substantially altering the physiological state of the individual. An example is drugs; but in excluding these, we must point out that some drugs are marketed and others can be prescribed. A number of drugs produce feelings of euphoria or reduce pain; some, like aspirin and alcohol, are marketed freely as goods, but are not usually subsidized by government. Some are taxed, limited to physicians' prescription, or prohibited. Among the reasons for such legal prohibitions is the possibility that persons using them may bring harm to others, either by their impaired capacity to function (like drunk driving) or by their need for resources to support an addiction. Nevertheless, substantial modification of personality by drugs and other physiological interventions has been permitted and subsidized in the case of mental patients given drug or electrical therapy, or psychosurgery.

Of greatest interest here are those types of policies that involve communication or the regulation of communication by government, or the timing and variety with which resources are supplied.[25] In principle, public authorities can lead people to experience greater subjective well-being by inducing them to use lower standards of comparison. If people are already low on a scale of objective well-being, then the authorities' task would be to limit their communication with (or awareness of) those more favored, perhaps by placing them in closed homogeneous groups. Aldous Huxley portrayed such a social order in *Brave New World* (1932), in which each member was assigned to a particular caste and taught from infancy to remain in his place. Such policies are forbidden in a free society in which individuals' choices of communication and association are unrestricted by government. They seem repugnant to us because of their inequity; as observers of an unequal distribution of objective conditions, we would see as unfair a manipulative policy by privileged groups who deceive the less fortunate into being happy with what they have.

These inequities would be reduced if all were asked to share in sacrifices, as in the face of a common emergency; but if information is to be curtailed by specific policies, some must know of it while others are kept ignorant. Such censorship is strongly criticized, either on the ground that it inhibits political opposition or on the basis of constitutional provisions. There are nevertheless social and legal boundaries, varying among nations, inside which communication is sometimes regulated, even though it is protected in the political and the private realms. Most relevant to subjective well-being are possible limitations on commercial communication; advertisements can encourage unattainable aspirations for material affluence and personal attractiveness. Proposals have been made to regulate advertising for children, on the ground that it cultivates inappropriate demands; but these have met strong opposition from the interests involved.

Public authorities can not only regulate others' communications, but can also send messages themselves. They can encourage people to separate men-

25. Government can be involved indirectly by subsidizing certain forms of communication, such as psychotherapy; but if participants enter voluntarily there are fewer ethical problems.

tally certain experiences from certain standards of comparison. Thus periods of special indulgence can be officially defined as exceptions to our normal expectations, in order to avoid raising those expectations; and people can be encouraged to separate their social statuses from one another, so as not always to compare their experiences with expectations based on their highest status. Occasionally it may be possible to "cool out" persons who might otherwise expect too much (Brickman and Campbell 1971: 292–94). When these sorts of things are done, however, they are usually done covertly or without explicit mention of the psychological mechanisms involved, especially when the communications are directed at a particular group.

What cannot be done by government can sometimes, however, be done by voluntary action. Groups of people who are critical of urban industrial society can create communes or communities aimed at technological simplicity, harmony with nature, and energy conservation, voluntarily cutting themselves off from comparison with those who have more material goods or reinterpreting that comparison. Dedicated religious communities have done similar things. Some of these types of private action might conceivably be encouraged by government.

A corresponding approach directed at those who have more material goods, but who enjoy them less because of higher standards of comparison, would be to encourage them to compare their state with standards lower than those of their own present milieu. People might be encouraged to remember past unhappy states to some degree so as to use them as standards (Brickman and Campbell 1971: 293). They might also be encouraged to compare themselves with others who are less fortunate; but this could lead to feelings of guilt or obligation and not merely satisfaction. Guilt or obligation feelings would not benefit the persons who experienced them, though they might eventually benefit others.

The temporal sequence in which people make use of (consume) goods is also a possible condition of subjective well-being. Brickman and Campbell (1971: 292) suggest that satisfaction can be maximized by making transitions gradual rather than sudden, whether they are upward or downward. The desirable rate of social mobility might thus be limited not only by the rate of decay of specific human capital, but also by individuals' capacity to adapt to new conditions.

The problem of habituation, noted by Solomon and Corbit (1974), would also suggest that we not plan for people to have determinate amounts or flows of material goods and facilities, but allow them opportunity for variety and novelty (Scitovsky 1976: Chap. 3). Their well-being may be seen as depending not simply on amounts or flows of goods, but on temporal sequences of consumption. Advocates of free markets praise them on the ground that consumers and producers can make efficient decentralized decisions; but even if we could give consumers exactly the flows of goods they now "demanded" to consume at given prices, they might enjoy them less over time. This would be because their preferences were not fixed, but continually in flux because of habituation, so that they would have to diversify consumption over time in ways that unchanging demand curves cannot determine.

Government influence over citizens' standards of comparison is limited not only by the legal framework of free communication, but also by the legal protection of individual privacy and families' internal activities. Persons who are members of functioning families are protected against the intervention of the state; only when a family is deemed to be providing inadequately for one of its members (as in the case of child abuse), or voluntarily allows a member to be cared for by outsiders (as in the commitment of a dependent elderly person to a nursing home) can outsiders, and the state as a regulator, make decisions about that person.

When the elderly are separated from their families and dependent on governmental aid, their condition is especially relevant for consideration of policies aimed at their subjective well-being. The elderly are considered deserving and worthy of public support, and they are often dependent without being fully supported by their families. It is important to study their subjective well-being because the usual criteria of production are less relevant to nonearners. Even their income, when unearned, is separate from job satisfaction and combined with more leisure than for full-time workers. A possible reduction in aspirations is a pervasive theme in analyses of the well-being of the elderly, introducing the psychological considerations we have mentioned here in contrast to purely economic ones. Thus significant applications of many of the questions discussed in this chapter can be found in analysis of policies for the elderly.

A final controversial problem concerning the promotion of subjective well-being is that satisfaction may be associated with conservatism (Buttel et al. 1977). A paradox in the effort to increase subjective well-being is that some actions increasing it in the short run may not maximize it over the longer run. This is not simply a problem of allocation of resources to known policies in the present rather than in the future; such a problem can be dealt with, in principle, through maximization over time. Rather, it is a question of our very awareness of possibilities for change. If we are blind to our own and others' present deprivations and to the potential for fundamental change, we may be happier than if we knew of them. Conversely, as we become aware of the possibility that things might be better for us and others we may well become less satisfied in the present. Dissatisfaction seems to be associated, even more than negative affect, with political demand.

Another sort of optimization problem then arises. How much of the time should we be concerned with changing the political and social order, and how much with making the most of it as it is? Furthermore, which subgroups of the population should be especially concerned with the future? In this perspective, even subjective well-being can be both an end-value and a contributory variable.

A continuing awareness, on the part of citizens and policymakers, of the contrasting directions in which the quests for subjective and objective well-being would take us can highlight important possible conflicts in our value systems.[26] It may lead us back to an objective, material standard of judgment because of the difficulties in seeking subjective well-being. It may remind us of the price that must be paid for the maintenance of a fully free society, and lead us to seek limits on freedom of communication that will not impair political dissent and change. It may also lead us to inquire just how the boundaries of privacy should be drawn. Thus a closer relationship between the study of subjective well-being and related policy choices can focus our attention on major long-run issues of choice, including the possible clarification of our collective values.

26. The desirability of including both objective and subjective variables in an indicator system is also pointed out by Zapf (1979).

7 Equity: Aggregation and Compensation

A third major class of end-values are those concerning equity.[1] Equity is an end-value and not merely a means to the general well-being. Even if we could transform the monetary standard of economic benefit into a standard counting each person more equally, such as one based on time, we would still not have dealt fully with the demands of equity. Whatever measure of value we may seek to aggregate, we must still acknowledge that it can be distributed in an undesirable or unfair way.

The status of equity as an end-value is complicated, however, by the variety of things that may be distributed. When end-values themselves are considered, their distribution is clearly a value that cannot be merely a means to other values. Questions of equity may, however, be raised for more specific values; the distribution of a particular good, or of citizen obligations, may be more or less equitable. If this good or obligation is not a general end-value, we may imagine that it can be traded off for a more inclusive or more nearly intrinsic value. An entitlement to housing could be replaced by money; military service, replaced by a tax; a wrong done to a person might be compensated by a monetary payment. We do not always allow such replacements,

1. From the perspective of policy indicators, equity is concerned with the distribution of valued things or conditions among persons. These things or conditions may range from income or happiness to access to hospitals or the obligation for military service. This feature of equity is nearly synonymous with distributive justice (Rescher 1966); but Barry (1983: 8–20) contends that justice concerns only the distribution of resources and not that of utility (subjective well-being), and Lucas (1980: 163) limits distributive justice to distributable goods. In a narrow sense, justice concerns matters or rightness independent of consequences; our treatment here is more in the spirit of the economic use of equity, as concerns distributions of things among equals and unequals, and we do consider how happiness should be distributed. As for the concept of justice, distribution does not exhaust it; some writers have connected justice with the keeping of promises, or with obligations to the state, which are not obviously distributive notions. Moreover, justice may require certain distributions in a sense stronger than that of desirability. The status of equity as an end-value differs from that of individual welfare; it does not rest on individuals' valuations of things that affect themselves alone, but involves our judgment of relations between individuals.

however, and equitable distribution of particular values that are not end-values is often treated as being of intrinsic value.

The distinction between undesirable and unfair distribution of values corresponds to two distinct aspects of equity. If a distribution of a value is merely undesirable (i.e., it fails to maximize some social welfare function), and if the responsibility for remedying it is deemed to rest on public-spirited citizens concerned with the general welfare, we shall call this *distributional* inequity. On the other hand, if a distribution is considered unfair or unjust, and if the responsibility for remedying it rests especially on a particular actor or group that is held to have acted wrongly in bringing it about, we shall refer to this condition (or claim) as one of *compensatory* inequity—"rectificatory" justice in Aristotle's (1942: 1131) classification.[2] We shall deal with distributional equity in the next three sections of this chapter, concerned with univariate distributions; and with compensatory equity in the last, concerned with bivariate and multivariate analyses.

Distributional equity, as reflected in policy indicators, is often associated with the conditions of the poor and needy in relation to those who are wealthier. The notion of need, though distinct from equity or distribution, overlaps with these concepts. If a minimum standard of health, housing, or safety (for example) is required by all,[3] then proposals can be made for policies that provide it. In a literal sense, these proposals would lead us to try to raise everyone at least to the minimum standard regardless of how many were currently above and below it. Some minimum standards, based on physiology or survival, would have this absolute character. Others, such as the poverty level, have changed historically and can reflect notions of equity in relation to what others have (Bulmer 1983: 357–59).

How, then, should well-being or other values be distributed? In answering this question, as in our discussion of end-values in the preceding two chapters, we must seek to put forward relatively consensual valuative criteria for equitable distribution, measurable by policy indicators. We expect less consensus on standards for equity than on other end-values, however, because questions of distribution tend to pit one group against another. The question as to what shares various groups should have in some value is likely to be open to struggle and bargaining, or perhaps to temporary compromise, but less to

2. Distributional equity is a forward-looking criterion that we may seek to maximize in utilitarian fashion; compensatory equity is more a backward-looking, nonteleological criterion (Lucas 1980: 13, 62). When compensation is called for, even though the actor at fault bears a special responsibility, public-spirited citizens may also feel responsible that justice be promoted.

3. One reasonable argument for such a standard would be that below certain levels of nutrition, or above certain concentrations of pollutants in the workplace people's well-being would decline precipitously. From this perspective these minimum levels or needs would not be end-values; rather, attaining them would be an efficient means to overall well-being. Need in this sense, or its fulfillment, would be a contributory variable.

agreement on general formulas that will last more than a few decades. The only formula that would seem to have *prima facie* preference is simple equality among persons. This is indeed the simplest definition of equity—that all persons should be equally well off, or equally happy.[4]

This claim of equity as equal distribution conflicts, however, with another widely held principle. Another aspect of equity suggests that people have a right to the fruits of their efforts if they have gained them in fair enterprise or competition. As Lucas (1980: 244–45) points out, if government acquires money that "just happened, it is arguable that it should be distributed among all members of society equally. But the gross national product does not just happen. People have to work, some of them quite hard, to produce it." We thus face two disparate, incommensurable end-values that are typically reconciled through political processes. A partial reconciliation is possible, however, because people's right to the fruits of their efforts is closely related to the economic notion that the aggregate net benefit is enhanced by incentives that bring with them possibilities of inequality. If we replaced the full right to one's earnings by an ethic of maximization, but maximized a sum of quality-adjusted life years (for example) rather than of monetary value, we might combine the two principles to some extent. This maximization would presumably then dictate that rules be made to preserve incentives.

The notion of distributional equity is most often introduced into policy analysis through the usage of economics, where vertical and horizontal equity are distinguished.[5] Typically, *horizontal equity* is defined as the equal treatment of equals, and *vertical equity* as the unequal treatment of unequals (Musgrave and Musgrave 1976: 216), the latter with the implication that they should be made more equal. These definitions leave problems as to what equal and unequal treatment mean, but even greater problems in identifying equals. We may, for example, treat all citizens of a nation-state as equals; but in a federal union, we may sometimes classify only members of each particular part as equal, with respect to action by that part. For other purposes we may consider equality among persons occupying a given type of job, as when we are studying inequities in salary; but we may further classify them according to their previous education and other variables before considering them really equal.

This diversity of definitions suggests not only imprecision, but also the part

4. In connection with educational policy we saw that equality of opportunity was valued; Coleman et al. (1966) studied this concept by examining educational outcomes. The same approach has been used in the study of opportunity for upward social mobility. In both cases we seem to exclude from our research any measure of whether opportunities, if they existed, were grasped—whether effort and will were invested. The study of equality of opportunity, in this literal sense, is difficult for social science, as effort and will are hard to measure. We return to this question below in connection with discrimination.

5. A different use of the term equity is found in philosophical writing (Lucas 1980: 241, 244).

played by social norms in establishing the criteria of similarity that are relevant for a particular purpose. Various societies, for example, may stress caste membership, religious observance, and economic effort as bases for equal treatment. At the very least, a common language and a perceptual framework are necessary for such a classification; and a common normative framework is probably also needed. This requirement, in turn, may mean that equity is more easily assessed within a single society, or social subsystem, than across systems. We then have to judge which criteria for equal treatment are relevant or essential (Perelman 1967: 22) and are tempted to base our judgment on the norms of the system rather than on universal ethical criteria.[6] Thus a test of membership may be relevant for equity.[7] The judgment whether to grant the rights of membership may be a first application of equity as a value, even though it extends only to the disputed social boundaries of the community or polity in question.

Equity (or justice) also has another widespread definition related to reciprocity (Barry 1979: 53–55). Notions that two people who enter exchange should give one another equal values, or that persons should receive equal pay for equal work, are of this kind. By extension we may include the notion of *compensatory equity*: that compensation should be made for wrongs against persons, or that punishments for wrongs against the community be commensurate with the magnitude of the wrong.[8] This notion has been formalized by Walster et al. (1978) in the requirement that the ratio of outcome to input be the same for different persons who are being compared. As we define it, it also includes the placement of blame on a specific actor responsible for a wrong, who then has a special obligation to compensate the victim of the wrong.

Yet even here there are important ambiguities. A crime or an exchange is implicitly viewed as an act committed at a particular time, with outputs or wrongs being judged relative to a preexisting condition. Some would claim, however, that even that prior condition could be unjust (Barry 1979: 74)—that "property is theft"; or that air or water pollution as a by-product of production

6. It is possible, however, for social measurement to ascertain the judgments of equity held in particular social systems so as to adjust policies to them. This was done in the design of the point system for discharge of soldiers after World War II (Stouffer et al. 1949: Chap. 11).

7. Rather than membership in a society (sharing its norms), presence "in a given justice-constituency," or in the domain of a political community, has also been proposed as the requisite membership for "doing justice" (J. Stone 1979: 101). For policy-relevant comparisons that transcend such boundaries, other end-values such as need are often invoked, but equity is also considered (Barry 1979: 76–77).

8. In public policy we are primarily concerned with compensation for wrongs, rather than compensation due from persons who have received benefits; the latter type of obligation can exist, however (Barry 1979: 70). In either case we are dealing with distributions of changes in well-being or possessions rather than with simple distributions of well-being.

or consumption, even if once considered normal, should now be considered a wrong requiring compensation to those who have been harmed.

We may ask not only whether equity is an important end-value, but also whether it is valued partly as a means. Rule-utilitarianism suggests that while we obey rules in particular instances, we can sometimes rise to a higher level of analysis by assuming the role of legislator (rather than judge), and ask whether the rule is justified (Rawls 1955). The principles of justification for the rule-utilitarian legislator are those of collective well-being. Alternatively, some may contend that a reason for rules of equity is that they foster the integration of the political community (Lucas 1980: 1), thus indirectly promoting its collective well-being. From these perspectives, equity is at least a contributory variable, whether or not it is an end-value.

Aggregation and the Transformation of Value Measures: Univariate Procedures

Most policy indicators measure characteristics of individual persons—health, income, housing, and victimization, for example. Indicator statistics based on such measures summarize these pieces of information by combining data for various persons. This summarization typically involves such familiar operations as addition, averaging, and computation of percentages—all of which combine by adding.[9] Yet these apparently noncontroversial operations involve value judgments. Insofar as an indicator statistic of this sort tells us how well off society is, it combines individual values into a statement of social value. Any possible additive combination of individual characteristics must specify a weighting of persons relative to one another; the most common weighting is equality. Perhaps the most important choice, however, which we typically make at the start, is to assign a weight of zero to persons outside our political community. Thus not merely the variables chosen, but also the ways in which information on individuals is combined, pose valuative questions.

Perhaps the clearest recognition of the valuative character of these sums and combinations appears in the economic literature on interpersonal comparison of utilities.[10] Throughout the nineteenth and early twentieth centuries

9. Other simple forms of social welfare functions involve measures of the dispersion of a distribution (Rescher 1966: 32) or use of its minimum value as in Rawls' difference principle (1971: 76–78).

10. For a summary of this literature see MacRae (1976a: Chap. 5). Equity can also be treated in economic terms without interpersonal comparison in the usual sense. A distribution of shares of two or more goods among persons may be judged to be fair if no recipient envies any other recipient's shares, that is, if he would not prefer the other's bundle of commodities to his own

it was widely believed among economists (following Bentham) that utility was a measurable variable, and that its sum over individuals characterized the welfare of society. Then a more sophisticated line of argument, dating from Pareto but given emphasis by Robbins (1937), denied the scientific meaning of interpersonal comparison of utilities. According to this argument, scientific meaning in economics was limited to variables that could be estimated through observation of economic behavior; and utility could not be inferred, in cardinal or numerical terms, from consumers' choices. Only an ordinal preference function could be so defined, and various individuals' ordinal preferences could not in principle be compared with one another on any scientific grounds (Arrow 1963: 9).

These critics' contention that interpersonal comparison of utilities was meaningless was a positivistic excess (MacRae 1976a: Chap. 3); but their conclusion that such comparison could not be made by science alone was correct. Insofar as we combine characteristics of persons and make valuative judgments about such a combination or sum, we have tacitly or explicitly made judgments about the relative importance of those individuals. These judgments usually take the form of assuming valuative equality between persons when they are in certain observable conditions or when they provide certain responses, which are used as yardsticks for comparing them (MacRae 1976a: 127).

The set of numbers or of attributes that an indicator variable provides for members of a population[11] can be expressed as a frequency distribution. Statisticians teach us how to summarize such distributions, by such measures as those of location and dispersion. Statisticians also warn us of the dangers of misinterpreting distributions on the basis of these summary statistics, especially when they are influenced by a few extreme values. Such misinterpretations involve our placing too much or too little emphasis on certain observations; they can thus involve judgments about the relative importance or value of the corresponding persons. What might be regarded as a mere question of statistics is thus a question of value if we are dealing with a policy indicator that is a value measure.

When we aggregate such numbers for individuals, we must ask whether they validly measure our values in a cardinal sense. The average income for a community may depend heavily on a few high incomes. If a few people have extreme values of the variable in question, we may hesitate to average them equally with others in a purported measure of general well-being. Rather, we

---

(Baumol 1982). This formulation takes into account possible differences in personal tastes and thus bases the notion of fairness on subjective preference rather than simply on quantities of goods.

11. We assume here that the indicator statistic in question is based on individuals, not on firms, places, incidents, or other bases of aggregation.

may tacitly or explicitly consider that a variable such as income is nonlinearly related to well-being, and transform it in some way before adding.

Such a transformation of a continuous variable can expand or contract one part of its range relative to another, thus stressing the part of the range that is relatively expanded. For a set of categories, a transformation can combine or collapse some together while leaving others unchanged. Thus diseases, crimes, occupations, or types of facilities available might be transformed to focus attention on serious diseases or crimes, onerous occupations, or essential facilities. The categories of special interest might be distinguished from one another while the rest were grouped together; or they might be given attention as a group by simply contrasting this group with all the rest together.[12]

We can illustrate these problems of transformation most easily with value variables whose social value varies monotonically with the corresponding numbers.[13] In the following discussion we shall take income as an example of such a quantitative variable that may be transformed.

Suppose that we have obtained an ethically valid measure of income[14] and a distribution of this measure over a population. We must now decide whether and how to transform it before combining the transformed values for persons into indicator statistics. We may present a distribution over income categories—often in intervals of unequal size, grouping together wider income ranges at higher incomes. We may collapse these categories so as to show only the poor and the nonpoor, making no distinctions among the nonpoor; or at the other extreme, we may present the proportion received by the highest category. If this last figure does not elicit enough concern, an analyst valuing equality may present the proportion of total income received by the highest category of the population, implying that that proportion should not diverge

12. In assessing the effect of transformations on the reader's definition of problems, as in data presentation generally, we must recognize that psychological questions of perception are involved and that a layman may respond differently from a trained statistician. Fienberg and Goodman (1973: 64–69) make this point with respect to bar diagram presentation of social indicator statistics.

13. We set aside variables such as blood pressure, for which unusually high and low values are both of concern.

14. We assume for the moment that individual or family income is a valid measure of well-being; we discuss the treatment of individuals' well-being within families further in the next section. The ethical validity of a measure of income primarily means that it measures an individual's opportunity to satisfy preferences, taking into account resources and obligations. By publishing statistics based on transformed sums of such numbers, however, we do not wish to give the impression that our problem is to maximize these sums in the very short run. Some persons may have incurred obligations through voluntary choice, or deprivations through punishment for crime (Nozick 1974: 154), and we do not necessarily wish to alleviate these deprivations as a matter of policy. Nor do we necessarily tax savings heavily. Rather, if we wish to maximize a sum of transformed incomes over the longer run, we must consider the effects on this maximization of maintaining social rules and incentives.

so much from the proportion of persons in the highest category. Similar statistics have been presented for the proportion of the world's energy consumed by the population of the United States.[15]

With the rise of benefit-cost analysis came the recognition that benefits and costs were not always distributed uniformly among members of the population—in terms either of value received or of recipients' income. Weisbrod (1968) recommended that anticipated impacts of public programs be assessed not in total, but separately by income categories, so that policymakers could better assess the equities or inequities involved. Hansen and Weisbrod (1969) then showed that for higher education in California, a policy that financed education from taxes effectively transferred income from poorer to less poor citizens. Stucker (1977) showed that a tax on gasoline may be regressive, and Dorfman (1977) also showed that certain environmental programs transfer resources from the poor to the less poor.[16]

Transformations of income have also been proposed so that we might estimate not only the total well-being corresponding to a given income, but also the incremental well-being produced by the results of a particular policy (in terms of the derivative of that function, for small changes). Rather than tabulating these differential effects by income categories, we can try to reexpress them in terms of well-being (welfare, utility) by mathematically transforming the net monetary gain by members of a given income group into a net welfare gain. We seek to do this by means of a function $W(Y)$, where W = welfare and Y = income, such that W increases with Y but does so at a decreasing rate as Y increases.[17] If $W(Y)$ has this property, then in order to maximize W through policies we shall have to devote relatively more resources to improving the condition of low-income groups than we should have done under ordinary benefit-cost analysis. A function of this sort may be postulated on the ground of one's own ethical intuitions, or estimated by surveys of others' ethical

15. This last type of statistic is an example of an index of concentration, of which the Gini index is another example; if we transformed income at the start, however, these indexes would usually be altered. We have not discussed blame and exploitation as interpretations of univariate distributions; but of all such distributions, that of income (or of property) is most often interpreted in terms of blame for past actions by the haves against the have-nots.

We can also concentrate attention (when stressing values other than equity) on qualities manifested in a small segment of a population near the top of a stratification system—the enjoyment of art, literature, or science, or the capacity for leadership.

16. Dorfman advocates separating the alleviation of external costs from income redistribution if possible, on the ground that redistribution is not a proper function for the Environmental Protection Agency (EPA). Some agency other than EPA might have both these functions, however, as the provision of specific public goods often affects both. We would then no longer be constrained by the "peremptory rule" (Braybrooke and Lindblom 1963: 150–58) that EPA's function limits the values we can properly trade off against one another.

17. Such a static functional relationship, of course, neglects the criticisms advanced in Chapter 6 concerning the role of temporal sequences and perception among the causes of subjective well-being.

intuitions or of individuals' judgments of equivalence of probability-weighted combinations of gains or losses. It might conceivably be chosen by a political community. This function would then take the place of the conventional monetary summation, which is itself based on certain ethical intuitions.

Various mathematical functions of this kind have been proposed for trade-offs between equity and other aggregative values.[18] These transformations give greater emphasis to the poor, but still preserve the initial rank ordering of income; unlike the collapsing of categories, these are strictly monotonic transformations. Other transformations ignore differences in income above a certain level and make distinctions only below it. Figure 7–1 shows several conceivable functions by which income (Y, the horizontal axis) may be transformed into a presumed measure of well-being or welfare (W, the vertical axis). For comparison, we show the identity transformation $W = Y$, which treats untransformed income as the measure of welfare; this transformation is represented by a diagonal straight line.

The remaining transformations shown in the figure place special emphasis on policies that produce changes in the region of lower incomes. One such way of transforming the scale of well-being would be to give highest priority to raising citizens to a minimum level of well-being such as a poverty line. If we believe that this minimum level is of first importance, we might transform the income scale as shown in the heavy solid line (step function) in Figure 7–1. Taking this transformation literally, we would account it of no value to raise someone's income if that raise did not carry him or her over the poverty line Y_0. Changes occurring entirely below the line, or entirely above it, would be counted as zero changes in welfare. The costs borne by nonpoor taxpayers would also be neglected as long as their taxes did not drop them below the poverty line. Such a viewpoint is obviously an extreme one for income, but we use similar reasoning whenever we classify persons simply as meeting or not meeting some minimum standard.

Another type of transformation of income might involve some increasing function of income with decreasing slope, such as its square root (Figure 7–1). Such a transformation, like the drawing of a poverty line, would place more emphasis on the poor, but it would not do so exclusively. A third approach might be to add the incomes of persons affected by policies, but to count all above the poverty level as effectively at that level, thereby attaching no importance to changes above it (dashed line in Figure 7–1).

Some writers using such transformations have simply postulated them,

18. In principle, the optimum for a collective good (Fig. 5–3) could be recalculated using well-being (utility) rather than price as the vertical axis; the calculation would be difficult because of the need to know incomes of all affected. Haveman and Margolis (1977: 19) suggest, however, that separate presentation of efficiency and equity impacts of policies, in contrast with their combination in a single value variable, allows decision makers to choose different weights for these two values.

Figure 7-1. Transformation of Income to Promote Equity

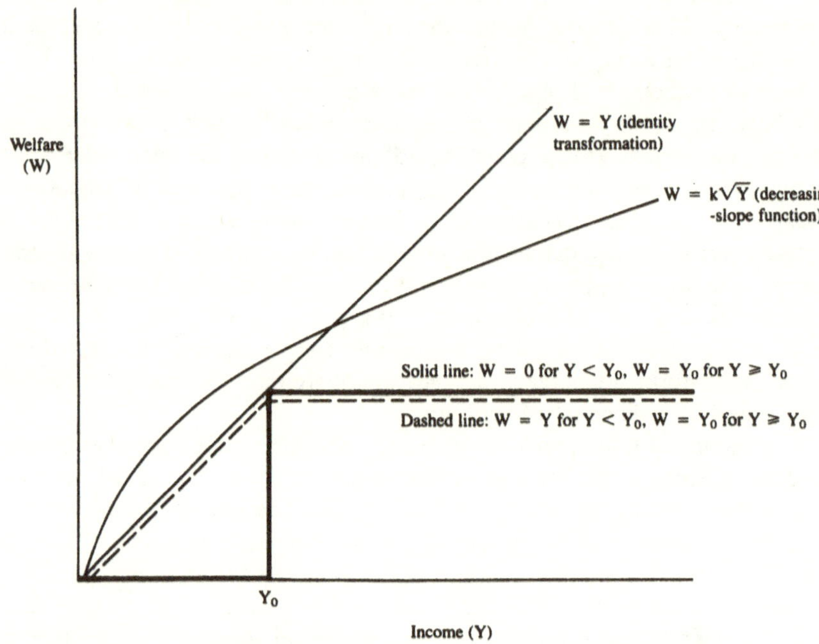

often furnishing calculations based on several transformations from which the reader can choose. The transformed function is usually considered to measure the utility or value to the person affected. Greene, Neenan, and Scott (1974: Chap. 3), for example, have evaluated benefits and costs to persons with various incomes, weighting these benefits and costs in terms of various functions of income. Freeman (1977: 253) lists seven such functions of income that might be used to incorporate the value of equity.

Other writers, however, have tried to infer the proper transformation of income into welfare by observation of individuals' choices. Such a transformation is simplest if first attempted for a single individual. If the individual's choices give rise to a determinate transformation, the remaining task is to specify precise rules (based on valuation rather than science) for combining one individual's preference function with another's. A particular case of this effort is the empirical study of individuals' valuation of money. An early study of this type was that of Mosteller and Nogee (1951), which made use of probabilistic combinations of possible outcomes.[19]

19. Subsequent researches on subjects' preferences among probability-weighted events have shown difficulties in the application of this method. It was first shown that subjects' preference or aversion for risk itself could introduce an extraneous influence into these choices. Then Kahneman and Tversky showed (1979) that other behavioral tendencies, such as a special valua-

An alternative approach to the transformation of income is suggested by the "leaky bucket metaphor" proposed by Okun (1975: 91–95). In discussing equity between income levels, he proposes that we consider the transfer of money from the rich to the poor in a leaky bucket so that some of it disappears in the course of the transfer. The amount of leakage that we are prepared to accept reflects the relative values that we attribute to marginal amounts of income at the two levels in question. Such a line of questioning, if it were carried out systematically and if the results were consistent, could provide a measure of the function or transformation that respondents used in estimating the well-being of others. The method estimates an aspect of respondents' social welfare functions rather than their own welfare; it would presumably be least biased if the hypothetical transfers did not involve the respondents as donors or recipients. A related study of respondents' judgments of equity by Alves and Rossi (1978) shows their judgments of fairness of earnings in various hypothetical family situations.

A final basis for transforming individuals' monetary benefits and costs into possibly more equitable terms makes use of their valuations of time. In Chapter 6 we mentioned the possible use of time together with subjective well-being as a basis for interpersonal comparisons, as a means of reducing the inequity that results from monetary comparisons. We here consider an alternative use of time as a basis for measuring demand, benefit, or cost—similar but not identical to the previous one.[20] The ethical justification for using time in this way derives from the fact that the distribution of potentially available time among persons is more uniform than that of money. If demand (and thus social benefit in policy assessments) is expressed in terms of money, then the priorities of policy will resemble those of the economy in directing goods and services especially to those who have the most money. If it is expressed in terms more nearly proportional to the amount of time people have, this bias may be reduced.

The substitution of time values for money values can be approached in two ways: by finding a suitable weighting factor to convert each individual's

---

tion of outcomes that are certain rather than uncertain, also led to choices that violated the rules of expected utility. These authors proposed another set of axioms for choice, based on departures from an initial reference point rather than on a variable such as overall income, or wealth. They also showed that the value function of money revealed by choices is steeper for losses than for gains—a phenomenon often observed in political responses to losses and gains. They proposed, in addition, that probability itself be replaced by a "decision weight" that is a nonlinear function of probability. This alternative formulation not only casts doubt on the initial axioms of measurement, but raises the question whether subjects' departure from a calculus of expected values should be accepted as an indication of value, or challenged as an error.

20. The estimation of quality-adjusted life years uses time as a basis of aggregating well-being over total time experienced. Here we consider marginal units of time and ask whether they could be exchanged (by changing their use) for money and vice versa.

monetary valuations into time units; or by considering in more specific terms the individual's economizing behavior with regard to time and his or her marginal valuation of time. The first method would allow us to make use of data on the monetary costs and benefits of a proposed project to individuals (if available) and transform them into more equitable form.

In the first approach, a leading candidate for the conversion factor is the wage rate *w*; this has been considered as the individual's tradeoff ratio between spending time in work and in other activities. If we could simplify this tradeoff by saying that each marginal unit of time was worth *w* dollars to a person, we could convert that person's monetary valuations of such units into time valuations by dividing by *w*.[21] This revaluation would decrease the increments of well-being received or paid by high earners relative to those for low earners.[22] To refine this estimate and apply it to persons not currently employed, we might question people about their monetary valuation of time or compare the values that they assign to commodities in terms of money and time.

The ethical validity of this approach depends, however, on our willingness to treat the marginal hours (or other units of time) that two persons consider trading against money, as of equal social value when they have the same wage rate. In some cases we might judge two such persons' time to be unequal in value, as the following example suggests. Suppose one (a retired person) works only part-time and has other compensating resources and few obligations; while another (an overworked breadwinner) works at two jobs, has many family obligations, and has little free time remaining. If asked to place a monetary value on time, the overworked breadwinner would presumably value it more highly than the retiree; we might then consider their own valuations more valid than their common wage rate. The result would be to assign

21. In a 1965 paper, Becker (1976: 92) assumed that the value of time could be set at the wage rate, in view of individuals' possibility of shifting time between work and other activities. Empirical studies have shown, however, that the values persons attach to travel time are only about twenty to thirty percent of their wage rates (Quarmby 1967). Becker (1977) suggests that such discrepancies may be due to the allocation of effort. Moreover, the use of wage rates requires estimation of equivalent wage rates for persons not in the labor force, such as children, retired persons, and homemakers; empirical studies of the values they place on time are needed if we are to include all persons affected by policies in our calculations.

22. In the simple case in which everyone worked full time, the wage rate would be a constant times the income per year (Y). If we equated each person's marginal or incremental valuation of time to his wage rate (dollars/hour) and measured welfare (W) in time units we could write

$$\frac{dW}{dY} = \frac{1}{kY}$$

where W and Y are rates of flow and kY is the wage rate. This would give rise to a logarithmic function for $W(Y)$, one of the functions often proposed for this relation when vertical equity or utility is considered. The function would of course have the same form if the value of time was a constant fraction of the wage rate.

to the overworked breadwinner a higher *equivalent wage rate* (money value of time) and thus to count the well-being (measured as time) that he gained from an increment of income as less. In terms of the efficiency of maximizing total well-being this might be justified: the retiree would have more time to combine with the money in "household production," and the contribution of a given amount of money to that person's welfare would be greater.[23] However, the retiree would have been better off at the start, by having leisure, as well as having a higher capacity to produce well-being from money; in this sense preferring him through policy would be inequitable. The use of individual monetary valuations of time as inverse weighting factors would direct resources to the poor who earn less money for a given unit of time, but also to persons who have more time at their disposal.[24] The use of time as a basis for assessing willingness to pay must thus be examined carefully before it is used for policy choice.

Intrafamily and Intrapersonal Aggregation

In discussing transformation of income we assumed we had valid data on individual incomes, which were to be transformed into measures that reflected individual well-being more equitably. But an ethically valid measure of income, if it is to reflect well-being, must take account of other resources and obligations (K. Fox 1974: 20–23);[25] and income that is transferred to others must thus be counted as benefiting those who use it in consumption rather than as benefiting the initial recipient. Resources include savings as well as income; equivalent income from ownership of home and capital goods as well as earned income (Weisbrod and Hansen 1977); capacity for obtaining help from others as well as own possessions; and the time dimension of these resources (when will they be exhausted or have to be paid

23. Morgan and Smith's (1969) index of "well-offness," also discussed by Sirageldin (1969: 100), effectively values time in proportion to the income that a person has to apply to that time. It first considers the division of income among family members, allocating smaller amounts to children than to adults and calculating the income "per equivalent adult." It then multiplies this income by leisure time per unit of nonsleeping time, to obtain a welfare measure per unit for a family. This measure would rank the well-being of single adults, if they spent equal work time, in proportion to their incomes. Breadwinners who had to support large families with a single income would rank lower in well-being. Retired persons and others receiving unearned income together with greater leisure time would rank correspondingly higher. This index assigns no value to work time as such.

24. This ethical "conflict situation" (MacRae 1976a: 93) exemplifies a long-standing criticism of utilitarianism—that it might dictate efficient production of subjective well-being by distributing it inequitably.

25. Becker (1976: 93, 257) has proposed notions of full income, including time valued in monetary terms; and social income, including nonmonetary contributions made by other persons.

back, becoming obligations?). Obligations include duties to care for children or dependent elderly relatives; job responsibilities and attendant costs; debts; and interpersonal obligations to kin or neighbors, such as repayment for past aid. The investigation of such resources in "means tests" has well-known difficulties; but if we are measuring well-being in principle, we must clearly define the concept of well-being.

The interpersonal components of resources and obligations remind us that for economic statistics the family is a "black box." Researchers can learn from tax records when two earners are in the same family (but not when they are simply living together), and about dependent children as exemptions; but except for the study of poverty, analyses of income distribution are not usually converted into terms that express well-being in terms of family composition. If lower-income families have more dependent children, for example, then effective income inequality is greater than the uncorrected statistics would suggest.

To prepare statistics on the effective or corrected income of individuals we would have to convert information on income received by some family members into individual statistics on income (social and monetary) received by each family member. For married couples we might assume that their total income was equally divided before performing the transformation.[26] If we extend our reasoning to children as individuals, however, a host of new problems arise. First, their entire "corrected income" can derive from transfers and services, largely within the family; and such an estimation can be quite uncertain. Second, there may well be significant differences within a family among children, as well as among adults, in the distribution of resources and in consumption of family collective goods such as housing. Third, if we assign to a child some fraction less than an adult share of family income, we should presumably form some ratio of income to needs in order to place the child of a wealthy family at the same level of well-being as his parents.

One particular instance of this problem is illustrated in the estimation of the poverty level for families. The definition of this level is important not only for policies affecting the poor, but also because it has been proposed by Watts (1977: 29) as the denominator of a "welfare ratio" by which the welfare of all families might be assessed.

If a poverty threshold (minimum income) can be defined for a single person, then equivalent thresholds for families of different size and composition may be estimated with the aid of consumption data. Watts (1977: 186–87) describes a method devised by Wetzler that equates the income thresholds for families of different sizes when their proportions of income spent on food are

26. For an application of this line of reasoning to the "marriage tax" issue see MacRae (1980c).

the same. A further order of refinement, however, would ask whether families differed from one another in their internal distribution of resources. Such differences can occur not only in extreme cases of child abuse and neglect, but also as a result of variations in the resources and obligations of family members; when both parents are employed, their income is greater than if only one is employed, but their time available for family responsibilities is less. As children become older, they enter school and some responsibility for their care is transferred to the school; older children become able to fulfill responsibilities in the home, and the degree to which they actually do so affects the welfare of other family members. Thus, if the benefits of detailed information were worth the costs, further details about the distribution of resources and responsibilities within families would throw light on their members' welfare.

Our concern with equity can involve not only equity among and within families, but also the distribution and summation of resources and well-being within a person's lifetime—*intrapersonal aggregation*. Indicator statistics typically describe the state of society at a given time. But insofar as they do, they fail to relate the well-being of each person at that time to his or her well-being earlier. Thus Braybrooke and Lindblom (1963: 182–85) discuss the possibility that over a time interval some members of a population may move down in health status while others move up. A victim of long-term illness may elicit more of our sympathy than a number of victims of short-term illness who together contribute similarly to instantaneous or annual statistics, and the same is true of chronic unemployment or poverty.

Milton Friedman considered a related possibility when he first proposed the negative income tax (Moynihan 1973: 50). Even though a progressive income tax had some justification, in Friedman's view it placed a special burden on persons (or families) whose income fluctuated and fell at times into the zero-tax region. If an equitable tax burden were considered to be a certain proportion of total income over a period longer than a year, then under a progressive tax law the fluctuating group would pay more than others who earned the same total with less fluctuation. The tax would thus treat existing taxpayers so as to equalize them in any given year, but would appear inequitable if a person's income were considered over a longer period.

We may similarly ask how much a retiree deserves as a pension, or an unemployed person deserves as compensation, in view of his or her previous income. In a contributory system, equity requires that retirees receive in proportion to the value of their contributions, at least in an actuarial sense.[27] But in a pension system funded by general revenue, should we forget about the past and set payments only in terms of retirees' present situations (Rein 1976:

27. Advocates of women's rights argue that annual payments rather than total expected payments should be equalized.

183)? Should the state give less to those who had more before? Or is there some way to measure a person's net contribution to society, for which recompense is due?

A philosophical issue underlying these questions is whether a person should be regarded as the same person at different stages of life. Related to this is the question of the extent to which a person's choices or preferences at an early stage of life should be allowed to affect his later welfare, or whether requirements such as compulsory insurance should be imposed to reduce this effect.

At the same time as we consider intrapersonal aggregation, we must consider how it is to be compared with interpersonal aggregation. If we are comparing the aggregate well-being of a population as it might result from two different policies, should we be indifferent as to whether that total is distributed over a larger or smaller number of living persons? We must thus ask whether, and under what circumstances, it is better for a population to have longer or shorter life. At first glance, this question may seem unnecessary; the premature loss of life is generally regarded as one of the great tragedies of the modern world—not only in the poorer nations where life is short, but even in the United States (Vaupel 1976). In addition, our consideration of the desirable length of life raises serious moral questions. Yet we do ask at times whether life, if it is full of suffering or maintained only by expensive life-support devices, should continue. These questions cannot easily be asked if a person is considering the prolongation of his own or another specific person's life; such a choice is ordinarily made by an individual rather than by government, and is circumscribed by laws and norms concerning suicide, homicide, euthanasia, and abortion. Even when impersonal and large-scale public policies affect the length of life, the consequences of allocating scarce resources can be tragic (Calabresi and Bobbitt 1978). But at the level of indicator statistics dealing with impersonal aggregate values, the question can more appropriately be asked.

Demographers have dealt with "dependent" proportions of the population, the "dependency ratio" being a measure of the proportion of a population that depends on production by other members. One reason for prolonging life is to reduce this ratio—retaining existing lives rather than replacing them with children—as long as the persons whose lives are prolonged are not dependent. But we need also to consider the well-being of dependent family members, the psychic support and satisfaction that they give to the more active family members, and the moral questions mentioned above; dependency is not always undesirable.

Considerations of equity also enter. As Wilensky (1975: 87–96) points out, the poor tend to die young, but they pay health taxes to support the more affluent who tend to live longer. The very fact of living longer or dying earlier may involve a different participation in the goods of life.

Intertemporal Aggregation and Equity: Discounting[28]

Indicator statistics presented as time series are not usually aggregated over time. At most, they are aggregated over a single interval of observation, as when the GNP sums production over a year. But in estimating the effects of policies we need to combine the values of effects that occur at different times. The conventional economic approach to this problem of aggregation, used in benefit-cost analysis, is to discount future economic values in order to express them as present values. When we consider nonmarketable values such as quality-adjusted life years (QALYs), however, we might ask whether it is equitable to count the lives of future generations, or of the same persons in future years, as of less value than years now. In some applications of QALYs, future life years have been discounted in the same way as economic values. This procedure has been followed because of logical difficulties that appear to arise when years of life are not discounted.

The general problem exemplified here—which also applies to the aggregation of subjective well-being over time—is that of consistent ethical argument (MacRae 1976a: Chap. 4). Claims have been made for different procedures of aggregation, with or without discounting, based in part on considerations of logical consistency; but like all ethical claims, they also rest on basic convictions as to what is right or good. There may indeed be irreconcilable ethical differences among participants in this argument as to the importance to be attached to posterity. But differences in conclusions may also be due to reconcilable logical and empirical differences, including the possibility that a participant may systematize his or her contradictory ethical convictions.

A major ethical argument against discounting life years is based on equity over time. Horizontal equity is defined as the equal treatment of equals. Presumably, this principle extends to persons in equal circumstances in the present and in the future.[29] Such equal treatment would seem to run counter to discounting of future life years. A similar position is taken by Ramsey (1928: 543) with respect to utility: "It is assumed that we do not discount later enjoyments in comparison with earlier ones, a practice which is ethically indefensible and arises merely from the weakness of the imagination."

I shall argue here that perhaps the various assumptions involved can be reconciled and that intertemporal equity may not have to be abandoned in these procedures.

28. This section is adapted from MacRae (1980b).

29. The generic term equity applies to intertemporal aggregation only in part, however, as future generations do not participate reciprocally in transactions with those now living (MacLean 1980; Lazear 1978). Moreover, the utilitarian approach may fail to include some aspects of justice (Veatch 1979: 202–05).

The potential inconsistency among our ethical convictions that relate to the future has been pointed out by Koopmans. He notes first that benefit-cost analysis involves discounting future monetary values, on the ground that "society can temporarily curtail the production of current consumption goods by transferring some factors of production to the formation of additional suitable capital goods, in such a way as to return a multiple (> 1) of the same unit bundle of consumption goods in the future. Efficient intertemporal equilibrium then demands that the present value of the goods returned to consumption be equal to that of the goods not now consumed. The quantity of the future bundle being larger, its per-unit present value must be correspondingly lower" (Koopmans 1979: 7).

He then considers the problem of attaching values to health and life; and though he expresses doubt about "a calculus of the value of human lives in large numbers," he goes on to consider a paradox involved in the application of such a calculus to problems with a long time span. He begins with the implication of horizontal equity that I have mentioned: "Suppose one accepts as an ethical principle that, in balancing risks to human life in the present and in the future, equal numbers of lives should receive equal weight. This would make the present value of future human life independent of the time at which it is lived" (Koopmans 1979: 9). But in view of the fact that "the present value of a standard bundle of goods in the future decreases as that future time recedes, . . . the present value of future life relative to that of future goods will be much higher than the value of present life in relation to present goods." He concludes that from the ethical rules, preferences, and behavioral tendencies that have been assumed, "sets of 'present values' formed at successive points in time need not, and generally will not, be consistent with each other" (Koopmans 1979: 10).

What does it mean to say that these sets of present values will not be consistent with each other? It means first, as Koopmans points out, that as we contemplate the present values of life and goods at a determinate future time, and allow the time at which we calculate to vary, the ratio of present values of those future quantities of life and goods will change. In terms of our decisions to allocate resources, we may initially be indifferent between allocating present resources that will produce given amounts of life and of goods at a given future time; but as the time of judgment becomes later we will decide to devote relatively more resources to producing those same future goods, and less to producing the same future amount of life. A central question here is whether we must insist that these relative valuations of life and goods be the same regardless of the time from which we view that future comparison. Conceivably this is not simply a matter of logical consistency, but also involves empirical questions.

This argument may be clarified by a second example in the literature.

Weinstein and Stason (1976), investigating the desirability of policies to deal with hypertension, measure the effects of such policies on patients in quality-adjusted life years and the input costs of the policies in monetary terms.[30] They then feel compelled to discount life years as they discount dollar returns, in order to make the resulting policy judgments consistent in Koopmans' sense. Weinstein and Stason (1977: 720) illustrate the problem by comparing several alternatives, each of which involves expending money and gaining years of life, as shown in Figure 7–2. For each of these alternatives the gain in life years is shown above the horizontal axis and the monetary cost (in constant dollars) below the axis; the scales of the two are incommensurable. Weinstein and Stason wish to compare Program A, which saves one year of life expectancy forty years from now at present cost of $10,000, with Program B, which saves one year of life expectancy now at the same cost. To make this comparison, they first consider a hypothetical program A_1 that saves one year of life forty years from now at a cost of $70,000 in forty years. This is considered equivalent to Program A because $70,000 in forty years has a present value (at approximately five per cent) of $10,000.

They next consider Program A_2, which translates the benefits and costs of program A_1 from the future into the present, and make a crucial assumption. "*Provided life years are valued the same in relation to dollars in the present as in the future*, Program A_2 should be considered to have the same long-run priority as Program A_1" (Weinstein and Stason 1977: 720; emphasis mine).[31] They point out that Program B should have higher priority than Program A_2, as it produces the same benefit at lower cost; and thus, under the assumptions made, that it should have priority over programs A_1 and A, which are equal in priority on their assumptions to A_2. By this set of comparisons they reason that saving a year of life in the future must have lower priority than saving a year now for the same present cost. Following this line of reasoning, Weinstein and Stason (1976: 27–28) discount future life years in the same way as future monetary values are discounted. They thus achieve consistency in decisions over time by abandoning intertemporal equity.

The arguments both of Koopmans and of Weinstein and Stason compare our possible preferences for different combinations of life and goods (or money); and Weinstein and Stason show that a certain sort of consistency over time

30. Ideally, we might wish to assess all the costs and benefits of health policies (or other policies) in terms of quality-adjusted life years, if the quality and quantity of life are what we seek for society. In actuality, however, analysts estimating quality-adjusted life years compare them with monetary costs, thus engaging in cost-effectiveness rather than benefit-cost analysis. (To assess the reduction in taxpayers' quality or quantity of life as a result of taxation seems rather impractical.) We thus face a two-attribute decision problem involving the ordering of pairs of numbers.

31. A similar assumption is made by Koopmans (1960: 293–94) in his postulate of stationarity. This axiom has been criticized by Ferejohn and Page (1978: 272–74).

Figure 7-2. Benefits in Years and Costs in Money, in the Present and Future

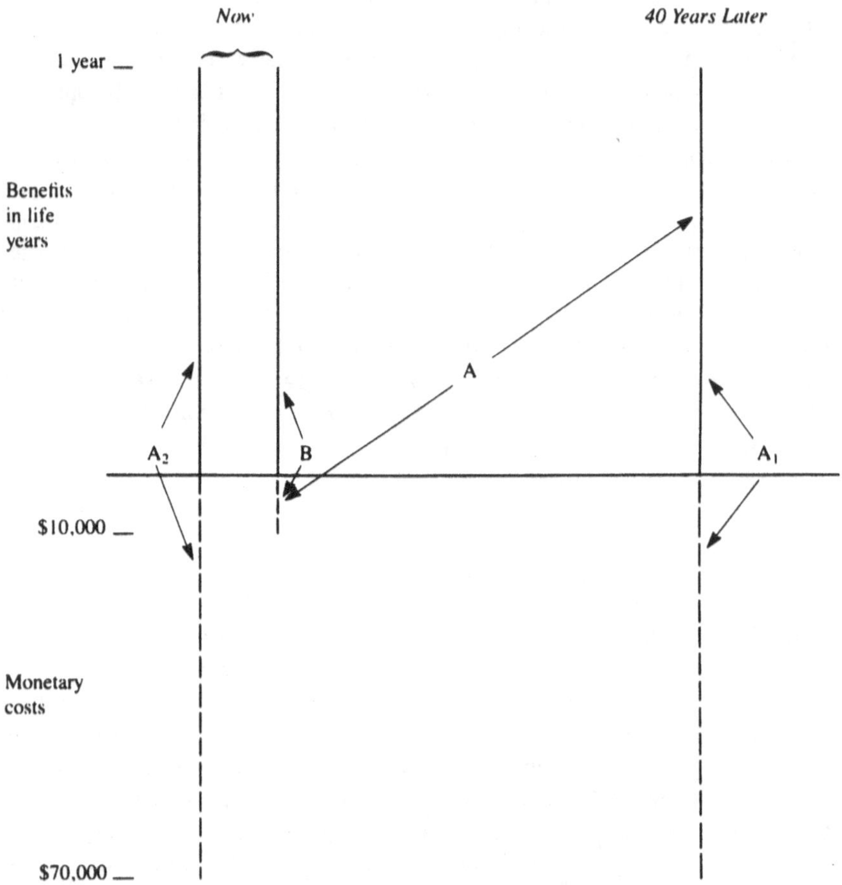

requires discounting of life. This consistency centers about our making the same judgment of relative value between life and goods, regardless of the time at which the judgment is made.

Let us now consider this paradox from the other side, stressing intertemporal equity. Koopmans has shown that if we assign the same value to life independently of the time at which it is lived, we must assign different relative present values to the same future amounts of life and goods, depending on when we make this assignment. Weinstein and Stason have shown that if we require that these relative valuations be the same whether they are made now or forty years later, we must discount life years. Suppose, however, that we give priority to intertemporal equity and that we do *not* require these judgments made at a different times to be identical; what would be the result?

If Weinstein and Stason had instead chosen to value life years the same in the future as in the present, and if they had continued to value future money the same in relation to future life years when this comparison was viewed now and forty years hence (A_2 equivalent to A_1), they would have encountered a clear contradiction in view of the reduced present value of future money. Suppose, then, that they had abandoned the assumption that the relative value of future life and money had to be the same when regarded from the present or from forty years hence. They would then have been led to recommend that our present resources be invested rather than spent on saving life now, so that these resources could be used to save more life years in the future. This alternative line of reasoning can also be illustrated with the aid of Figure 1. Program A_2, under this alternative assumption, would be inferior to B or to A, which would then be considered equivalent to B because years are not discounted. Thus, instead of spending \$70,000 now to save a year of life, we would invest only \$10,000 now and wait forty years to save that year. Arriving at that later time, we might find it preferable to postpone our spending till a still later time, and so on. The paradox might thus take the form of continual postponement of consumption (i.e, postponement of use of money to prolong lives), with apparently insufficient attention to present life-saving.[32]

We might also examine what would happen if we were considering only two points in time and made the same choice repeatedly. Under the assumption that life years are not discounted, we would then transfer resources from present to future life saving until the latter, at the margin, was far less productive in years per dollar than the former. For the forty-year interval considered, the equilibrium between present and future life saving would occur only when we devoted so much of our resources to future life saving that only one-seventh as many lives could be saved per marginal dollar then as now.

We are now invoking empirical considerations as well as purely logical ones. We have acknowledged that the options available to us at the margin, in terms of our capacity to spend money to prolong life, may vary with the amount we have already spent, according to the law of diminishing returns. This consideration leads to some limitation of the apparently inordinate tendency to transfer resources to the future. But there is another important empirical consideration that needs to be introduced: does the existence of an interest rate entail economic growth and the possibility that future expenditures will be less productive of saved life years than present ones?

If so, counterarguments against discounting of life years are possible, based on possible different relations between money and life at different

32. This problem of postponement is cited by Olson and Bailey (1981: 13) as a reason for discounting, or positive preference of the present over the future, in national policy choices. We might, however, examine empirically the yield of life-prolonging procedures per dollar at different historical times.

times; we shall simply suggest their nature (MacRae 1980b). (1) The technology of converting money into life years at the margin, as suggested above, may be less effective in the future because of diminishing returns, and this change may be linked to economic growth and the social discount rate. (2) If we consider the quality aspect of QALYs, the increase in quality per dollar in the future may decrease because of the form of the function W(Y); diminishing returns may result from this source as well in a growing economy (Dasgupta et al. 1972: 154). (3) In a steady-state economy the desirable social discount rate would be zero.

This issue is far from resolved. Most of the technical and mathematical analysis on the subject favors discounting (Keeler and Cretin 1982); but an increasing line of qualitative argument, stressing equity and the special case of nontradable goods, has questioned this procedure (Goodin 1982a; Parfit 1983). The study of this issue is of the utmost importance, however, if length of life or measures of subjective well-being are to be aggregated over time in the choice of policies.

Distributions Among Social Categories: Value, Causation, and Blame

Some of the most important, and probably the most controversial, uses of policy indicators involve distributions of value measures among various population groups. The indicator displays used to show them are cross-tabulations or other portrayals of association between a value measure and some social category in which individuals are placed, such as race, ethnicity, sex, age, area of residence, occupation, or income group. The interpretation of such a display, however, can involve complex argumentative exchanges as to how it should be used to define an inequity, an obligation, or a policy problem.

Such a display might show that American school children of different racial and ethnic groups differed in achievement test scores (Coleman et al. 1966) or in school years completed (Duncan 1967). It might show that women were not represented on university faculties in the same proportions as men. It might show that federal aid was being directed more, per capita, to one region than another. Or it might show that unskilled workers in Sweden were more likely than the rest of the population to live in flats rather than in single family houses (Sweden, NCBS 1981: 64).[33]

In preparing or interpreting such a display we face a set of questions that

33. A variety of statistical indexes have been proposed to measure inequality and segregation. Such indexes need to be examined not only for their statistical properties but for their ethical validity.

are partly similar, and partly dissimilar, to those we faced in developing causal policy models. These questions revolve about two major themes: value on the one hand, and causation and blame on the other. Let us consider them in turn.

The social categories we use in these displays reflect hypotheses. These hypotheses state that some variable related to people in these categories—the past treatment that they have received or their own capacities or activities—has caused them to differ on the value measure. At this point we shall not examine causal relations in detail, but simply note that the value measure is the dependent variable in the models being examined. If a difference is found, the question at stake is whether that difference is inequitable. This question can, however, have the two different senses we have defined: that of simple distributional equity, of general concern to public-spirited citizens; and that of compensatory equity, in which someone deemed at fault (often a group favored on the value measure) is deemed to have a special obligation to right past wrongs. These two senses of the question lead to differences in several aspects of our analysis, which we shall now explore.

a. Tradeoffs and optimization versus blame and compensation. Inequity involves an improper difference with respect to the distribution of a value. We must thus first ask whether the measure being used in the display is a suitable measure of the value in question—whether it is ethically valid as well as reliable. Living in a flat, for example, may not be an indication of hardship for some business and professional families in Manhattan. Earned income alone might not provide an adequate measure of deprivation among the elderly, if they received proportionately more unearned income than other groups. As in the case of any value measure used in policy analysis, we need to know whether the results found are based on valid and reliable measurement.[34]

More important, however, is the question whether the value measure in question is a general one. From the perspective of value reconciliation that we have presented (Chap. 2), we might wish that measures used to describe inequity in a population would be the most general possible; but claims of compensatory inequity are often more specific. Consider, for example, the claims of women and of blacks in the job market. These are often expressed in similar terms. Yet the material living conditions of women are largely comparable with those of men when they share homes with them, while those of blacks are considerably inferior to those of whites. If we were to deal with a general equity criterion, the distribution of overall well-being, we would give priority to the conditions of groups lowest on the scale. If we consider well-being at an

34. The fairness of electoral laws also involves ethical validity. A provision that legislative districts be equal in population prevents some discriminatory strategies in districting, but not all. A provision that multi-ethnic cities arrange ward-based districts so as to increase the chance of electing minority representatives may increase the power of minorities, but under some conditions minorities might obtain even greater influence by coalition formulation under at-large elections.

individual rather than group level, we should deal with those who have the least resources regardless of their social category; an extremely poor white male would be worse off than a well-to-do black woman. Yet for purposes of compensatory equity, we might compare each person's condition with what it might have been, rather than with the condition of others in different social categories.

Claims of inequity are sometimes made by groups who are better off than the average, on the basis of specific rather than general deprivations. Women professionals have been concerned with fair pay for their group, and have been criticized for neglecting similar problems of working-class women. Such professionals may constitute a reference group or a political coalition, concerned with justice and rights for themselves and not the application of a general principle of distribution. Specific group claims may also rest on considerations other than well-being; an economically advantaged group such as the Basques in Spain may strive for recognition, in a nonmaterial sense (Allardt 1979: 46).

The public, however, often invokes considerations of general well-being in judging such claims. A strike for higher wages by well-paid airline pilots attracts less sympathy than one by low-paid workers. Similar general considerations enter when government taxes various groups. If government were to demand arbitrarily more tax payment from some citizens than others, such a practice would be regarded as inequitable; but when those asked to pay most are those with the most resources, a proportional income tax or even a progressive one is now widely regarded as fair.

When compensatory equity is in question, the value at stake is not general well-being but a specific aspect or amount of some value that has been improperly denied or taken away. The previous state of well-being (or the state that would have been expected if a past wrong had not been done) is taken as a baseline and the question of equity revolves about unfair changes from that state. Specific and presumably impartial rules are invoked. A wealthy person whose property has been damaged by someone else can thus legally seek restitution or monetary compensation. But even though the specific values sought can be expressed in terms of the more general end-value of economic benefit, the claim of compensatory inequity is based not on one's overall well-being but on specific unwarranted deprivations.

When claims for compensatory equity are made, they can involve comparison between the allegedly wronged party and others whose treatment provides a standard for comparison. An example is an employee who has been denied a promotion that similar others have received. The choice of a reference group (e.g., persons with similar training and experience) (Merton and Rossi 1968) is then no longer a mere sociological fact but an action requiring justification. The claim is directed at the employer, who should have treated similar persons equally but allegedly did not, distinguishing among them in terms of irrelevant categories such as race or sex. Sometimes the criterion is narrowed to include only voluntary and intentional action (e.g., *de jure* rather than *de facto* racial segregation). Moreover, not inequalities in general, but only those resulting from actions by the blamed party, are relevant to compensation by that party. The value of compensatory equity is thus quite different from that of distributional equity and more complex in its application.

The issue of compensatory equity becomes one of blame and responsibility rather than simply one of optimal action. This distinction has been made by Coase (1960) in contrasting economic and legal approaches to the remedy for an external cost imposed by one party on another. From the perspective of economic efficiency, the problem is to allocate factors of production optimally; this can be done either by requiring the actor causing the externality to compensate the victim, or by requiring the victim to bribe the other actor to reduce the externality. These two "economically equivalent" policies would be regarded as grossly different by the law, which in considering equity would require the one who causes harm to be the one who pays. Certain policy alternatives are then preferred, not because of their aggregate results alone, but because they restore a social balance through payment or fulfillment of an obligation.[35] Moreover, compensatory equity can involve retribution—simply depriving the blamed party of some value.

Distributional equity is a teleological value; we can seek to maximize it by calculating the consequences of policies. Those aspects of compensatory equity involving blame, although teleological, can be anti-utilitarian; retribution for crime, in the sense of imposing harm on the perpetrator to match the harm to the victim, can be pursued regardless of the aggregate consequences. In our choices of policies toward crime, these feelings may be mixed with utilitarian criteria concerning deterrence or incapacitation. We excluded retribution from the list of end-values we wished to examine further, after our value analysis of social indicator concepts in Chapter 1; but it is supported by the public, especially when combined with other values. If it is seen as involving only harm, it may meet with opposition; if deterrence or restitution to the victim is also emphasized, it is likely to have more support.

b. Definition of the problem. We pointed out in Chapter 1 that a major function of policy indicators is to help people to define the problems that they face collectively. In defining problems, however, compensatory equity leads to special lines of reasoning.

First, assume we are at the stage of problem definition and are interpreting indicator statistics so as to decide whether a problem exists. When we deal

35. The disparity between perspectives of simple maximization and of compensatory equity is suggested in Ryan, *Blaming the Victim* (1971).

with compensatory equity, there is often a presumption that policies should reverse the causal processes that initially gave rise to the problem by dealing with the person or group at fault. This is not necessary for policies in general or even for distributional equity; policies can deal with symptoms rather than root causes, and can seek different ways to alleviate a condition, besides reversing the processes that initially caused that condition. A special emphasis on blame and responsibility usually[36] characterizes problem definition when compensatory equity is involved.

When compensatory equity enters into the definition of a problem, a rhetorical change occurs. In our earlier discussion of problem definition based on indicator statistics (Chap. 1), we might have taken the previous level of a policy indicator as a basis for comparison with its present level, but no question of equity would arise between ourselves last year and ourselves now. Similarly, we could take a planning goal as a standard for comparison; again, equity would not enter. But when we judge the level of well-being of one human group by comparing with that of another group, we can regard the well-being of that other group not only as a simple goal but also as evidence of unfairness. Less developed nations have long taken the standard of living of the developed nations as a basis of comparison. From Lenin's theory of imperialism to more recent theories of center and periphery, disadvantaged nations and regions have been able to argue that the situation is not their fault and not simply up to them to remedy; rather, they view it as evidence of exploitation and unfairness, requiring compensation.[37] Similar transformations of the argument have taken place for women and minority groups.

A second feature of our definition of the problem concerns whether inequity should be defined and remedied at the individual level or at the level of social categories. Legal processes are designed to ascertain facts about individual cases and act on them; but they also create precedents leading other cases to be decided similarly.

Public policies cannot usually be designed to do justice to every individual affected. They address general categories of action or of eligibility; but the policy analyst has some freedom to redefine the problem and propose such categories in view of their effectiveness or political feasibility. The analyst is not compelled, outside the realm of equity, to address social categories because of obligations owed to or by them. One example of an analytic problem definition that avoided problems of compensatory equity was that of Lyndon Johnson's war on poverty. Though actually concerned with the plight of minorities, the designers of this policy nevertheless framed it to deal with a

36. It is also possible to compensate without placing blame specifically, as by insurance or governmental restitution.

37. The question whether differences in occupation and income are unfair or merely undesirable may be reflected in the greater use of such variables as bases of social indicator displays in Europe than in the United States (Ramsøy 1974: 47).

group that apparently had less claim to compensation—the poor—and as a "war," not an obligation based on past wrongs.

But public policy as it is actually administered can lead to inequities even when they are not brought about by law or explicitly intended by administrators. "Street-level bureaucrats," under pressure to make quick decisions, often feel forced to develop simplifications to aid them in judging their clients (Lipsky 1980: 111–16). These simplifications can involve judging people in terms of race or sex—in terms of perceived group tendencies that are used as screening devices for individual cases. This may be done explicitly, as in *de jure* segregation, or semiconsciously, as in institutional discrimination (Alvarez et al. 1979). If it can be shown that such a categorization has been employed in the past to the unfair disadvantage of a group, then a claim may be made that that categorization should be used in compensation as well.

Compensatory equity, therefore, is not in general an end-value that can be maximized in policy choice in the same way as economic benefit or subjective well-being. Rather, it represents an indefinitely large class of claims that can be made for equal treatment under a set of rules. We can devise formulas measuring departures from equality (in the size of legislative districts, in the distribution of the races in schools), but they deal only with aspects of the larger notion of equity. Unlike the measures we have considered for other end-values, we cannot expect that they will come near expressing the entire notion validly. More than any of the other end-values we have considered, equity has a meaning that is continually being shaped in public debate (Alvarez et al. 1979: 24–30).

The assessment of blame differs somewhat from that of causation. It includes deciding whether an action was voluntary, whether due notice has been given and proper procedures followed, and whether the person affected has been disadvantaged involuntarily or has accepted the conditions in question. These are categories of law and ethics more than of social science.

c. Compensatory equity and causal inference. Special questions arise about causation when compensatory equity is at stake. In assessing causation we are concerned, as in our previous policy models, with properly specified causal models, including control of possibly confounding variables. We also wish to estimate the size of the effects involved so as to know whether they are of practical significance. There are nevertheless important differences between the reasoning involved here and that of Chapters 3 and 4 concerning models.

When compensatory equity is involved, the inferences we draw from nonexperimental causal models depend more than usual on the question of whether our models have been properly specified and whether adequate statistical controls have been used. Duncan (1967), for example, analyzes the effect of race on years of school completed, controlling for the education and occupation of the family head, the number of siblings, and whether the family is intact. What remains after these controls is taken as a residual definition of the effect of "color."[38] The magnitude of the effect is considerably reduced by these controls, but a difference of one to two school years remains. A defender of the rights of blacks might claim that some of the variance attributed to the controls was in fact due to differential treatment of the races; a defender of the existing system might ask whether the controls had been measured accurately enough or whether others might have been added.

In interpretations of these causal relations, a question of burden of proof arises. This type of question also arises in other sorts of policy debate—indeed, it is essential when the information is not fully conclusive—but here it takes an especially adversarial form in which accusations of special pleading are more frequent.

The problems of assessing causality can be illustrated somewhat more generally by the simplified model shown in Fig. 7–3, dealing with movement of persons into desired social positions.[39] We are concerned with the interplay between the desired equal treatment across "irrelevant" social categories and differential treatment according to "relevant" characteristics, in a society where both are deemed legitimate. The positions into which people may move can include those of students in colleges or universities, or of employees in jobs. In either case we start with a social category such as race, ethnicity, sex, age, or geographic location, which is assumed to be irrelevant; it may, however, be associated[40] with a relevant characteristic such as capacities for study or work, or action in seeking the position. Membership in a social category may influence relevant variables through these associations (path *A*), and may influence irrelevant aspects of treatment by others (path *B*). Relevant characteristics may influence gatekeepers' treatment of the person (admission or employment) through path *C*, thus contributing to a relevant aspect of this treatment. The individual's eventual well-being or possession of some resulting value may then be affected both by his own relevant characteristics (path *D*) and by the action of gatekeepers (path *E*).

Our task in detecting inequity is to see whether the effect through paths *B* and *E* together—from the initial category to irrelevant treatment to well-being—is of practical significance. This requires controlling for the relevant

38. A similar procedure has been used more recently in *Social Indicators of Equality for Minorities and Women* (U.S. Commission on Civil Rights 1978: 53). Annual earnings are adjusted for a group's level of education, level of job prestige, income level of state of residence, weeks worked, hours recently worked per week, and age. The remaining differences in earnings "are considered . . . to be the cost of being female or minority, or both" and construed as "inequity of income." The identification of such residual associations with discrimination is criticized by Cole (1979: 28–31).

39. Numerous other forms of possible discrimination require similar model construction, e.g., access to housing, loans, and insurance. More complex models have also been proposed for analysis of longer-term discrimination in organizations (Alvarez et al. 1979).

40. If such an association exists, the claim of equity often requires that the associated irrelevant category not be used as a screening device or a surrogate for the relevant one.

Figure 7-3. Schematic Causal Model of Effects on Admissions or Employment Policy for Various Social Categories

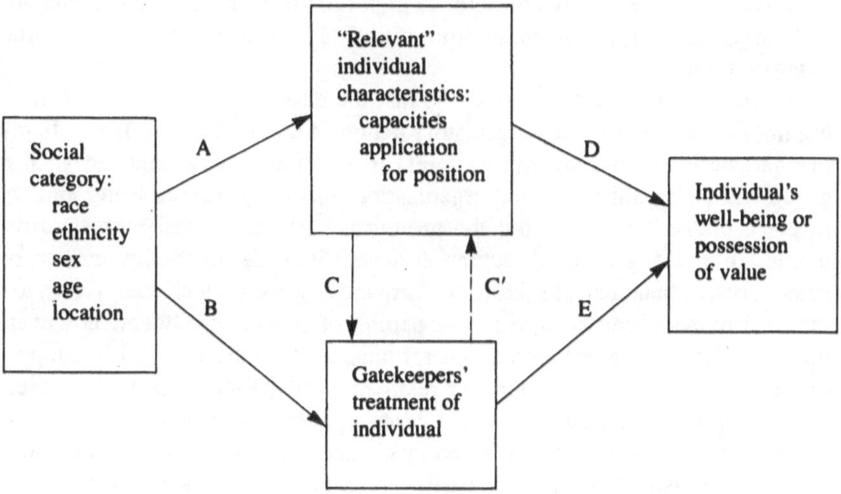

characteristics in terms of capacities and of application for the position (Hoffmann and Reed 1981).

One form of counterargument has been that another causal arrow C' accompanies C but has the reverse direction—that gatekeepers' treatment of the individual causes differences in motivation and skills or in application for the position (e.g., discouragement of females or minority members from seeking advanced training; tracking in schools). This path is shown as a dashed line, because it can interact with the social category. If a path C' exists but is not included in the analysis, statistical control of relevant characteristics may leave little residual variation but a source of discrimination will be ignored.

A second counterargument centers about the measurement of the relevant individual characteristics. If this measurement is unreliable, then statistically controlling for it will leave residual variation that appears to pass through paths B and E but does not result from discrimination. If the measurement is biased with respect to the initial categories (as in the case of a culturally biased test), then controlling for it will obscure some of the effect of paths B and E and thus of unequal treatment in terms of the category alone. A credential used in assessment of individual characteristics, which can be acquired by those favored on irrelevant characteristics, will have the same kind of effect as a culturally biased test. The purchase of relevant human capital (as by attending private school) may also lead to characteristics valued by the gatekeeper even if resulting from irrelevant advantages in the past. Finally, the very definition of the relevant functions of the organization may be debated; for example, should a leading university be concerned only with distinction in research, or should it devote more resources to instruction and community service? Such a decision is important if members of particular social categories (e.g., women and minorities) tend to be needed more for the latter functions.

A third type of counterargument is that the disadvantaged group is suffering not simply from the discriminatory action of others, but also from effects of characteristics that are not its fault. In the transactions between such a group and an organization, the organization may deny responsibility and try to place blame elsewhere; but the group may still claim that redress is required. In a stronger form, such an argument is made for the irrelevance of some types of handicaps to future performance. Some such characteristics are indeed irrelevant and require the dispelling of prejudice. Others, however, may be relevant. Nevertheless, general human rights are invoked in support of mainstream school treatment for retarded pupils, even when their capacities are limited. An argument in this case might be that general natural abilities are not one's fault, but are given by chance and thus ought not to be punished by exclusion from general participation in the public schools.[41]

Some of the complexities of argument in this area are illustrated by the debate as to whether women scientists are victims of discrimination in appointments and promotions to positions in academic faculties, as well as in other forms of recognition. Cole's *Fair Science* (1979) makes use of measures of relevant variables—quantity of publication and numbers of citations received (a measure of quality)—as statistical controls, and he contends that after such control the differences attributable to gender are small as regards reputation (119) and the rank of the department in which scientists are employed (69), but higher for academic rank (70). Reviewers have suggested, however, that the proper functions of the university faculty member go beyond research and publication (Andersen 1981; H. C. White 1982); that women have been denied equal access to the "means of production" in the form of research support and facilities; and that it is only the quantity of publications, not their quality, that distinguishes the sexes (Mason 1980).

These are examples of generic arguments concerning compensatory equity. Cole's initial work followed the line of Duncan's early analysis, controlling presumably relevant variables and seeking the residual. Had the residual been larger, defenders of the existing allocation of rewards in science might have sought further and more accurate controls. Because it was small, the critics considered whether relevance was properly measured (what is the goal of the

41. This argument was also supported by one based on consequences: large-scale summative evaluations showing that "children in special classes did no better on average than comparable children whose schools kept them in regular classes" (Cronbach 1982: 14).

university?); whether employers had previously denied resources to women (path C' in Fig. 7–3); and whether quality alone ought to be considered, a tacit claim that factors limiting quantity of publication were not women's fault and that quantity of publication should thus be considered irrelevant.

Policy indicators are relevant to compensatory equity, and have been used extensively in this connection. Indicators of equity seem, however, to be immersed in controversy, far more than other indicators. This immersion, I have tried to show, results in part from the complexity of ethical argument surrounding compensatory equity. Social scientists and policy analysts designing indicators in this field thus face more difficult choices than those who devise and use policy indicators simply for maximization of well-being. They must first choose between an avowedly partisan role and a role in which they can hope to maintain some consensus in their technical communities. In a partisan role they will simply try to serve the rhetoric of their side, though without "lying with statistics." In a more consensual scientific role they may try to restrict their participation to presenting "facts"—even though their choice of facts can have important effects on readers' definition of the problem. We must hope that experts in this field can systematize the types of arguments used in it, and thus propose standards as to what information should be presented so as to give a reasonably complete account of relevant causes within a framework of justification.[42]

42. The collective enterprise of Alvarez et al. (1979) represents an effort to take "steps necessary to analyze and act on claims and counterclaims" in the area of institutional discrimination, and to form a relevant technical community or "invisible college" dealing with these models.

8 Indicators of Social Integration

Most of the indicator variables we have discussed, and all the end-values, refer to individuals. Yet there are also contextual indicators that refer to social entities larger than the individual. Among these indicators of the social context one type has been widely proposed for defining problems: it measures a quality such as collective morale, loyalty, or altruism; and the absence of such a quality may be measured by indicators of divorce, turnover in membership, or anomie. This type of variable refers to the integration of social systems ranging from families to organizations, communities, and nations. Qualities of this sort are not end-values; there is much debate as to whether and under what conditions they are desirable. They are likely to be contributory variables, though the models that tell just how they contribute are incomplete. Nevertheless there is wide support for measuring them, both as problem definers and as supplements to an otherwise individualistic approach to indicators. They are widely used and proposed, but need to be examined within a common framework, which the concept of integration can provide.

By *integration* we refer to a widely shared configuration of sentiments ("attachments"; to be discussed in detail below) held by system members toward one another and toward the system in question. We include measures of aid and cooperation as indicators of these sentiments, hypothesizing that sentiments of attachment conduce to aid and cooperation.[1] We devote a spe-

1. This definition of integration focuses on one aspect of this multiply-defined concept. Parsons (1951: 77, 97) outlines a series of aspects of the integration of ego with alter: attachment, in which one person may be the source of an organized system of gratifications to another; loyalty, in which attachment is organized in terms of a cultural pattern of shared expressive symbols; solidarity, in which loyalty is institutionalized and acquires moral sanction and obligation in terms of the common value system of the society; and loyalty to a collectivity as an object of attachment. The capacity to work together is similar to functional integration (Landecker 1951) or "organic solidarity" (Durkheim 1947 [1893]). We do not, however, propose to measure functional integration directly because of the variation among social systems of a given type in the tasks (functions) that they choose or face. We shall also deemphasize cultural integration. Our definition overlaps with that of "cohesion" (Schachter 1968). Individuals may also give attachment or support to

cial chapter to this concept because its logical status requires clarification. The prevalent individualistic approach usually proposes economic or social-psychological indicator models in which social contributory variables refer to individuals. Variables referring to the system as a whole have been recognized to be important and neglected (House 1981: 445; Juster 1981: 28).[2] Yet these variables have not usually been grouped together or given a common treatment in the literature on social indicators. We must show, therefore, that these properties of social systems, which can affect individual well-being, can be organized within a common framework.

Social integration is by no means always valued, nor should it be. We expect that it can be a means to well-being under some circumstances; the integration of a firm can increase material production and thus the well-being of consumers; the integration of a family or community can conduce to subjective well-being as well as to the coordination of its members' activities. But integration can also hinder social and political change, facilitate the performance of useless or harmful tasks by a system, or exact a price from members (Coser 1974). Insofar as we value it as only a means, not an end, it is not an end-value; and if its contribution to end-values is not clearly positive or negative, it is not necessarily a value concept.

The importance of social integration for policy indicator statistics cannot rest on its clearly signaling a consensual value for use in collective problem definition. We do not, in fact, advocate its use in public statistics as strongly as we do the use of economic benefit, subjective well-being, or equity. However, the concept of integration does allow us to draw together in a common framework a number of more specific variables whose status in indicator systems is ambiguous: divorce, job turnover and morale, mobility between subsystems, charity, crime, participation and alienation.[3] All these variables can reflect aspects of the integration of social systems or subsystems. If we can describe the integration of the various systems to which these measures refer by placing them in a common perspective, we may clarify some of the tasks of indicator development.

The integration of social systems is a particularly social (as contrasted with economic) source of coordination of people's activities. In an economic perspective, exchange is the great coordinator; people cooperate voluntarily

---

particular others without the incorporation of these relationships in integrated social systems; these relations may be important for well-being but we do not emphasize them here because of our interest in system properties.

2. Some writers seek to include them by dealing with holistic, system-level values; but we do not. Rockwell (1983: 2) argues that organizational indicators (properties of a special type of system) are relevant both because of their influences on individual well-being and because of their influences on social change. We consider only their relevance to individual well-being and equity.

3. Land (1975a: 20) chooses two of these variables, divorce and alienation, to illustrate the diversity of valuations given to indicators.

through offering different things to one another and exchanging them. In these exchanges, individuals deal with other individuals; but they also face contextual features such as firms and market prices—entities seen as unitary by participants though each is a symbol summarizing numerous particular actions and transactions.[4] Underlying market exchange, moreover, are further contextual features of each participant's situation. The functioning of a market requires adherence to underlying norms of truthfulness and respect for property, without which the market could not exist (Parsons 1951: 98–99). In addition, the productivity of a firm can be affected by the atmosphere of the workplace. Both norms and atmosphere are characteristics of social systems, not of individuals. Social integration is a particular sort of contextual variable, characteristic of social systems.

The type of integration with which we are concerned here is social; but it impinges on political integration as well. Easton (1965: Chaps. 11–13) has distinguished among three political objects—each socially defined and more than an individual—to which a citizen may give support: the political community, the regime, and the authorities. All of these center about a politically organized system, which often coincides with a society, and about its potential "authoritative allocations of values" (Easton 1965: 350). Social integration affects the capacity of a political system to make decisions that further its members' well-being, and not merely the capacity of individuals and groups to aid one another through private action.

The integration of a community or a society can affect the legitimacy accorded to, and thus the effectiveness of, a corresponding set of political institutions. Attachment to a community or society can be expressed by attachment to governmental symbols—the flag, the presidency, the Constitution. The connection of some citizens with one another is mediated by the actions of government. Thus we must eventually deal with political aspects of integration when we attempt to describe social integration.

Social Integration

In measuring social integration we may begin with observations based on pairs of persons. One person may give another social support that enhances that other person's objective or subjective well-being. If objective well-being is enhanced this support may take the form of money, food, housing, or other goods; donors, when they furnish such goods, may reduce their own resources. Similarly, when one person aids or cooperates with another without the transfer of money or goods, perhaps enhancing the recip-

4. Such symbols—as perceived by members of social systems who communicate about them—were referred to by Durkheim as "collective representations" (Parsons 1937: 360).

ient's subjective well-being, the donor may expend time and effort. But members of families, organizations, communities, and nation-states often aid their dependent or deserving fellow members, or provide for their fellow members generally, without asking for a specific compensation in exchange (Wellman 1982: 78). When aid entails such an apparent sacrifice, some motive other than self-interest is required to explain it.[5]

This aid is encouraged by the special sentiments (Gusfield 1975: 10) that members of social systems may have toward one another and toward their social systems. We shall call these sentiments *attachments,* treating their presence as a variable which may assume positive, neutral, or negative values. They are typically measured by asking questions of people. Attachments to a system are (if positive) favorable attitudes of members toward other members and toward the system as a whole; we expect such an attitude to lead the respondent to aid and cooperate with the others toward whom it is directed. These may be attitudes of friendship or loyalty to persons, or of devotion to the entire system and its symbols (Kasarda and Janowitz 1974); or if negative, of hostility and disloyalty. As we shall see, they may derive in part from social norms.[6]

The meaning we give to attachment is thus an altruistic one: it is a sentiment which, if positive, predisposes a person to help the persons or the system toward which it is directed, irrespective of any compensating gain in the donor's well-being. When directed toward an organization that provides pay or profit, this implies a willingness of the person expressing the attachment to aid or cooperate over and above what would be expected on the ground of these monetary returns alone.

The sentiment of attachment resembles the sentiments that Durkheim (1947 [1893]) characterized as involved in "mechanical solidarity," except that our definition centers about sentiments expected to produce mutual aid and Durkheim's stresses another manifestation of them—moral feelings about crime. Not all crimes, as he points out, impair the well-being of the group in the same measure as they elicit moral indignation. He notes, however (105) that mechanical solidarity involves attachments: "not only are all the members of the group individually attracted to one another because they resemble

5. Altruistic action is typically described in economics in terms of an increase in an actor's utility relative to other acts that he might have chosen—a requirement, by definition, for any action. We need not, however, require that the well-being of the actor be increased by an altruistic act; this is an empirical question that should not be answered by definition (MacRae 1976a: 138–44).

6. These sentiments do not fully determine the extent of aid and cooperation, which also depend on other variables such as resources, skills, and understanding on the part of the donor or of representatives of the system; and the need or other qualities attributed to the recipient. We define attachment in relation to giving of aid; but in actual sociometric or network analyses, the ties studied can concern reciprocal relations or the seeking of aid.

one another, but also because they are joined to . . . the society that they form by their union."

An integrated social system or group is thus something more than a social network with numerous interpersonal ties. Especially for larger systems, as we shall see, it involves attachments to the system rather than to its individual members. A person who is attached to a community as a whole may be willing to devote resources to actions intended to benefit the community, without consideration of the specific members who gain from them. A Gestalt is formed, such that the donors' motivations all come to include recipients' well-being in similar ways, in part because each donor knows that the other donors have similar sentiments. This Gestalt, which can be an aspect of a norm, transforms interpersonal (individualistic) attachments into system characteristics perceived by the participants.[7] When this occurs we should expect the membership of the group to be defined not only by cohesive interpersonal ties, but also by similarity among members in their attachments to other objects and symbols (Burt 1978). The formation of this sort of integration is of course a matter of degree.

The relations between the attachments of individuals and the integration of social systems can be clarified with the aid of sociometric matrixes such as those shown in Table 8–1; each entry in such a matrix represents a particular donor-recipient pair that could be the basis for an attachment. This table shows four possible configurations of attachments in social systems[8] each of which consists of six members. The major part of each matrix (a 6 x 6 square array) shows the attachments between all possible pairs of persons; the entry a_{ij} in row i and column j shows the attachment of i to j, with the self-attachment of each person being represented by the diagonal element a_{ii}. Each of these entries is designated as +, 0, or −, indicating whether that particular relationship is positive, neutral, or negative.[9] Self-attachments are shown as positive to simplify the appearance of integrated subsets.

7. This simultaneous patterning of donor motivations and recipient distributions may be revealed by direct factor analysis of sociometric data (MacRae 1960), or by blockmodels (White et al. 1976). It is also possible for structural properties of a social system to be seen by observers even when some participants are not aware of them.

8. We assume that we know the boundaries of the system, though this can be a difficult empirical problem (Laumann et al. 1983). A social system need not be defined by attachment alone; indeed, it must be defined independently of attachment if its integration is to be logically distinct from its existence. People may interact regularly (and thus form an integrated system in senses other than attachment) because of exchange, obligation, or ascribed social position (as by being born into a system), in which case they may be members without necessarily having positive sentiments toward the system.

9. Data used in such matrixes are usually obtained by questioning the donors of attachment i. We assume that valid measures have been obtained. Cell entries might be numbers rather than these three categories if the questions used permitted it. A set of equations for "interdependent utility functions" might be expressed in similar matrix form if the entry a_{ij} were assigned a value

Table 8–1. Sociometric Structures in Matrix Form

A. Fully integrated

i: \ j:	1	2	3	4	5	6	S*
1	+	+	+	+	+	+	+
2	+	+	+	+	+	+	+
3	+	+	+	+	+	+	+
4	+	+	+	+	+	+	+
5	+	+	+	+	+	+	+
6	+	+	+	+	+	+	+
G**	+	+	+	+	+	+	

B. Differential participation

i: \ j:	1	2	3	4	5	6
1	+	+	+	+	+	+
2	+	+	+	+	+	0
3	+	0	+	0	+	+
4	0	+	0	+	0	+
5	0	0	+	0	+	0
6	0	0	0	0	0	+
G**	+	+	+	0	0	0

C. Hierarchy

i: \ j:	1	2	3	4	5	6
1	+	+	+	0	0	0
2	+	+	+	0	0	0
3	+	+	+	0	0	0
4	+	+	+	+	+	+
5	+	+	+	+	+	+
6	+	+	+	+	+	+

D. Two-factional conflict

i: \ j:	1	2	3	4	5	6	F_1	F_2 ***
1	+	+	+	−	−	−	+	−
2	+	+	+	−	−	−	+	−
3	+	+	+	−	−	−	+	−
4	−	−	−	+	+	+	−	+
5	−	−	−	+	+	+	−	+
6	−	−	−	+	+	+	−	+

*S designates the system as a whole or its symbols.
**G designates the government of the system.
***F_1, F_2 designate factions 1 and 2 respectively or their symbols.

In such a matrix it may be possible to measure integration by the density of attachments, provided that the matrix represents a generally positively connected social system.[10] Table 8–1A shows a fully integrated social system in which each member has a positive attachment to every other. In addition, each member has a positive attachment to *S*, the system as a whole or its symbols. The government *G* of the system provides benefits (we can no longer speak of attachments in the same sense) to all members as well. We might expect in

corresponding to the partial derivative of *i*'s utility with respect to *j*'s. Measures of the entries in such a matrix could also be obtained, in principle, by asking persons to identify others from whom they might receive support; this type of definition of attachment is used by Henderson et al. (1981: 31).

10. Negative ties should contribute negatively to density. In general the strength of ties would also be relevant, but this is not reflected in our trichotomous matrix entries. The interconnectedness of subsystems may also be relevant to the integration of an entire social system. Measures related to density have been used to define cliques, in which a high degree of integration corresponds to the existence of a primary group (Burt 1980: 97–100). For simplicity of presentation we juxtapose government (more often a property of large systems) with a small sociomatrix. In larger systems we expect a lower density of direct interpersonal attachments, with more importance of symbols and indirect attachments.

this case (though data of this sort do not prove it) that the six members would regard the system as an entity in its own right.

Table 8–1B shows a pattern of attachments that we may call differential participation; person 1 gives the highest number of attachments (six including a self-attachment), and the number of attachments given steadily declines as we go downward until person 6 reports only a self-attachment. Person 6, though receiving three attachments from others, gives none to other members. The government G of this group provides benefits only to the top three members (a relatively integrated subset) but not to the rest. In this case we have allowed the members to specify different numbers of attachments; such an approach is more useful for our purposes than the type of question that asks for one's "five best friends," which cannot distinguish five strong attachments from five weak ones. Note also that we must not only identify the members correctly at the start, but also arrange the rows and columns in the proper order to see such a pattern of differential participation easily.[11]

Table 8–1C shows an asymmetric configuration: members 4, 5, and 6 have positive attachments to members 1, 2, and 3; but members 1, 2, and 3 are neutral to 4, 5, and 6. Such relations are characteristic of a prestige hierarchy, in which unreciprocated attachments are directed "upward." Measures of integration should not be expected to reveal hierarchy.

Finally, Table 8–1D shows a configuration with negative attachments in a systematic pattern. Members 1, 2, and 3 have positive attachments to one another but negative ones to members 4, 5, and 6 and members 4, 5, and 6 similarly like one another but dislike 1, 2, and 3. In addition, each subgroup shows positive attachment to its own symbols (1, 2, and 3 now being designated as F_1 or faction 1, and 4, 5, and 6 as F_2). We refer to this pattern as two-factional conflict or hostility. From the viewpoint of overall well-being, it is important to know whether an integrated social system such as F_1 or F_2 harms outsiders or not. If the −'s were changed to 0's, we would have two subgroups that were indifferent to one another and one another's symbols; this might be called simply segmentation. It might then be more appropriate to say that the two subgroups were internally integrated but the entire set of six was not.

Diagrams of this sort suggest some of the types of indexes of integration that we might use if sociometric data were available for the subsystems we wished to study and characterize. We could describe the overall density (Berkowitz 1982: 5–46) or prevalence of attachments in the system, at one

11. Various types of structure might be sought in a matrix such as that of Table 8–1B. We might try to permute the rows and columns so as to bring entries near the main diagonal (Coleman and MacRae 1960); we might look for particular network paths connecting one member with another; or we might simply try to summarize the general amount of attachment among members in a single index.

remove or more;[12] if there were a metric that permitted subtracting the −'s from +'s, we could express the algebraic sum as a proportion of the total number of entries in the table. We could examine the prevalence of subsets that are nonparticipant or ignored, or that are antagonistic. We might also examine the extent of reciprocation of attachments, and study the question whether reciprocity enhances the overall density of attachments. We could try to identify subsystems and their degree of segmentation from data of this sort; but alternatively, we could specify subsystems—e.g., families, organizations, and communities—on the basis of our prior knowledge and ask the same questions about their members without facing the difficult task of identifying them from voluminous sociometric data. We could also simplify the data analysis by asking similar questions about respondents' attachments to elites or to possibly mediating subsystems within larger systems, though this would not necessarily give us a full account of respondents' social ties.

The indexes we have mentioned so far depend only on the attachments of individuals to one another, as is conventional in sociometric or network analysis. If, however, we include system symbols such as S, F_1, and F_2, or collective actors such as G, it is harder to propose indexes of integration that are comparable across systems. The difficulties of identifying comparable common symbols and establishing common metrics (for attachment to persons and to symbols) complicate the problems of measurement. For practical purposes, some of the most useful measures of integration have been based on proportions of members who consider their organization or community to be "best." Not only is it difficult to choose common metrics for attachments to persons and to symbols; the importance of attachments to the system as a whole, relative to those to individuals, tends to increase with system size. Thus although attachments to individuals and to the system can be combined in a study of "structural equivalence" (Burt 1978), as by factor analysis, the two types of attachment may have to be presented in separate statistics.

Even if we could summarize a matrix of attachments so as to yield statistics characterizing the system, the interpretation of these statistics would depend on the definition of attachment that gave rise to the data. A relation between two persons can involve a complex mixture of emotions, perceptions, and activities. In addition, it can be mediated by social structures and channels of influence. For example, the effective relation of a taxpayer to a migrant laborer may be mediated by government, and thus be effectively positive or negative without any direct personal attachment or contact.

We wish to define attachment so that it is relevant to one person's action

12. If we were to measure integration by density of attachments, we would still have to demonstrate that it affected well-being as a system (contextual) property over and above the effect of individual processes (Lincoln and Zeitz 1980).

that affects another's well-being, but not to make that relation tautological by using the effect on well-being in the definition. We could, for example, ask one person whether he likes another, or likes to associate with him; the response may reflect a desire to receive help as well as to give it. More to the point, we might ask whether the respondent would help (or has helped) the other in specified ways, with the possibility of negative responses as well. The overall disposition of one person to aid another may also depend, however, on the normatively defined relation between the two persons; for this reason we must recognize that normative integration can also contribute to well-being, over and above simple attachment.

Normative integration has been defined (Landecker 1951: 333) as consistency "between cultural standards and the conduct of persons. It measures the degree to which the standards of the group constitute effective norms for the behavior of the members." These standards are of various levels of generality; some, concerning equity or charity, can encourage altruistic aid.[13] Other norms can define proper behavior in a variety of social roles. Depending on the substance of the norm, there may be gainers and losers from its existence; the norm (law) establishing one person's property rights also deprives others of use of that property.

The normative aspect of the relation between two people may or may not affect the disposition of one to aid the other; normative integration can include a wider range of sentiments and activities than attachment. It may concern their styles of greeting, their dress, proper topics of conversation, or modes of supervision on the job. Conformity to these norms may symbolize attachment to the system without contributing directly to others' well-being. The application by i of his internalized norms to j may indeed result in applying negative sanctions or in doing things "for j's own good" that are not appreciated by j, rather than in increasing j's well-being. There can thus be "crimes without victims," departures from normative integration that do not directly harm other group members. But there are also norms that support aid and cooperation; we shall call their effects on particular interpersonal relations *normative*

13. A norm involves the expectation, within a social system, that certain behaviors will be produced in certain circumstances; an internalized feeling that these behaviors are right; and sanctions rewarding conformity or controlling deviance. All these features may be present in varying degrees (MacRae 1976a: Chap. 8).

Aid to others may seem to result from unilateral attitudes on the part of the givers. Nevertheless, aid is encouraged by a shared feeling that giver and recipient are both members of a group. Such feelings may be supported by the members' sharing attitudes, conforming to common norms (Naroll 1983), or (in the case of rights to the benefits of the welfare state) believing that common citizenship entails special contributions by the more fortunate to provide for common benefits. Aid may also be part of a process of social exchange (Blau 1964), but this is not necessarily related to attachment.

The degree of emphasis on norms in social structure has lessened somewhat in the recent interpretations of network analysts (Wellman 1982: 64).

ties, analogous to attachments in being directed from one person to another. The overall effect on the recipient will then include that normative component—if there is one in the relationship—together with any additional nonnormative contribution. Thus if a teacher is disposed to help a pupil because it is an expected part of her social role, this is a normative tie; if she goes beyond this because she likes the pupil, this is a nonnormative attachment. The latter is more likely to be reflected by sociometric questions.

Normative integration is a matter of degree in several senses. First, a greater or lesser proportion of members' activities may be governed by norms. Second, a given realm of behavior may be regulated to varying degrees by norms; a single permitted option may exist, or there may be freedom to choose any of several. Third, a system may provide more or less guidance as to the options available; norms may vary in their degree of institutionalization. Fourth, even when a norm defines the range of permitted behavior in a given system, various members may adhere to it in various degrees. Their deviance from a norm (nonadherence) detracts from normative integration. When the norm in question relates to aid for others, deviance will thus also detract from this immediate effect of attachment; crimes *with* victims are an example. The extent to which parents care for their children, employees work efficiently, and citizens pay taxes or carry out military service are indicators of normative integration in social systems that have the corresponding norms.

Thus in a society that institutionalizes the traditional male-dominated family, the total predisposition of a husband to aid his wife includes the effect of the dominant male role (authority), perhaps leading to less aid than in another type of family, as well as whatever duties of support and protection that role entails; and it includes in addition the effect of whatever special degree of fondness (nonnormative attachment) the husband holds for his wife.[14] Similarly, the overall tendency of an employee to aid the organization for which he works might be composed of the general normatively defined expectations for employee behavior—perhaps changing as organizations became less authoritarian—plus the contribution of specific norms of his own organization, plus his particular personal attachment, or lack of it, to the organization.[15]

The result might be that the husband who cares for his wife will dominate her less; and the employee who is personally attached to his organization will work for it with more dedication than his pay and the prevailing norms of work would suggest. An outside observer sensitive to the effects of the norms might see the combined result of norm and personal attachment; but the participants might be more conscious of any departure from the prevailing norm,

14. The normative definition of the husband's role then becomes a baseline from which the nonnormative component of attachment is judged. Parsons (1954: 394) makes a similar distinction for the assessment of performance in a system of social stratification.

15. We refer at this point only to the integration of the organization, not to the solidarity of the work force.

taking the norm for granted and responding only in terms of these departures when asked questions containing options such as "a great deal" or "very little." As a result a perceptive observer might see the overall contribution of the interpersonal relation to the well-being of the wife, or of the person-organization relation to production by the organization, while the participants might be aware of only the nonnormative aspect.

The simplest example of integration of a social system would concern two people, considered in isolation from others. If each wished to aid the other (attachment) each would presumably try to enhance the other's well-being and to cooperate in tasks. Moreover, the attachment of one might increase that of the other, transforming the two separate attitudes into a system property. In measuring this property, however, we would probably have to rely on averaging the two members' reported attitudes toward one another, since there might not be clearly expressed attitudes toward the group as a whole.

We cannot so easily speak of the normative integration of such a pair; they may well regulate some of their interactions by rules, including judgments as to what is right conduct, but are less likely to formulate and codify the pair-specific aspects of these rules than are the members of a larger social system. A married couple may be guided by the general norms of their society that define their roles, but for two friends these norms are less precise. Our measurement of the degree of integration for an isolated pair is often likely, therefore, to be limited to measurements of strength of attachment.

Integration may still not maximize the well-being of the pair, however. Altruism (attachment by one party) may sometimes be carried too far, leading one member to sacrifice himself or herself to the other more than the other gains. A pair having the relation of master and slave might live together in harmony, each accepting this relation and judging it right (normative integration); yet an observer might judge this situation to produce less total well-being than would otherwise be possible, or to be inequitable. Thus integration is not synonymous with members' well-being or with the realization of other end-values, even in an isolated social system.

As we consider larger social systems, the meanings of attachment and normative integration become more specific and detailed. Let us first consider a larger social system in isolation. Attachment in a larger system tends to include sentiments addressed to the system as a whole rather than to specific members—to the family as a larger entity, to the town as a place to live, to the company as a place to work, to the nation; and to symbols such as the family name, the local baseball team, or the national anthem. It also comes to depend more on mediated attachments: attachment to small groups can contribute to attachment to a larger system, and the integration of elite members can both affect and reflect the integration of those whom they represent. Attachment to a given organization or community is often measured by asking whether a respondent thinks it is better than others; but an ethnocentric response (Hillery

1982: 32) may reflect negative attitudes toward outsiders as well as pride in one's own system.

If we were to measure only attachment to the symbols of the system as a whole, however, we might ignore distributional effects that could be revealed by the ultimate or mediated effects of attachments. From a survey of those who felt attachment to a political community we could not directly tell who received the benefits from the community's government. The mediating effects of organizations and of government, receiving these attachments and transforming them into policies, might aid some citizens while ignoring or harming others. Some members might be integrated with one another through these mediating effects, while others might remain outside these reciprocal ties and receive less aid or cooperation—perhaps even while being loyal to the system. Whatever indicator of integration we choose for such a system, we must be careful not to ignore the distributional effects of policies if they exclude some persons from benefits. Indeed, one of the meanings most commonly attached to nonintegration (as for members of ethnic minorities) is the exclusion of some groups from participation and from social support. Although an index of integration can usually be increased by excluding some persons from the system,[16] we cannot at will exclude citizens or residents of a government's territory from our calculations of the public good.

Normative integration also becomes more complex in larger social systems. A norm can there be expressed in universal terms, referring to all fathers, all employees, or all guests, rather than merely to specific persons. A normative system of this kind may lack some of the spontaneity and flexibility of personal understanding that is found in smaller systems.

In a larger group with considerable normative integration, some members can suffer from the existence of a norm in a way that could not be true in an integrated pair. If a pair is normatively integrated, both members have to accept the norms governing their relation. In a community or nation, however, a large majority may be united in adherence to norms regarding the duties of certain categories of persons to others, but some may wish to escape from the restraints of these norms. The flight of youth from the small town to the city, and the creation of a counterculture are examples of this desire to escape. Regardless of which set of norms were the dominant basis of normative integration, the group members who did not adhere to these norms would experience less well-being as a result of being different. Lack of normative integration can thus have effects that are broader than those implied by the indicator category of "crime and public safety," in which deviant actions harm conformists. Normative integration can decrease the well-being of deviants and not simply increase that of the conforming majority, since some types of deviance do not

16. Some advocates of abortion claim that a smaller family, or one without unwanted children, is likely to be more integrated (Steiner 1981: 51).

reduce others' safety or well-being. Moreover, although integration can aid collective action, a certain degree of conflict can be a fruitful stimulus to collective action in political systems.[17]

The effects of normative integration thus depend on the scope of the system of norms. A social system may treat alternative behaviors in certain domains of conduct as matters of taste and may even encourage diversity in them. Attachment to other persons, together with conformity to a norm of diversity, would then dictate the encouragement of diverse paths to well-being.

Finally, the normative integration of a subsystem can have effects on a larger system whose members' well-being we wish to promote. The subsystem and its members will be affected by the larger system's norms (what does the society define as proper family behavior?) and its allocation of resources (how are the resources of a community affected by the state of the economy and by central government policy?). But the subsystem will affect outsiders as well as its own members. It may produce goods that are consumed by outsiders; compete for resources; participate in decisions that affect outsiders; socialize potential members of the larger system; welcome outsiders or exclude them. The contribution of subsystem integration to overall well-being is thus complicated by distributional considerations outside as well as inside the subsystem. For example, a highly integrated community may successfully lobby its national government for funds that benefit the community so little that they ought not to be taken from taxpayers elsewhere. A highly integrated suburban community (not in the racial sense) may try to exclude others who seek entry. A highly integrated political subgroup may achieve internal consensus on its political goals, but be intransigent and uncompromising externally, hindering consensus in the larger system. A subsystem may also contribute positively to the well-being of the larger system—though we need to specify the conditions under which this can occur.

When we seek to measure the effect of integration of a social subsystem on well-being, we must consider not only the relations among given subsystems but also the possibility that persons may move their activities and attachments from one subsystem to another. A person chooses to divide his time and resources between activities in groups and alone, between work and family, or between family and community, in terms of his resources, tastes, attachments, and constraints; but these choices may or may not maximize the well-being of all who are affected. For example, some family members may spend recreation time separately from others. Children may further their longer-run well-being by attending school rather than using that time to aid other family members. With regard to the well-being of outsiders we may thus ask: How much time should the world-class concert pianist take from practice to spend

17. Thus an important indicator in cross-national comparison is the extent of political freedom (Taylor et al. 1980: 126–29).

with her husband? Should the top business executive be sure to arrive home every night when dinner is ready—or should he prepare it himself? Subsystem integration may thus conflict with the general well-being. It is a contributory variable but not clearly a value variable.

In the remainder of this chapter we shall consider in turn the integration of families, organizations, social communities, and politically organized communities. For each type of system we shall develop suggestions for appropriate indicators by first considering measures of the prevalence of membership in that type of subsystem; then the effects of membership and integration on the general well-being (as they affect both members and outsiders); and finally indicators of integration, including the contributions of normative ties.

Family and Household Integration

The family provides useful initial examples of indicators of social integration because it is small and its integration affects the well-being of its members directly. Among families, we begin with the simplest case, a husband and wife without children in the home—a couple.[18] By extension, much that we say about the couple could be said about an unmarried couple, or a pair of persons of the same sex living together.

The couple. The couple in modern society is partially separate from the larger society in two senses. It is usually physically separate in residence; and, because it is without children, it is not usually producing resources collectively for the society. Thus its activities are not scrutinized or tested externally to the same degree as those of work organizations. In economic terms, the household is a consuming unit; and even when its members engage in household production this is largely for their own consumption. Its two internal sociometric attachments contribute to well-being through social interaction (including sex) and the more material aspects of household production. We are thus mainly concerned with the contribution that its integration makes to its own members' well-being.

The couple is of course highly affected by the larger economy and society. In modern societies its members may work for pay outside the family; the demands of work and the pay received can affect the couple's store of resources, which they allocate in part because of their attachments to one another. Moreover, the couple's definitions of permissible behavior are influenced by the society's norms. The resources that society and the economy make available, including products and jobs, also affect their relationship.

In assessing the contribution of any type of subsystem to well-being, we

18. In actuality there will usually be kinship relations outside the household, including those with parents and siblings or with grown children.

must first examine its prevalence, which together with its effect per person determines its overall effect. The proportion of adults who are members of couples (as of other types of families) is customarily reported. From the viewpoint of well-being, we need to give special attention to persons who are not members of couples or comparable groups and who might become members; they may be separate through choice, but may also suffer lack of material and social resources.

We must next ask the extent to which membership in a given subsystem, and its integration, contribute to well-being, both in sum and in distribution; for the couple, we ask these questions at the start only for the members' well-being. Campbell et al. (1976: 52–53) show that married persons tend to report greater subjective well-being than those who have never been married, and that those who are married without children in the home have the highest well-being; though there may be selectivity in this comparison, it suggests that marriage contributes to well-being.

Among those who are married, integration (attachment) leads to higher subjective well-being for the couple. Goode (1982: 80) summarizes findings in the field: "If both partners are inclined toward 'altruistic behavior,' i.e., if both give way to each other in conflicts; if both have similar definitions of the other's proper role behavior; or if both find much value in interaction with each other, . . . we can predict that their marital relations will give them more happiness." This summary includes aspects of normative integration as well as attachment.

Even though agreement on role behavior promotes happiness, a given set of roles and norms may still be criticized on the ground of inequity. Thus conventional marriage has involved gains by men at women's expense, as Durkheim has shown. He compared the suicide rates of men and women in countries that differed from one another in the difficulty of divorce. The data showed that "in countries where divorce is forbidden or only recently permitted, the suicide rate of married women exceeds that of unmarried women, while in the same countries, the suicide rate of unmarried men exceeds that of married ones. Inversely, where divorce rates are high, married women have a lower suicide rate than unmarried women, and husbands have a higher rate than unmarried men" (Tiryakian 1981: 1028–30). Durkheim (1951 [1897]: 275–76) concluded that although monogamous marriage "is often represented as a sacrifice made by man of his polygamous instincts," man benefits more from marriage than woman, and by accepting monogamy, "it was she who made a sacrifice." These findings may relate particularly to families with children, which we discuss below.

Although we have described the couple as relatively isolated from the rest of society, we can also ask under what conditions it contributes, positively or negatively, to the well-being of others. Any direct contribution would depend on activities undertaken by the couple together and affecting others. If their shared activities do not affect others, society can ignore these activities. The family may also be a "haven in a heartless world" (Lasch 1977), contributing positively to its members' outside activities by providing respite from them.

An increasing concern in recent decades has been the relation between couple integration and the participation of both members in the job market. This relation depends heavily on the expectations brought to bear on it by the couple and those around them. If they start with traditional expectation that the husband's employment takes precedence and that the wife has disproportionate duties in the home, the wife's employment can lead to conflicts for her, intensified by the presence of children (Coser 1974: Chap. 6). To a lesser degree there are conflicts for the husband: Pfeffer and Ross (1982) find that men's careers, though furthered by marriage, are hindered by the wife's employment. At the other extreme, the increasing number of dual-profession couples who live apart because of the location of their jobs suggests that a relaxation of expectations can be tolerated; fewer activities will be normatively regulated by the couple in this case, even though the attachment may be strong.

We propose the following bases for indicators of couple integration (we shall reexamine them below for families with children):

1. Survey questioning of each member about sentiments toward, or disposition to help and cooperate with, the other. This type of information is the most valid, the most direct, the most subjective, and the most expensive in that both members must be asked. These questions may also deal with the allocation of money or cooperation in specific activities such as household chores, either of which may include normative aspects. The responses must be combined to form an index for each couple—perhaps by averaging.

Data of this sort are not conventionally part of national statistical systems, as they may seem inappropriate for government to gather. They have been gathered by private organizations (U.S. Dept. of Commerce 1980: 40; A. Campbell 1981: 243), sometimes with government support, and provide information on trends over time. Local government becomes concerned when the degree of integration is so low that domestic violence occurs. This condition, however, is reported unsystematically on the basis of complaints; and while the incidence of such cases can be tabulated, it is likely to be unreliable (like many indicators of deviance reported to the authorities) as a measure of conditions in the population. Some styles of intrafamily conflict have been measured by a questionnaire scale (Straus 1979).

2. A less direct measure of couple integration—not strictly a measure of sentiment but a predictor of it—is provided by the time that they spend together. We might assume that a highly integrated couple wish, as a result of their sentiments, to be together often; or conversely, that having spent much time together they have come to care more for one another. These two causal relations (sometimes assumed for other social subsystems as well) require

closer examination, especially when other variables affect the time spent together; Marsden and Campbell (1984) show some limitations of duration and frequency as measures of the strength of a social tie. Nevertheless, the use of time provides an approximate indicator of integration. Estimates of time spent may be based on questioning of individuals rather than couples. Time budget data on family interaction might thus be gathered periodically under private auspices, to supplement earlier studies (Szalai 1972; Robinson 1977).

The amount of leisure time that a husband spends at home has been suggested by Varga (1972) as an indicator of marital integration. Drawing on a cross-national time budget study, he shows a significant negative association between the divorce rates in nine countries and the amount of leisure time spent at home in those countries by employed married men with children.[19] Leisure time away from home, he points out, shows a closer association with divorce than does time spent away from home during holidays, which is more likely to be spent with the family. What is needed is a direct measure of time spent with the family; but even free time spent away from home with the spouse is positively associated with divorce rates across countries. This mode of analysis seems promising, but we need more complete models tested with larger numbers of individual cases.

3. A still less direct measure is provided by the rate of breakup of couples; its analogue for an organization or community is the rate of voluntary departure. The principal existing indicator that purports to reflect couple integration (or family integration more generally) is the divorce or separation rate. This rate is not, however, an indicator of either the prevalence of membership in families generally or of their integration. A particular divorce reduces the number of persons who are members of families;[20] but remarriage increases this number, and to know the rate of change in prevalence we need to combine them. The net effect of these two processes, for couples at least, may also be an increase rather than a decrease in integration or in well-being; divorce may indicate that an unsuccessful trial has been terminated. A. Campbell et al. (1976: 330–32) present data suggesting that remarried men have more satisfactory relations with their spouses than those first married, but this effect may result from differential selection (L. White 1979). Thus studies of the

19. These data actually refer to families with children but we use them to illustrate a possible measure of couple integration. The presence of children may affect this use of leisure time, however. Varga (1972: 368) also shows that national averages of time spent in conversation with family members are inversely associated with national divorce rates, especially when conversation includes all family members together.

20. Divorced and separated persons tend to have a lower level of subjective well-being than do the never married (Campbell et al. 1976: 52–53), reminding us that subjective well-being is often affected by sequences of conditions rather than by conditions themselves. We thus need to study and report sequences of events for particular persons and families (Cherlin 1981: Chap. 3).

integration and well-being of couples are necessary to interpret divorce and separation statistics.

Alternatively, we might make use of marital histories to derive statistics on duration of marriage. The interpretation of these statistics would be complicated by historical and cohort effects, as well as by the fact that some marriages have not yet reached their eventual duration; but further analysis can take these factors into account.[21] We expect that termination of marriages after a short time will involve less cost in well-being than termination of longer marriages. If we tabulate rates of termination of marriages (from separation, divorce, or death) by duration of marriage, we can derive average duration from them, by analogy with life expectancy.

If we use the time spent by a person in a social system—either hours per week or years' duration—as an indicator of attachment and integration, we shall face a problem of interpretation not only for the family but also for organizations and communities. The time spent may be either a cause or a result of integration, and it is affected by other variables as well. Especially when it is interpreted as a result of attachment, we must be sure that this allocation of time is voluntary. When we ask whether duration of interactions causes attachment (as in the case of length of community residence), we must separate the effects of attitude change from those of selection, in that some who are less attached leave the system. We must eventually seek indicators based on more adequate causal models.

Families with dependents. As we move from the couple to the family including children or other members, many more alternative structures become possible (Goode 1982: 8–9); more members can be involved and relationships vary. Relations with persons outside the household may also become more important; but at the start we shall again refer only to members in a single household. We shall also emphasize families in which the dependents are children rather than elderly persons or handicapped adults, and families centering about one or two adults who provide resources rather than three or more.

Again we must summarize the prevalence of these relationships. We previously suggested reporting the proportion of adults who are members of families, stressing that nonmembers were of special interest. Some of these might be potential members of families such as we are now considering, who might enter families as dependents: elderly persons living alone or in institutions; children who have been separated from their families, as by placement in institutions, or who have left home without adequate preparation for independence; single parent families lacking resources. Statistics of this kind are

21. The interpretation of this indicator will also depend on statistical controls for longevity and of effects (if any) of the presence of children.

often taken as assessments of the need for aid or social services; but whatever policy may be addressed to them, they are indicators of numbers of persons possibly lacking social support.

Next, to aid our choice of relevant indicators, we must assess the possible effects of membership and integration in families on well-being. This is complicated not only by the diversity of family structures, but also by the difficulty of predicting long-run effects on children's well-being and on society (including its integration) through these children. The distribution of well-being is also affected by the family's influences on the eventual social placement of its children, through the passing on of advantages. The well-being of young children, in the long and short run, seems especially dependent on continuity of adult care; thus indicators of children's cumulative experience are desirable (Bumpass and Rindfuss 1979; Watts and Hernandez 1982: 38; Furstenburg et al. 1983). A family supporting elderly dependents in the home may have similar concerns, but is less likely to try to influence elderly dependents' long-term future or personality development.

In spite of our concern with long-run effects on children, we must also study their short-run well-being. Not only their health and intellectual development but their present subjective well-being should be studied (Watts and Hernandez 1982: 16–17), even though in a free society their well-being within the family is not usually a topic for public policy except in the case of child abuse.

Our two hypothesized effects of attachment—aid and cooperation—largely overlap in the family because most common tasks are undertaken for the benefit of the members; household production is for the members' consumption. An exception might be the tasks of bringing up children; although adults dealing only with one another may cooperate to increase their own well-being, in child rearing they may sacrifice some of their well-being for the children's sake, with effects that last longer than the family and affect the society as well.[22]

We then turn to possible indicators of family integration. Whatever measures we use, we should subdivide the population of families according to their membership, and perhaps in relation to resources and obligations: income, numbers of productive adults, time obligated to work, and numbers of preschool children are relevant. Comparisons of degrees of integration can then be made more clearly within particular subcategories of families. We propose the following indicators, paralleling those proposed for the couple:

1. Again the most valid but most expensive approach would be based on survey interviewing of all members of a family about sentiments toward, or disposition to help or cooperate with, each other. The particular pairwise relations, or relations to the family as a whole, then need to be summarized or

22. Punishment of children may also result in a sacrifice of their present well-being for the sake of longer-run goals.

averaged if we wish an indicator of the integration of a family. We might try to measure parents' attachment to one another; adult-child relations, including aid from older children; and sibling relations. There are, however, problems in measuring young children's attitudes.

2. As before, for each member we can ask how much time is spent in family activities and for what other members' benefit, if any, they are intended.[23] This includes the activities we have mentioned in connection with couple integration; household production (including preparation of meals, shopping, and cleaning) now benefits children as well as adults. But now child care and the interaction between adults and children play an important part; time budgets for child care have been studied extensively (e.g., P. Stone 1972). Older children can contribute to household tasks; and the presence of children of all ages (as of adults) affects the need for resources.

Devotion by older children to family tasks is expected to decrease with age, however. In a society where children are eventually expected to leave the parental household and become independent, they may have to give priority to learning, work, and self-support. Social norms and peer pressure can encourage activities that compete with aid to the family. Adolescents and young adults will then necessarily become less integrated in their parental families; in constructing and interpreting indexes of integration we must consider the children's ages.

So far we have dealt only with reported attachment as an aspect of family integration. For completeness we should study normative ties as well; a larger family group might have and inculcate norms of its own. An extended family including several households is more likely to do so than a separate household (Goode 1982: 102). If we were to try to measure the degree of normative regulation in a family, however, we would have to take account of variation among families in the nature of their norms. Conformity to general norms of the society, community, or social class could be perceived by an observer in the home (if access were granted) or in a school or playground; but these similarities would relate to integration in a larger social system, not necessarily to the special norms of the particular family. We thus do not propose indicators of normative integration for particular families.

3. Third (in addition to the marital dissolution statistics mentioned for couples), we propose measurement of the duration of experience of a child in a single parent family (Bumpass and Rindfuss 1979). This measure is expected to relate to the child's personality development, though this effect may also depend on the child's age and other factors. The effects of divorce or separation on children differ from the effects of the death of a parent (Goode

23. A father who leaves his family so that they will become eligible for welfare support may be performing an altruistic (though illegal) act that is unlikely to be recorded. Activities by one family member may also benefit others without explicit intent, if such activities are carried out for purely normative reasons such as ceremony or tradition.

1982: 167). Our knowledge of the effects of parent-child relationships in this respect is limited (Blechman 1982); but a review by Haskins et al. (1983) suggests that children of divorce suffer cognitive and emotional disadvantages.

As we try to measure the integration of families, we must recognize that any definition restricting the family to a residential unit or household is arbitrary. The system of interaction within the residential unit can be enlarged through contact with an extended family outside the household. These relations with persons outside the household are created by children's maturation and leaving, by divorce or separation, and by the creation of new families with in-law relations. Caplow (1982: 199) points out that "much of the 'community solidarity' of Middletown is really kinship solidarity. When there are relatives living in town, especially close ones—parents, brothers and sisters, or grown children—they are seen, talked to, and visited more frequently than are friends or neighbors. When Middletown people need assistance or advice, they are apt to turn to their relatives."

Relations with kin outside the household thus bring with them possibilities of aid, In some societies the father may leave his family for a period of years, going to another country to earn money for their support. Alimony and child support payments are also a source of resources. Distant kin can sometimes be called on for aid in emergencies.[24] These positive relations counterbalance the conflicts that result from the one-sided relations of some kin outside the household with the mother's or father's side of the family; cross-culturally, marital stability may be associated with the involvement of the family in a single dominant set of blood kin as well as in other external networks (Scanzoni 1965). If it were possible to perform repeated surveys, under private auspices, to trace the prevalence of aid through such kinship ties, we might know more about alternatives to governmental care for the needy and about the effects of governmental aid programs on aid by kin.

Organizational Integration

An *organization* is a social subsystem created for a particular purpose (Blau and Scott 1962: 5). Among these purposes are the production of goods, services, and regulation, carried out by private firms and public bureaucracies; direct benefit to the members themselves, as in recreational and fraternal associations; and political influence, as in pressure and interest groups.

People may become members of organizations because of monetary return from participation, because of sanctions for nonparticipation, because of their

24. A number of possible indicators of support from extended kin have been proposed (Watts and Hernandez 1982: 43–44).

valuation of the purposes of the organization or their activities in it—or from mixtures of these motives. We are mainly concerned here with productive organizations, in which members work and expect monetary return; roles in these organizations are widespread, claim much of their members' time, and affect their activities in other subsystems such as the family. We thus begin by assuming that in a productive organization all members are paid for their work.[25] Whatever contribution these organizations make to the general well-being comes largely from their effects on outsiders (e.g., customers or clients). In contrast to the family or a recreational association, they are expected to enhance their members' well-being more through pay than through activities in the organization itself.

When people are paid for their work, we might expect from an economic perspective that they will work only for pay and without special attachment to their organizations. Nevertheless, both workers and managers have other motives for working and rewards from it. Among them is satisfaction with the organization (Caplow 1976: 128) or the job. In measuring integration we need to focus attention on those nonmonetary motives that conduce to aid or cooperation; and because the focus of work organizations is on tasks, we give special emphasis to cooperation on these tasks. Workers may, however, aid one another, or persons outside the organization, on matters unrelated to their tasks.[26] Quinn (1976: 36) points out that "quality of work" is actually an ambiguous term, which can be viewed valuatively in different ways from the perspectives of employers, employees, and the community. Employers, he suggests, are primarily interested in productivity, workers in their well-being, and communities in external effects that may impose costs on them.

The firm—a private productive organization—was earlier characterized by economists as brought together by the maximizing calculations of employers and employees in their exchange of work and money. Later, March and Simon (1958: 140–41) saw it as composed of satisficers; and Roethlisberger and Dickson (1946) added a sociological perspective by showing that interpersonal relations influence output. Selznick (1957) called attention to sociological aspects of the transition from an "organization" to an "institution," involving commitment by members to shared values and goals. More recently, Leibenstein (1976) has criticized the analyses of firms by his fellow economists, pointing out the prevalence of inefficiencies within the firm ("X-inefficiency") distinct from the allocation of factors of production and associated with such variables as organization and effort.

25. We omit unpaid clients such as students or prisoners who are subject to the organization's authority, even though they are sometimes considered part of the organization (Caplow 1976: 30); and voluntary productive organizations, such as a volunteer fire department or the committee organizing a church charity sale.

26. Aid to outsiders cannot, however, derive directly from organizational integration unless it is the goal of the organization.

Roethlisberger and Dickson called attention to the importance of workers' informal organization for its effects in increasing production; conceivably similar effects result from formal worker participation. At present, however, a more important form of organization growing from membership in the firm is the labor union. Unions are concerned with their members' well-being in relation to the job, as it is affected by wages and working conditions. In providing this support to their members, primarily through collective action, their actions can run counter to the benefits that production provides to consumers; if wage increases raise the price of the product, workers' well-being is enhanced at the expense of consumer surpluses. At the same time, unions have institutionalized labor-management conflict about work, promoting it, yet often limiting and channeling it. Unions join across firms in political interest groups; as such they must be treated below, together with employers' associations, in connection with the political communities in which they operate.

We shall deal especially with organizations having hierarchical structure—a nearly universal feature of productive organizations but not a logically necessary one. As we attempt to measure the integration of such an organization, we encounter an important distinction between the two possible effects of attachment that we have suggested: members' mutual aid and their cooperation on organizational tasks. Etzioni (1975: 281) distinguishes between "peer cohesion" and "hierarchical cohesion" and points out that cohesion among the members of an organization does not necessarily promote compliance with organizational directives. Thus members of a labor union, or an informal group of managers or workers, may aid one another without necessarily increasing production or profit. They may be cooperating on a common task that is not an official task of the organization. We shall assume here that organizational integration is measured with respect to the task set by the organization's leaders; if we judge that this task does not have desirable effects, then we may conclude that organizational integration is undesirable (as many Americans judged for the military during the Vietnam war).

In characterizing the integration of a productive organization we again proceed from the sociometric perspective. Each member, whether owner, manager, or employee, may have particular degrees of attachment to the organization as a whole and to its other members—as well as to subsystems such as branches or agencies, or superimposed systems such as unions or internal factions. In hierarchical organizations, lower-ranking members' positive or negative attachment can be directed upward in a way that is impossible in more egalitarian systems—to the head of the organization, often associated with the symbols of the organization as a whole, and to other superiors. "For the rank and file, the responsible manager is a living symbol of the organization's collective identity" (Caplow 1976: 31). The upward and holistic aspects of attachment (or loyalty) are thus particularly important in organizations. Loyalty in this sense can often be increased by the leader's actions, including allocation of pay and recognition, communication, responsiveness, and successful external action to gain resources and accomplish tasks. The leader is also expected to show attachment to the organization and its values,[27] and the attachments of higher-ranking members are especially important. High morale tends to go with high rank (Caplow 1976: 129), but disaffection or conflict at the top can be especially damaging.

Members' upward attachments are affected by peer attachments such as those studied by Roethlisberger and Dickson; informal groups can contribute to morale in the military (Shils and Janowitz 1948) and in the firm. Indeed, Ouchi and Johnson (1978: 293) suggest that the Japanese style of management (as reflected in certain U.S. firms) relies more heavily than does the American style on attachments like those of the family or community, as it involves "low task specialization, . . . low turnover, and primary or wholistic relations among employees."

Informal or participatory relations of workers to the firm can, however, work in other directions than simply to serve management's goals. Such relations can take the form of union membership. They can also be incorporated into alterations of bureaucratic authority in the direction of worker participation, possibly modifying the organization's goals. Preference for kin can be introduced. In each of these cases we need to ask the difficult question whether such changes in the balance between production and worker satisfaction contribute to the general well-being.

In addition to simple attachments to an organization there are normative ties. Caplow (1976: 67) reminds us that "every living organization has some aura of sanctity in the eyes of its members." Biderman and Drury (1976: 224–31) also stress that employees' motivations are by no means limited to self-interest—that employees are concerned with accomplishing worthwhile tasks and that their loyalties affect their actions. Some normative ties are based on intraorganizational norms such as the commitments described by Selznick (1957). Members of an organization, if they can contrast their own organization with others, are likely to be aware of these norms and may be able to report the corresponding ties in response to survey questions. Attachment based on societal norms concerning work and organizations, however, is less likely to be reported; these norms are likely to be assumed as baselines relative to which one's own attachment is judged.

The contributions of integration of productive organizations depend on the prevalence of membership in them. This is given approximately by the em-

27. His attachment to individual employees can be contingent on his resources and their performance, but attachment to organizational goals is essential. He must "defend the faith," that is, "take the group and its work more seriously than anyone else" (Caplow 1976: 97). Especially in public bureaucracies, his loyalty must reciprocate his subordinates': "Loyalty is hilly, and it has to go down if it is going to go up" (Allison 1971: 166). One result can be excessive commitment to suborganization goals at the expense of more general values.

ployment rate in general or for particular population subgroups. The employment/population ratio in the United States for persons sixteen years old and over was 60.0 per cent in 1979 (U.S. Dept. of Commerce 1980: 397). The rate of participation in the labor force (including the unemployed), was 63.1 percent for the same age group, having increased about 4 percent over the previous two decades. Over this same period the participation rate of white females increased by 14 percent (indicating the rates of change of such statistics) while that of males sixty-five years old and over decreased by 13 percent. Employment can affect well-being in a variety of ways, as we have seen (Chap. 5).

We next ask, as a further guide to choice of indicators, how organizational integration can affect well-being. The major economic effects work through production and sale of outputs (in the private sector) and through pay. We would like to know whether a given policy that increased integration led to a net increase in the value added to the product, and how this additional value was divided among management, labor, stockholders, and consumers. At the same time we assess the subjective value of work to workers and examine the contribution (positive or negative) of integration to workers' well-being.

The indicators we propose are as follows:

1. Indicators based on members' sentiments toward fellow members and the entire organization, and on their disposition to aid or cooperate with one another. If we had unlimited resources for data collection, we might try to characterize the sociometric structure of each organization; but because of cost and resistance by management and employees, this can probably be done only in illustrative studies (Lincoln and Miller 1979). It is both cheaper and less controversial to seek statistics that provide averages over numerous organizations. Such studies do not threaten particular organizations or employees as organization-centered studies may; national samples of home interviews provide this possibility, but deal less directly with organizations as social systems.

We wish to measure such sentiments independently of what would be produced by pay alone. One way to set aside effects such as that of pay is to ask a respondent to compare his or her firm with others, using categories such as "one of the best," "above average," and the like (*Psychology Today* 1982). Similar questions can be used in asking respondents to compare job types and communities.[28]

28. Questions of this type can also be asked about commitment to work in general and to the respondent's particular type of job. One measure of a person's commitment to work is his response to the question, "If you didn't have to work to make a living, do you think you would work anyway?" (Veroff et al. 1981: 256). Combining this with a question on job commitment (271), Veroff et al. showed that among men and women who were committed to work, job commitment decreased by nine percent in each group between 1957 and 1976 (283). Such a finding, even for only these two years, performs an important function as an indicator series. It may reflect a de-

A direct and behavioral indication of lack of integration in organizations is the incidence of strikes; and unlike family violence, these are public and recorded. Reliable statistics on pilferage and embezzlement by employees are harder to obtain.

2. Statistics on the amount of time spent at work. Since this time is constrained by the employment contract and by pay, we should not expect mere time spent to be as valid an index of attachment as it would be for a system with voluntary membership. Moreover, the time people spend on the job is influenced by numerous variables other than attachment to the organization, including need for earnings, the normative definition of "full time," the state of the job market, the technology permitting flexibility in coordinating labor with other factors of production, and competition with other roles. One scale of work involvement, reflecting this sort of tradeoff explicitly, asks about relative commitment to job and family (Bailyn and Schein 1976: 163). Wilensky (1961: 43) has nevertheless suggested that the long hours worked by men who control their schedules result from a higher degree of involvement—perhaps in careers or jobs as well as organizations.

Although overall time spent at work may reflect organizational integration to some extent, it reflects too many other variables to be a useful indicator of integration. It may simply provide a finer measure of the prevalence of employment as it is affected by part-time work. Only if we could distinguish those expenditures of time that went beyond the requirements of the job or fell short of them could we use time data to indicate integration. Possibly voluntary overtime, or absenteeism or lateness, would have this property.

3. Statistics on job turnover. The separation of a person from a job is analogous to the termination of a marriage, and interpreting it poses similar problems. In the creation of either relationship there is a preliminary period of trial, followed by the establishment of a more lasting relationship (marriage or employment). The early stages of the formal relationship may be in part an extension of the period of trial, so that turnover may be less costly in that period than later. Thus job turnover or marital dissolution can contribute to greater integration in the next job or marriage; but at the same time it can reflect a lack of integration in the organization that one left.

If we wish to use data on job turnover to reflect organizational integration, we must try to control for causes of turnover other than low integration. We should measure rates of leaving rather than of hiring, even though the two are interrelated. Overall separation or leaving rates also reflect economic conditions as well as integration. Voluntary departure rates (quit rates) may be more valid, but even they can reflect other variables such as wages and the presence of other job opportunities. Conceivably a highly integrated organi-

---

cline in general commitment to jobs, and thus in commitment to cooperation in work, that could not be detected by questions comparing one firm with another.

zation could have a high turnover rate for new arrivals if it did not choose them carefully or gave them only probationary status. For jobs as for marriages, we should thus compile statistics on duration of membership prior to separation. Terminations after many years on the job, but prior to normal retirement, may tell more about integration than terminations concentrated in a period of trial at the start.

Overall managerial succession rates have been shown by Pfeffer and Moore (1980) to be negatively associated with organizational integration, even though this type of turnover may be involuntary as well as voluntary. The firing of lower-ranking employees can also reflect low integration, but such firing depends more on economic conditions than does change of managers.

Three additional types of organizations, other than the pure "productive organizations" we have been discussing can give rise to special problems of indicator design.

First are organizations that control groups of clients over extended periods, such as prisons or schools; their integration depends not only on their employees but on the students or prisoners, who must be considered members of the system. It is difficult to speak of the integration of a prison or a concentration camp; thus Hillery (1982: 73–90) contends that total institutions are not communities. If members give aid to one another, it will be within their own subgroup—guards or prisoners—with rare interactions between the two. The task of managers and guards is incarceration, which is not a task shared by prisoners. Rehabilitation and production might conceivably be shared goals, but are unlikely to be dominant ones.

In compulsory schools (even when the compulsion comes from parents), a partially similar relation arises. Pupils may establish their independence through peer groups that stress values other than official goals of the school (Coleman 1961). Integration into these groups is not necessarily a contribution to fulfillment of the school's official tasks (Chap. 4). Some possible indicators of integration about official tasks are attendance, lateness, vandalism, and violence in classrooms, as well as voluntary teacher turnover.

A second type of group, simpler to analyze than work organizations, includes "expressive" associations (Gordon and Babchuk 1959) for members' gratification through activities such as recreation and consumption. Although these groups do not face outside scrutiny as productive groups do, they contribute directly to well-being. Their prevalence and integration may be of interest, if these groups are deemed to be an appropriate concern of public policy.

A third type of group is a voluntary association with political goals, influencing the decisions of communities or larger governments. Such organizations can be integrated internally to various degrees; but even more than for firms, we need to place them in the context of the larger system they seek to influence. We shall therefore consider them briefly in the next section.

The Integration of Communities and Political Systems

The final type of subsystem with which we are concerned—communities and the larger nation-states of which they are parts—differs in important ways from families and organizations. It consists of a set of individuals and households, and often a set of organizations, interacting with one another in a common spatial area. Such an area may correspond to an economic system whose component parts are drawn together by the location of raw materials, transportation facilities, jobs, consumption items and other amenities, as well as by the presence of other households and organizations. Such an area may correspond to a governmental jurisdiction. Some subunits such as neighborhoods also group persons and families together through their choices of residence. We are concerned, however, with the tendencies to aid and cooperate in such groupings—first apart from government and then through government.

We choose spatial communities for analysis not because all communities are based on locality—technology is increasingly freeing them from narrow spatial limits (Wellman 1982: 63)—but because local areas are convenient bases for comparable time series. Regardless of whether the spatial community is a local area within a city, perhaps served by a community newspaper (Janowitz 1967), or a wider metropolitan area throughout which support networks extend (Wellman 1979), we can still characterize the integration of local areas and trace its changes over time.

We thus pass over numerous social systems that are not based on locale (Bernard 1973: 180). Some social groups that provide much meaning and support to their members arise from work and are separate from the place of residence. A considerable amount of shared feeling is directed to the mass media, and sentiments are expressed by mail, telephone, and computer linkage as well. The shared feeling and aid implied by the term community are not usually found, however, in groups that communicate only through the media. Probably the most important groups that are not identified with particular areas are political groups; but these are ingredients of larger politically organized spatial communities.

In dealing with the family and the firm we have separated the economic functions of consumption and production. In analyzing community integration we shall separate social relations from those of political decision: a neighborhood, though neither having a government nor organized to influence government, can still be a source of aid and cooperation.

The social community. Although it is convenient to deal with indicators for legal and political communities such as incorporated cities and towns, we can clarify the meanings of community by beginning with definable social groupings that do not have governments—neighborhoods, unincorporated areas, groups of residents of housing projects, possibly extended kin groups not occupying single households. At the same time, we can deal with that aspect of legal communities or jurisdictions (Parsons 1960: 258ff.) that does not involve formal governmental action.

The prevalence of membership in informal residential communities is difficult to measure statistically. Their very informality reflects the fact that their boundaries and membership are often ill defined. Even in regard to legal or political communities, most people might be treated as members of some such unit; but the meaning of membership is so varied and in some cases so slight that tabulation of its prevalence seems useless. Only for those who are completely nonmembers of the community where they live—vagrants or social isolates—would it be of interest, if feasible, to tabulate their numbers. Similar attention is desirable for institutionally segregated members of stigmatized groups (Sarason 1974: Chap. 7).

It is more important, though even more difficult, to characterize the types and integration of social ties within a given spatial area. As we consider larger units, with thousands of members, we can no longer examine all the possible pairwise relations among them. We may be led to sample individuals, and by questioning them about their relations with others try to learn about the "personal community" (Wellman 1982) or "interpersonal environment" (Kadushin 1982: 149) of each. Some indication of the integration of this environment can be obtained by asking each respondent about contacts among the persons named (Wellman 1982: 73; Kadushin 1982: 156n.). If clear definition of the strength of ties can be obtained, corresponding to a specified degree of attachment, we may be able to compare the prevalence of attachments in different communities. Whether these indexes of prevalence would validly reflect the integration of the communities in question would remain a problem for analysis.

It seems plausible that community integration can contribute to well-being, but the verification of this causal relation requires careful research. There is some evidence connecting social support with psychological well-being (Turner 1981), and the integration of interpersonal environments with lowered stress reactions (Kadushin 1982); but studies in this area have been criticized on methodological grounds (Thoits 1982). Henderson et al. (1981: 163), analyzing a longitudinal study, conclude that the development of neurotic symptoms (including depression) is associated not with lack of social relationships but with lesser perceived adequacy of these relationships. For our purposes, it would be necessary not only to show an effect but also to demonstrate that it resulted from a system property.

Because the prevalence of charity has been used as an indicator of community integration (Angell 1951), it is plausible to look for special contributions of integration to the well-being of the poor or needy. If integration conduces to special aid to the needy, it might result in higher levels of well-being, both economic and subjective, for those who were worst off. We must ask, however, which needy persons are considered authentic members of the community and thus entitled to support; if integration excludes some of them, its effects will be correspondingly limited.

In proposing indicators of community integration, let us begin in the same ways as for families and organizations. We might therefore propose:

1. Sociometric indexes based on reports of interpersonal attachment. In principle this study of interpersonal bonds is important first as a means of delineating subsystems within a geographical area or ties that cross its boundaries. Some of the various indexes of attachment to other persons include questions about acquaintance, friendship, and visiting (A. Campbell 1981: 244), or records of behavior such as intermarriage. Data on these aspects of attachment can reveal not only their general prevalence but also cleavages in relation to other variables such as ethnicity or social class.

We recognize immediately, however, that genuine sociometric studies of communities (with 100 percent samples) cannot be carried out extensively for national indicator purposes because of their cost; with sample size equal to average system size times number of systems sampled, they may be even more expensive than for organizations. We are led, therefore, to three possible courses of action in addition to the study of personal communities: to sample individuals rather than subsystems, reporting their average attachment to their communities but not to other individuals; to study a small number of communities intensively (Laumann and Pappi 1976); or to economize as the participants themselves must do, and study relations among organizations or elites, or the relations of individuals with these mediating entities. On each of these approaches we would hypothesize that a community or neighborhood with specified boundaries exists and then test whether the system(s) found fit that hypothesis.

Various indexes have been proposed that measure an individual's attachment not to other subsystem members, but to the community as a whole. Even a community without legal organization may still have shared symbols to which members can show attachment. The name of the neighborhood, some of its major institutions, and its informal leaders, can be among them (Janowitz 1978: 287). Rossi (1972) suggests several such indicators of "solidarity": perception of the locality as a collectivity; affective involvement in the locality as a collectivity; interest and involvement in local events; "social climate" measures; use of residential localities as reference groups; segmental solidarity; and attachment to the residential locality, also used by Kasarda and Janowitz (1974).[29]

A community can also be integrated by the interactions of families and of

29. Rossi (1972) also defines integration differently from our usage, including ties of exchange; as indexes he proposes the prevalence of market relations, voluntary formal associations, kinship, and friendship.

organizations. Hillery (1982: 49) defines the folk village, for example, as "a localized system of cooperating families." In more developed societies, a community can contain organizations that undertake collective action—resembling governmental organizations but without the legitimate use of force. It can include charitable and recreational associations, and churches that symbolize a religious or ethnic aspect of shared allegiance. These organizations can add to, or channel, individual or family attitudes to aid one another or act together. Productive organizations may also contribute incidentally to community integration; schools by bringing families together, firms by contributing to community causes or by linking employees to one another, and political pressure groups by performing a local integrative function apart from their influence on government.

Individuals may also interact with organizations, not as members but as customers, clients, students, or citizens subject to regulation. These relations with "corporate actors" (Coleman 1974) need not generate strong attachments to the community, especially when they are impersonal; but when the organizations represent the community as a whole (e.g., charities), they may contribute to attachment to the community. Conceivably, community members' interactions with private firms (if such interactions become habitual and personalized) may also contribute to individuals' attachment to the social community.

A further available source of indexes of attachment derives from publicly recorded statistics on extremes of negative attachment—in the community, crime, especially if directed against others.

Indicators of charity and crime were the bases of Angell's (1951) study of moral integration of American cities. Both indicators appear to reflect the inclination of members of a community to help or harm other members. If there are general tendencies in those directions, these indicators should reflect them. Angell's "welfare effort index" (based on Community Fund data) and crime index (based on FBI statistics) were negatively associated ($r = -.43$). The two were combined in a "moral integration" index, which was negatively associated with population heterogeneity and population mobility; a homogeneous, nonmobile population apparently was conducive to "moral integration." A repeat study (Angell 1974), however, failed to show a close association between the two component indicators and cast doubt on the generality of the variable being measured.

2. Measures of amounts of time spent and activities in the community. The integration of a community can be affected by numerous variables—contributory variables at a second remove. These include social homogeneity, though this can also reflect exclusiveness and discrimination (Parsons 1960: 254). Integration can also be affected by the variety of activities shared by members, from mere residence in a "bedroom community" to shared work group membership. Similarly, the overlap between children's schooling and place of resi-

dence can be a factor, reminding us of the cohesive and exclusive aspects of the neighborhood school. Thus studies of time spent by individuals in the community, or in the company of fellow members of the community, can be relevant just as for time spent in the family. These studies can probably best be done under private auspices.

3. Length of residence in the community (like duration of marriage or employment) is again relevant as both a cause and a result of integration. As in the case of families and organizations, we can study mobility or turnover rates among residential communities, and classify moves according to length of residence up to the time of the move. We may thus gain comparable statistics on families, organizations, and communities, possibly related to social accounts. For communities as well as for organizations, it is important to separate those motives for mobility that depend on integration from those that do not. Speare et al. (1982) have shown that the presence of strong social bonds as well as duration of residence are negatively associated with migration from a community.

The political community. We next add government to the features we are considering for the social community (Parsons 1960: 252), and ask what additional meanings integration acquires. These meanings will differ depending on the size of the system; as size increases, members' attachments will go proportionately more to general symbols rather than to other individual members. In a large political community or nation-state there will also still be personal attachments to its subsystems (neighborhoods in a city; ethnic groups; local governments in a national system), which may or may not contribute to the integration of the larger system.

Indeed, a major problem that arises again in our consideration of the formal political unit is that its boundaries may not be defined in relation to integration. From the new nation uniting warring tribal groups, to the city in which recently arrived ethnic groups are in conflict with older residents, to the older nation in which classes or linguistic groups have acquired a new militancy, political units face internal conflict. The power of governmental coercion, and the influence of the resources that government can allocate, may suppress or mitigate this conflict; but such forces cannot so easily produce integration in a social system that was not brought together voluntarily, like the family, or defined by its integration, like a sociometric clique.

We here treat city and nation-state together, in spite of their great differences.[30] Central national governments conduct foreign affairs, support armed forces, and usually have greater taxing power and central control of justice. Yet it is similarities, and similar indicators, that we seek.

These political communities tend to be larger than neighborhoods; as they become larger, aid and cooperation (as well as conflict) among their members

30. Weber (1978 [1956]: 902–04) reserves the term "political community" for the state.

tend increasingly to work through governmental action, and thus through the interaction of political groups. Aid, if it is given at the system level rather than through families or localities, is likely to be no longer voluntary but partly coerced by the majority. The terms, amount, and beneficiaries of aid become matters of conflict as well as of agreement—not only because some taxpayers are giving unwillingly, but because of disagreements about the deservingness of recipients, the effectiveness of policies, and the relative worth of other uses of resources. The attachments that further this aid may then be to parts of the community rather than to the community as a whole.

In both the large social community and the political community we encounter the question as to how subsystems contribute to the integration of the larger system (Gross 1966b: 198–200). In a large social community (such as New York or Chicago considered in this aspect), does the concentration of particular ethnic or class groups in local areas lead to overall integration or to conflict? The answer may depend on whether the separation is voluntary and on the interests and cultural relations of the groups.

Similar questions arise in the large political community for the political interest and pressure groups it contains. These groups, which need not be locally based, can contribute to integration or conflict in the larger community (Janowitz 1980). Political conflict, as we have suggested, is a way of producing aid and cooperation (at least within the majority) through public policy. The contribution made by any particular social subsystem or political group to the integration of the larger system is an empirical question (Grodzins 1956). We need therefore to measure not simply the prevalence of membership and participation in such groups, but also the extent of cooperation, segmentation, or conflict among them; segmentation or conflict among groups can detract from the integration of the larger system. Lipset (1960: 248–52) has stressed the capacity of members of a potential group to communicate with one another as a condition for the group's mobilization for political conflict, in contrast to cross-cutting ties that may limit conflict. An alternative approach, however, stresses the possibility of harmonious intergroup relations nurtured by political relations between their leaders or representatives, even when members of one group have little contact with members of another (Lijphart 1975; J. Steiner 1974). Thus in a large political community, where capacity to act collectively affects well-being, weak social ties may be as important as stronger attachments (Granovetter 1982).

The internal integration of political groups is therefore of interest only at a second remove. The main effect of these groups' activity is in the production of majorities supporting particular policies. A single group will presumably be more influential if more highly integrated, but a system of groups that are each highly integrated internally may give rise to unresolvable conflict more than if the groups were less integrated. In short, the integration of interest groups bears such an indirect relation to well-being that the value of measuring it seems doubtful. Political scientists study the cohesion of political parties, but time series of indicators of this condition do not seem directly useful in defining problems for public debate.

In following our systematic approach to the integration of political communities, we may apparently pass over the prevalence of membership in them since nearly everyone is a citizen of some nation. However, there are persons who, though formally members of political communities, are not granted their protection when they reside elsewhere—migrant laborers and illegal aliens being important examples. Even those residing within their own government's territory may not be granted full citizenship rights, and the degree of their protection may differ between national and local government. A person's effective membership may ultimately depend on his participation in, and recognition by, government. A member of a neglected minority in a large nation-state, unaffiliated with any major political coalition, may be effectively less of a citizen than others even when all have equal legal rights. Participation is an approximate indicator of actual exercise of the rights of citizenship, but because of its diverse causes, it falls short of being an accurate measure of integration.

The contribution of integration to well-being is more difficult to assess for political communities than for any of the other types of systems we have considered. Because this effect works through public policies rather than through voluntary action, observers will be in less agreement on its beneficial effects than they would on those of aid through the family or the social community. One source of this disagreement is the possibility that the existing order, about which integration centers, should be replaced by another. Political immobility, bred either by excessive normative integration or by excessive conflict, can hinder collective support of major changes that might aid citizens. System properties other than integration might have to be measured to indicate capacities for change.

If we proceed to indicators of integration of political communities, we can find analogues to those we have considered for other social systems.

1. Surveys of attachment by individuals to major governmental symbols and institutions.[31] We now forego measurement of direct attachment to other persons in any but the smallest systems (e.g., towns), and focus on specifically governmental symbols of the community. As in the case of organizations, we may consider alienation from government to be a value assumed by the variable attachment.

We can ask individuals their attitudes toward the symbols of various groups in the political community, and analyze the structure of their attitudes by

31. An analogous measure is the distribution of language use in a multilingual society. Deutsch (1953: 104–11) uses the distribution of language as one among several indexes of the mobilization of national populations.

methods such as factor analysis. Alternatively, if we can find persons who represent the various groups involved, we can ask their attitudes toward other groups or their representatives and analyze them in "elite sociometry." One variant on this theme is the analysis of numbers of events relating one nation to another, as indicators of pairwise relations between nations (Azar and Ben-Dak 1975).

Just as we considered published data on events of violence in families, organizations, and nonpolitical communities, we can consider instances of political violence as evidence of lack of integration of a political community. The converse type of action, compliance with citizen obligations, could also provide indexes of integration.

2. We can ask how much time is spent in activities related to the political community—not merely residence in it, which will be continual for most people, but activities related to integration. These might include participation and attention to public issues; but such participation would have to be weighted by the degree to which that particular activity contributed to system integration. A revolutionary could spend most of his time in political action within a system, but this might not contribute directly to that system's integration.

We can also measure units of activity rather than time spent; Deutsch (1956) uses the ratio of foreign to domestic mail, controlled for demographic variables, as an index of international communication. Jacob and Toscano (1964) use comparable methods for studying integration in international and urban systems.

3. Finally, turnover or out-migration rates could be measured. For cities, such rates could reflect attitudes toward the government over and above the social community, but it is unclear whether they would reflect political integration. For nations these rates would be small and could reflect high integration supported by coercion; but they could also reflect economic conditions, which would have to be controlled statistically if we sought to measure integration.

We thus face the question as to what indicators of the capacity of a political community for aid and cooperation will be useful to that community—or to higher-level communities. For scientific research aimed at general long-range remedies, these indicators may provide useful research data. To a decision maker outside the community, such as the centralized French prefectoral system that can abridge the powers of municipalities, they may be a guide for action. But for defining problems in the same political system that is being described by the measurements, these indicators must be viewed with caution. Indicators of system integration are likely in this case to be objects of controversy. Since they can then be invoked in opposing ways by those in power and by dissenters, they seem particularly unlikely to be effective in defining common problems or in resolving conflicts.

Social scientists, as members of an institutionalized occupation, have often hoped to do good in society from a relatively noncontroversial platform. They have disagreed as to how far they should enter into political controversy, some stressing a positivistic restriction to matters of fact and others taking cautious or bolder steps toward advocacy. Thus political scientists and others have tried to argue for values that are not in the forefront of controversy: examples related to integration are conflict resolution and the reduction of violence. It is doubtful, however, that even these are sufficiently consensual or free of hidden controversies to be proposed uncritically as policy indicators in one's own political system.

I therefore suggest that what is required here are partisan indicators, designed and provided by sources that do not claim the support of consensual political values. These sources cannot be governmental. Those who produce the statistics should abide by professional standards, so that the uncommitted can scrutinize and use them; but conflicting groups cannot be expected always to present the same indicators. One group may deal with the exclusion of the poor, another with the need for national loyalty. More importantly, one group may stress political integration while another might stress individual rights or the need for change.

Such a counsel of pessimism and partisanship is relevant to some political indicators that have been widely proposed. Participation, for example, appears at present to be an indisputable good; but as recently as the 1950s a dominant school of political sociology feared the undisciplined participation of the masses, giving the rise of Hitler as a negative example of such participation. Not the simple fact of participation, but its quality, was at stake. Similarly, alienation may seem undesirable; but insofar as the usage of this concept implies that the fault might lie with the object (government) from which citizens are alienated, this usage involves an implicit and controversial political judgment. Indicators of the integration of political communities should thus be recognized as explicit components of political debate, and not masked as science or fact or even linked to consensual values.

Within this framework of controversy it might still be appropriate to measure attachment to the political community, the regime, the authorities (Easton 1965: Chaps. 11–13). We most often measure only the last two of these, in the form of support for major governmental institutions such as the presidency, Congress, the judiciary, and their incumbents. Such statistics (gathered nongovernmentally) are presented regularly (Lipset and Schneider 1983: 48–49), but interpreted differently by those who wish to restore confidence and those who wish major change. Other aspects of general supportive attitudes are conformity with obligations to the system, such as obedience to the laws.

In contrast, I would consider those indexes whose meaning can be altered by the quality of involvement—participation, allegiance to groups and parties—to depend for their validity on conditions that need to be specified. They do not adequately measure integration (let alone well-being) unless we specify more as to the quality of involvement and the relation of such involvement to the entire polity. Possibly after further statement of these conditions the indexes in this domain will be more soundly based.

Finally I suggest as partisan indexes measures of the degree of responsiveness versus deliberation, intensity of conflict, and the distribution of governmental outputs. Conceivably opposed parties might agree on measures of these things, but they might also propose different indexes and argue about their meanings.

Conclusion

We have proposed a general direction for development of indicators of social integration, with parallels between indicators for families, organizations, and social and political communities. These proposals are summarized in Table 8–2. For each type of subsystem we have proposed three major types of indicators of integration: (1) measures of individuals' attachment to each other and to the system, including behavioral data on extreme instances of conflict or negative attachment; (2) proportion of time spent in the system; and (3) turnover or leaving rates. To some extent, parallel indicators are possible in the four types of systems. Series of this type may allow us to see whether the various types of integration vary together, or whether one type can vary to compensate for changes in another.

Indicators of type (3) resemble some of the categories used in proposed systems of social accounting (Land and McMillen 1981). They deal with the movement of individuals into and through particular social statuses, that is, membership in the four types of subsystems considered. In addition, indicators of type (2) consider the degree of involvement of a person in that system, over and above formal membership. This feature could conceivably also be included in a social accounting system; but at present such a system seems far from practicable. Such comprehensive data systems do not seem to be prerequisites for the indexes mentioned.

I have discussed policy models for value variables (Chaps. 5–7). For system integration, I shall mention some such models in order to show that this is a manipulable variable that can be affected by policies. Since system integration is not a value variable, however, we face particular difficulties in suggesting policies to affect it; we may wish under some conditions to increase integration and under others to decrease it. I shall speak, on the whole, about how it may be increased; but this selection may reflect a judgment about the present situation, unsupported either by data presented or by the reader's values. Moreover, the appropriate directions for policy may vary depending on the type of subsystem we consider. There are advantages in considering disparate subsystems in comparable ways, as we have done in this chapter, but

Table 8–2. Types of Indicators of Social Integration

		Type of Social System			
	Type of indicator	Family & Household	Productive Organization	Social Community	Political Community
(1)	Surveys of attachment	Spouses, adult-child, siblings attachment	Peer and hierarchical cohesion	Support for community symbols	Support for major governmental symbols, institutions; intergroup sociometry
	Behavioral data on positive or negative extremes of attachment	Spouse abuse, child abuse	Strikes	Charity, crime; isolation of stigmatized persons	Compliance with public obligations; internal political violence
(2)	Proportion of time spent in system	Time budgets: time with spouse, children	Unpaid overtime, absenteeism	Time spent in community or with community members	Time and resources devoted to political participation
(3)	Turnover rates	Divorce, separation statistics by duration of marriage	Voluntary leaving rates by length of employment	Out-migration rates by length of residence	Out-migration rates

these similarities do not guarantee that similar policies are appropriate for all of them.

Consider first what policies—in a broad sense—might be adopted to increase family integration. We cannot take this as an unqualified value, if only because children as they mature are expected to leave the family residence; and even if they maintain attachment to their parents and siblings, they will become less available to aid and be aided by them. Recent court decisions have increased their independence. An additional limitation on any external intervention in family affairs comes from the protected status of the family:

government does not have the right to peer into the workings of the family or to modify such workings directly without an important reason, such as serious harm done to one member by another.

Governmental policies can, however, furnish or subsidize services that families can choose voluntarily: family counseling, family planning, child care, assistance in caring for elderly dependents. A nongovernmental organization (the Catholic Church) can require of members who plan to be married in the Church that they provide information and obtain counseling in advance. Public school courses in marriage and family life, or sex education, can contribute to informed choices in relations leading up to marriage and family membership.

The social integration of the firm is also protected from government intervention, but in a different way. Government cannot enter the firm directly and interfere with its management in the name of integration, again except in cases of inequity or of harm such as dangerous working conditions. Interventions of this sort might be opposed not only by management, but also by unions if aimed at harmony in the firm at the expense of worker organization and activity. Management itself may try to increase workers' integration into the firm, especially if such integration furthers management goals; among the internal policies that may increase morale are adequate pay, just resolution of conflicts, satisfactory working conditions, reorganization of tasks and decisions, and communication of the significance of the job in relation to the product and its value to consumers.

In considering units with their own governments—communities and nation-states—we may more readily consider policies to increase these units' internal integration. One, frequently adopted, is to stress unity against a common enemy; but this simply suggests that conflict in a larger system is the price of integration in a smaller. Unity can also be achieved in the face of a common problem—for recovery from a catastrophe or putting a man on the moon—but in the case of social problems, some segment of the community will more often be called on to sacrifice while another benefits.

Another means of promoting unity is the celebration of shared values through ceremonies. A community street fair, a Fourth of July fireworks display, or the Easter service of a church in which community members share membership, can have this effect. More generally, the building and maintenance of institutions (Janowitz 1978: 399–406) can combine particular occasions and interchanges in a longer time perspective and contribute to community integration.

The integration of members into a social system depends in part on what they receive from it. Thus what at first is sheer individualistic or economic reward may eventually come to foster allegiance to a particular seller, or to a firm, or to a governmental unit. For a governmental unit, it is all the more important that policies and leaders' acts provide symbolic rewards to citizens;

the appearance of effectiveness, of progress toward common goals, and of bases of identification between leaders and the public, can all foster integration in the short run. For this effect to persist, however, the promises of the short run need to be fulfilled by effective governmental action in the longer run (Lipset 1960: 77–83).

In the very creation of political communities there are opportunities to foster integration. These have been studied with respect to the new nations created in the post-World War II period as colonies gained their independence. The ethnic or religious homogeneity of the population included within a government's boundaries can affect a new state's integration. Leadership that responds to the people's expectations—without creating unrealizable expectations—can also contribute to integration.

When the principle of population homogeneity is applied to communities and urban areas, however, it can favor one group of citizens while excluding another. Racial and ethnic integration may reduce the political or social integration that previously existed in an area in the process of allowing new groups to enter previously homogeneous communities so as to realize those groups' rights as citizens. The integration of the nation then takes precedence to some extent over the integration of homogeneous subsystems. Neither the exclusion of new groups seeking entrance, nor the expulsion of citizens already present, can be justified in the name of social integration when overriding values of a larger political community are at stake.

The building of communities, as of institutions, requires resources and skill; thus the presence of persons who are perceptive and dedicated to this social construction is important. Thus, if policies encourage the movement of such persons from one system to another, the result may not always be beneficial; J. Wilson (1975: 38) has noted such costs in the migration of middle-class blacks to the suburbs.

In summary, social integration is a significant general concept encompassing a number of variables that have appeared in indicator systems. Time series of these variables are of interest to the public even though they may be given diverse interpretations. These varied interpretations result in considerable measure from popular models of causation, and in part from the fact that some consider integration to be valuable for its own sake. Public policies may therefore sometimes be based on such indicators; but insofar as social science is called on to justify them, it must proceed cautiously. In spite of the prevalence of these measures, the models that connect social integration with well-being largely remain to be developed.

Part IV.
The Implementation of Policy Indicators

9 The Political Context: System Needs, Biases, and Users

The choice of policy indicators involves an interplay among values, causal models, and methods of measurement. Permeating this process of choice, however, are many practical considerations about which we have said little so far. The effectiveness with which these choices are made, carried out, and used depends on the political system—its needs, capacities, and resistances—and on the organization of the expert communities that are involved. In this final Part we must consider some of these more practical problems—examining first some political conditions affecting the choice and measurement of policy indicators (Chap. 9), and then the desirable organization of scientific and technical communities (Chap. 10). We summarize the problems of choosing policy indicators in Chapter 11.

The choice of policy indicators is a political question and not simply a scientific one. It depends on the capacities and needs of the political system for which they are chosen. In any such system, moreover, if indicators are to be related to values and policies there are significant political stakes in their choice (Morss and Rich 1980: Chap. 2). To declare a particular problem to be important by measuring its manifestations over time, or to declare another unimportant by discontinuing a series, can affect many careers as well as the interests of groups and organizations. Similar stakes are affected by the choice of a particular measure for a given concept.

The introduction of a new indicator series, to be prepared regularly by a government agency and used in a democratic political community, can be slow and difficult. A free society, in which citizens can organize politically, allows opposition as well as support to develop. Citizens have the right to question the priority to be given to new issues, or the political authority that an expert group may claim. The needed support from both experts and interested groups must be developed over time. Different groups of experts may propose rival concepts. Even the development of consensus within a single expert group on choice of specific statistical measures requires time. Not only the decision to establish the series, but also its implementation, encounters similar problems. Conceivably this process can be speeded by use of the prior experience of other similar political communities, but this possibility needs to be demonstrated.

Series must be terminated as well as begun. The social indicator movement, like the previous producers of indicator statistics, has stressed continuity and duration as essential qualities of indicator series. This has been a needed correction to a widespread emphasis in social research on cross-sectional studies that lack time perspective and thus provide less valid assessment of causation. Short-term as well as long-term series are needed for public decisions, however; these series must vary not only in intervals of data collection but also in overall duration. The very possibility of gradual change in the fundamental values of society can also require that policy indicator series be changed. We must thus examine some of the political and social mechanisms that affect changes in indicator series.

Politics is also involved in the dissemination of statistics. We have defined indicators and public statistics in a broadly inclusive way, in terms of statistics available to the public. Policy indicators are used in both governmental and nongovernmental statistics. Our earlier suggestions about the choice of indicator concepts, if they are to be put into effect, require us to consider a variety of ways of making statistics "available." There are many publics, varying in education and interests, to whom this information may or may not be effectively available even when it is in the public domain. The necessary decisions about the availability of information involve not only its content but also what organizations (financed from what sources) should collect information and how it should be disseminated, what incentives should be provided for research on related policy models, and how its results should be made available. Some of these decisions are policy choices by governmental units, but others are made by private organizations and individuals—research organizations and the governmental officials who seek their help, foundations, universities, professional and disciplinary associations, and individual professionals and scholars. Although these diverse decisions are interrelated, we shall emphasize decisions by government in this chapter and by expert communities in the next.

The ultimate test of whether we have done well in designing such information systems must rest on an assessment of the decisions that have made use of them. We would like to be able to monitor this use, perhaps by meta-indicators that tell whether a policy information system is serving its purposes. An assessment checklist for statistical programs, including usefulness and feasibility, has been proposed (President's Commission 1971: I, 190–94), yet assessment in terms of outcomes is extremely difficult (Downs 1967: 204–06). Studies of utilization of policy-related information, even using rather inclusive definitions of utilization, have often yielded disappointing results for both social indicators and applied research.

We must therefore distinguish among the various types of political systems that use information, and the different conditions they impose on it. We must consider the specific sources of pressure that can alter the measurement and meaning of indicators, as well as the general political pressures that influence the choice of variables and measures used. We must examine the roles and institutions in society that can facilitate the use of indicator information, including public leadership, the educational system, and the media. We must also consider the possible shifts of power that information can bring, both as sources of political motives and as advantages or disadvantages of information systems. These questions will be our concern in the remainder of this chapter.

Differences among Political Systems

In designing an information system for collective decisions, we must first identify the political system or political community[1] to be served by decisions based on it. Then we must identify the decision centers within that system that affect the generation and use of information—e.g., the legislature, the executive, or the courts. Sometimes decision centers in two or more levels of government are involved, as when census data are generated nationally but adapted for local purposes.

Before entering the political system (if we are outsiders), we must recognize that political systems vary greatly, both in the information they need and can use, and in the influences that they bring to bear on information systems. They differ, for example, in size, centralization, and capacity to use technical information. These differences affect the initiation of statistical series for particular indicators, their maintenance and continuing quality, their use, and their termination.

We must be alert to the different information needs of different systems. A national indicator system cannot be transferred to a locality by a simple reduction in scale; nor can the management information system of a bureaucracy be copied unthinkingly by a democratic community. These differences are due in part to the benefits and costs—outcome values—that we have stressed throughout; but they also reflect what is politically feasible.

1. For greatest generality we should speak of a "political system" and of the "affected population" to be served; such a usage would apply to hierarchical organizations whose decisions affect nonparticipants and nonmembers, and to decision-making systems whose populations are deeply divided and feel little sense of community (such as the United Nations). We treat organizations here because of the work done on management information systems. By using the term political community, however, we wish to suggest an aspect of a functioning democratic order by implying that most of those affected are participants. The term is defined by Easton (1965: 177–79), but not with this implication.

A political community that needs statistical information about itself is likely to have at least a certain minimum size; a small community whose members all know one another well has relatively little need for statistics on its own well-being (Clark 1973: 17). The citizens of a large nation, however, cannot sense one another's well-being so directly; they must more often supplement their personal knowledge of one another by indirect sources of information such as statistics.

Political systems also differ in their members' capacity to generate, understand, and use statistics. Prospective types of information such as new indicator series, data banks, and management information systems do not exist in isolation; rather, they add to a preexisting body of knowledge, information, and experience (Lindblom and Cohen 1979). The value realized from the use of new information sources is therefore an incremental value, depending not only on what is added but also on its relation to the old in intelligibility and compatibility. Indeed, in all human information processing "perception is guided by mental structures existing prior to the act" (Dunn 1974: 32). This relation requires special attention when there is a cultural disparity between the source of new information and the intended user, as often occurs when members of one culture provide information to another in order to aid national or local development.

Preexisting information can aid members of a political community to understand and interpret statistics. If preexisting capacities are concentrated among elites, the added information can accentuate their dominance; in the planning of statistical systems we must consider such effects. Prior information can also enable citizens to act more independently of official statistics and to criticize them. Some such countervailing information can come from expert groups or cultural elites other than those proposing to add new statistics. Thus natural scientists can occupy the status of laymen relative to economists (Koopmans 1979). Lawyers have acted as critics of the work of policy analysts and statisticians by making claims to rigorous thought and language without claiming scientific competence as such. Literature and the arts also interpret human conditions. These alternative perspectives can affect the design of statistical systems through public debate.

If a political community is to use indicators, those in it who produce and use the statistics must have some degree of skill and objectivity, and its citizens and leaders must provide support for statistical quality. When indicator statistics lead to embarrassing questions and controversial conclusions, it might sometimes seem easier to suppress them. If we believe in knowing the truth about ourselves, we must be prepared to tolerate and even support the messenger who brings bad tidings. Not every political community is willing to do this; but if we could ascertain this willingness before designing indicator systems, we might be more selective in our choice of communities to serve.

Stability is also a necessary condition for the collection and use of indicator statistics. Continuing data series require continuing political support; in a pluralistic democracy with diverse possible majorities, this means a broad consensus. Zapf (1972: 249) has thus characterized the political orientation of the social indicators movement as "liberal incrementalism." Insofar as indicator systems merely define problems in terms of small departures from a previous situation, they can involve the bias that problems and their remedies are seen as temporary and the previous situation as essentially satisfactory (Sharpe 1978: 309–10). But an incremental approach to the use of indicators, presuming some stability in the framework for decision, can be employed in socialist systems as well (Osipov and Andreenkov 1979).

In designing information systems for democratic political communities we can gain perspective by considering types of political systems other than national government that use information. Two important types of systems for comparison are bureaucracies and local governments. Much work has been devoted to the development of management information systems for organizations, including the use of modern information technology. It may be easier to design an information system for an organization than for a political community, especially if the organization is hierarchical and smaller in size. Goals may be better defined, channels of communication more determinate, and information needs easier to ascertain. We do not wish to say, of course, that democratic political communities should be made more hierarchical to serve the needs of information systems; on the contrary, we must guard against this possibility.[2] Nevertheless, the system designer can often benefit by studying simpler problems before turning to more complex ones.

Two examples will illustrate possible contributions from the literature on management information systems. Yugoslav organizations with workers' self-management encourage a higher degree of worker participation in design and use of information systems than is customary in hierarchical organizations (Rajković 1980). This participatory approach can be compared with possible developments in community information systems. Secondly, information systems can evolve over time. Starbuck (1975) notes that organizations considering change and development of their information systems can find it inefficient to change them in discrete steps; rather, processes permitting gradual evolution may lead to higher overall efficiency of the systems over time. These problems of participation and evolution are vital to policy indicator systems for political communities as well as for bureaucracies.

Local governments provide another important contrast to national indicator systems; we shall treat them in more detail in the next section. They differ from national systems in the indicator variables they use, as well as in their

2. Emery (1975) suggests, however, that organizations may modify their structure to make more effective use of integrated information systems. Democratic governments include bureaucratic agencies with internal management information systems (Hunt 1971: 450–54).

goals (Bradburn 1973: 26); studying them may provide perspective on the information needs of both types of systems. We might expect design to be easier at the municipal level because of smaller size and greater consensus; but larger systems are likely to provide more resources, since benefits are presumably proportional to numbers of persons affected while costs rise less rapidly with size. The skilled personnel necessary for preparation and use of information may also be more available in larger systems.

National, local, and organizational decision systems can be distinguished in terms of two dimensions: degree of participation and size. We may express these dimensions in a fourfold table such as Table 9–1. The four cells in this table simplify a two-dimensional space in which the reader may wish to add further examples; the examples shown may not fully represent the corresponding cells, but are important in a practical sense. So far we have mentioned the local community, the bureaucratic organization, and the democratic state. The cell we have not yet discussed is that on the lower right, designating a large system with a low level of participation. We give as an example the nondemocratic state; the problems in design of information systems for such states may teach us lessons for other systems. Nondemocratic states differ from bureaucracies in other ways than size, however; they control the lives of their populations far more than an organization that is merely an employer. Aspects other than participation may also distinguish the nondemocratic state from the democratic: the role of the private sector in information systems may be less in a nondemocratic state, depending on whether the state controls the economy; and the entrepreneurial and competitive role of the media may be absent. Another example that would belong in this cell is a relatively autonomous governmental organization affecting a large political system, such as the Federal Reserve Board in its use of economic statistics.

Participation requires that information systems be widely intelligible if they are to be a part of broad public judgments of policies rather than an obstacle to public understanding. The designer of indicator statistics, or the political system itself, must also ensure that information is thoroughly disseminated—genuinely made available, with little cost to the user for obtaining and understanding it—to diverse population groups. In addition, a more democratic system must find ways to engage the public in the design process itself; such engagement may take the form of periodic protest against an indicator or its use (de Neufville 1975: 80–83), and it is important that such issues be debated.

Larger political systems are likely to have more complex information needs; but these needs depend on the degree of centralization or decentralization of functions. In the United States, educational statistics have a special relevance to the state and local rather than the national level. Because the size of a political community affects its degree and type of participation, the two dimensions are not independent; large national democratic systems tend to have

Table 9–1. Typology of Information-Using Political Systems

		Degree of Participation	
		High (Participatory)	Low (Hierarchical)
Size	Small	Local community	Bureaucratic organization
	Large	Democratic state	Nondemocratic state

well-developed systems of interest groups (Peterson 1981: Chap. 6). These groups, together with political parties, use and transform information for the mobilization of voters; the information is thus often interpreted through symbols such as those of injustice, suffering, discontent, and blame. The need for political mobilization may accentuate the temporal fluctuation of public concern with particular problems. Downs (1972) has noted a repetitive rise and fall of attention to issues; this fluctuation may result from the difficulty of sustaining mobilization, or the temporary character of the events (such as elections or perceived crises) that call forth that mobilization.

Decisions made outside of government are also important for the shaping of information systems. The market makes some of these decisions, as some information needed for public decisions is sold by "information brokers" (Alonso and Starr 1982). Scientific and technical communities, through their decentralized decisions, choose lines of research that generate information, and theories and conceptual priorities that shape the terms of their advising and contracting for government. Nongovernmental initiatives sometimes pave the way for governmental activity, as in the case of the planning for the National Assessment of Educational Progress (Bradburn 1973: 27). Decision processes of this kind can exist within political communities that leave some independence to the private sector, but are more difficult within work organizations or in highly centralized states that control the economy. Even in these latter cases, however, informal expert or technical groups can propose decisions.

Policy Indicators in Local Government

The social indicators movement has been largely concerned with indicators of national conditions. Its emphasis on modeling social change rather than on policy has led to a concern with a relatively closed system—the national society—just as economic indicators deal with the national economy. Numerous policy-related indicators have been developed and used, however, in state and local government, and a substantial literature deals with them. We can compare such local indicators with national policy indicators, both to show that different systems have different needs and resources, and to illustrate some of the limits on implementation of information systems. We shall deal particularly with urban information systems, which have been studied in some detail.

The field of urban information systems "has weak identity" because such a system is "an amalgam of different professional fields" (Kraemer and King 1977: 80). Much of its literature has been technical, dealing with computer applications, and optimistic. Like the early literature on social indicators, this literature promises that its innovations will do much to improve public decisions. Downs (1967) warned early, however, that substantive payoffs were hard to assess and that power shifts might be more important determinants of implementation. Subsequent studies of actual efforts at implementation of computer-based municipal information systems have also shown that this approach was oversold (Danziger 1977; Kraemer 1977). Moreover, even when the federal government has tried to aid cities in diffusion of these systems, it has achieved only limited success (Kraemer and Perry 1979). Thus a technical or programmatic approach to information systems (including that of our preceding chapters) has great need of empirical observation of the problems of implementing and using them.

The Urban Institute has carried out a substantial program of design of municipal indicators, related to measurement and use but less to computer systems. This program has taken two approaches, one dealing with the measurement of productivity (Hatry and Fisk 1972), and another with that of citizen satisfaction (Webb and Hatry 1973). The former, dealing with objective measures of inputs, outputs, and costs, has been disseminated to a number of larger cities (Hatry 1978). The latter, dealing with subjective indicators of satisfaction, has been used regularly in only a handful of cities. Some questions have been raised as to whether responses concerning satisfaction with a particular service really reflect service delivery, but with judicious use by politically sensitive city managers such surveys can guide policy changes (Stipak 1979, 1980).

There are also municipal information and indicator systems in place that are less ambitious in developing new data or computerizing them, and that deal with somewhat different variables from those proposed for national social indicator systems. They deal with spatial distributions of persons and facilities in considerable detail. They cover functions that are performed at the local level because of diseconomies of larger scale, such as transportation, removal of refuse and garbage, fire protection, and water supply; and functions decentralized by law, such as education and policing. They include information about sources of revenue and types of expenditure,[3] which we did not stress in Chapter 1 or 5. Measurement of the economic production within a local community, an important source of revenue and employment, is aided by adaptation of national statistics.

Municipal indicators can differ from national either because of the different needs of the two sorts of systems, and the corresponding benefits from information, or because of disparities in resources in relation to the costs of generating information. On the cost side, communities lack resources and are thus dependent on the Census and other national sources for information (Mindlin and Levy 1980). Local governments also draw on administrative records of their agencies for data; new surveys, unconnected with the tasks of existing agencies, can be difficult to develop.

Some of these differences may be apparent rather than real, because at the national level there are policy indicators that are not classified as social indicators or economic indicators. There are, of course, national statistics on taxes and expenditures, and they are important for the federal government's definition of problems and choice of policies. These are not economic indicators (measures of the functioning of the economy) except insofar as government is a secondary contributor, along with the private sector, to macroeconomic conditions; however, they are important contributory variables that are used for policy purposes.

Some of the differences may result from the scarcity of skilled personnel for generating and interpreting data on the local level, relative to the national; this scarcity exists for policy analysis as well (Szanton 1981). Private initiatives from local universities, however, can sometimes provide indicators to citizens or stimulate interest on the part of local citizens and government. The decennial *Local Community Fact Books* provided for Chicago by sociologists at the University of Chicago from 1930 through 1960, based largely on census data, were funded in part by local users and were in considerable demand (Choldin 1980: 269–70). A computerized knowledge base system was developed to guide school-closing policy in Birmingham, Michigan, through the initiative of a Wayne State University faculty member who began developing the system in a course on large-scale systems analysis (Erlandson 1981).

The major remaining differences between the concepts used in municipal and national indicator systems seem to lie in the governmental functions considered, the local emphasis on input measures (services rendered), and the

3. A list of major municipal priorities and concerns, which approximates a list of potential indicator topics, is given in the chapter titles of Brecher and Horton (1982): the setting—population and economy; securing and managing resources—state aid, taxes, economic development, financing, and labor relations; delivering services—police, fire, sanitation, health, education, income maintenance, and mass transit. Clark (1973) also emphasizes fiscal indicators.

spatial orientation of local government. All these differences can be related to the fact that municipal governments are at the lowest level in a federal hierarchy.

Functions such as sanitation, transportation, and fire protection are carried out at the local level because they are local collective goods; they provide economies of scale or externalities at the local level but not beyond a certain scale. Some such local functions overlap with national concerns; health, for example, is a subject of national indicators even though hospitals are local. Health is more a concern of national policy than is education, and there is no counterpart for health to the strength of local school boards in policymaking.[4] Health, transportation, and safety are general indicator concepts, but they have specific local aspects.

Local programs take distinctive forms in different localities. A general governmental function that is of interest at the national level will often be delegated to states and localities, which perform it in somewhat different ways. This decentralization can lead to diverse measures of performance, which can be standardized only with difficulty; a local measure of crime, for example, can be affected by the degree of professionalization of the police (Biderman 1966a: 125–26).

National standardization of local policy indicators, to the extent that it can be accomplished, thus has costs as well as benefits. Among its advantages is the more convenient use of data and technical skills available from the national level. A standardized set of statistics on local programs in a given social goal area might permit comparison among localities in terms of performance and productivity (Clark 1973: 19). Nevertheless, because somewhat different things are being done in different places, local diversity is concealed by such standardized measurements. For this reason Scott et al. (1973), in proposing international indicators of local development, leave open the possibility of distinctive noncomparable local indicators.[5] Because of the better fit that may exist between special local indicators and local citizens' perceptions of what they are getting for their money, local specificity may have advantages. For this reason, input statistics may also have a special significance at the local level; citizens may have a better feeling for their meaning and a greater capacity to act on them by holding local officials accountable.

The special concern of local governments with spatial data—comparing conditions in local neighborhoods (Simutis 1980)—and spatial planning results in part from the fact that they have control of that space to a greater

4. Although federal control of public education is strongly resisted, states have developed indicators in the form of uniform statewide achievement tests. The minimum competency testing movement, though primarily producing tests of individual students, has also produced aggregate indicator statistics.

5. Argyris (1980: 18) has noted a similar difference between management information systems depending on whether they deal with local questions or more distant ones.

degree than do higher levels of government.[6] The United States federal government is limited in planning national spatial development because of the interposition of the states as power centers. This limitation combines with a widespread opposition to national planning to make the federal government's inefficacy in this area even more pronounced. States have somewhat greater power over their municipalities but still must deal with them, especially if they are politically influential; a state dominated by a single large city, for example (like New York or Illinois), cannot plan for that city unless the city is in great need of resources. Even small towns have some degree of autonomy from the states. But within the municipality, no smaller governmental units exist.[7] Thus zoning, transportation planning, property taxation, and other functions require continual attention to spatial distributions of indicators. The increasing concern with equitable distribution of municipal services has led to further indicators of spatial distribution, and comparison of educational achievement among community areas has added more. Some such comparisons are made nationally in the census and in reports of federal expenditure, but the policies for which they are relevant involve distribution of resources, not direct control of space.

Local governments' control of space suggests that they may be unusually powerful; but this power is balanced by their special susceptibility to elite influence (McConnell 1966). Their orientation toward the economy as merely part of their "setting" (Brecher and Horton 1982) reflects the fact that, unlike the national government, they cannot pretend to manage the economy; like smaller nations, they are at the mercy of larger trends.[8] Economic development, from the viewpoint of a state or local government, often means seeking to attract industry from elsewhere rather than increasing the national product. This competition subjects them further to the wishes of persons and industries controlling mobile resources (Peterson 1981), who thus have an additional "vote" beyond what they possess as citizens.

Finally, we must compare local information systems not simply with actual national systems but with our own recommendations. We have illustrated the importance of specific functions performed in the localities, and of costs and corresponding economies resulting from the use of census data and local agencies' administrative records. We may well ask whether these determining

6. A more centralized national government such as that of France, in contrast, can engage in regional spatial planning. In the United States, geocoding at the federal level can facilitate local use of spatial information (Barraclough 1971).

7. The Swedish municipal reform of 1962 actually consolidated many small communities throughout the country and changed their boundaries—a degree of change unimaginable in the United States (Niemi 1966). Neighborhoods in the United States can also be mobilized on issues that affect them.

8. Angrist et al. (1976: 194) point out that "proper distinction between the controllable and uncontrollable factors at the municipal level is critical to an effective linkage between urban policy analysis and social indicators."

factors leave room for the types of general indicators we have recommended in Part III. Benefit-cost analysis is performed for local projects, especially if federal funds are involved; equity is increasingly measured and these measures can aid conflict resolution; but subjective well-being and community integration have not been high priorities. Satisfaction with community services may come to be measured more often;[9] and if so, municipal experience may provide guidance for state and national use.

Indicators of subjective well-being and community integration seem, for the present, to require development apart from immediate governmental decision needs. The time may come when the public will be concerned with these variables and will envision policies aimed to change them; but at best such indicators are in their infancy, as they have not yet been widely related to major problems of policy and administration.

Measurement and the Reactions of Information Producers

Within a political community with particular needs and a particular decision structure, we must consider the conditions affecting feasibility of specific proposals for data collection and presentation. The technical standards for data collection and presentation are well known to applied statisticians and to the relevant disciplines and professions. There are, however, important differences between measurement of policy indicators and measurement in basic research. The choice of policy indicators and of the concepts behind them does not rest entirely on disciplinary theory; and the very prospect of practical use of information can lead to bias.

The choice of indicator concepts must combine technical and policy considerations in an iterative process (de Neufville 1978: 182). The detailed specification of measures also requires successive interactions between experts and the public so that measurement will permit problem definition in terms of widely held values; this sort of interaction has occurred for the measurement of unemployment (de Neufville 1975: 203–06). Although models and theory cannot be the sole criterion for choice of measures, they can play an important part in linking measures with concepts and in clarifying the meaning of both as they are debated. The stakes in choice of measures are significant because "the particular formulation of an indicator [statistic] can affect the likelihood of the adoption of a policy, . . . the probability that a specific agency will take responsibility for it, and the composition of the group which will benefit" (de Neufville 1978: 176–77, 179, 181–82).

9. The support of providers of services seems to be required for the use of such indicators of satisfaction.

In the process of policy indicator design and use, the statistics produced can be biased by political influences at several stages. They can initially be influenced by reactions of their producers, such as respondents and organizations. These reactions, while not necessarily working through governmental decisions, can affect the questions asked, the responses to them, and the subsequent reports; we consider them political in a broad sense involving the calculated counteraction of the intentions of others. In addition to producers, the groups and organizations potentially affected can exert pressure on the use and interpretation of particular indicator statistics, as well as on their content and even their existence. These two sources of influence—producers and affected groups—overlap somewhat; but we shall deal in this section with the direct effects of individual producers in biasing the statistics, and in the next with collective political pressures from those affected. We face the task of systematic analysis of "bias-free record-keeping" (D. Campbell 1971: 29).

When we say that a statistical measurement procedure yields a biased result, we imply the existence of a true value from which the observed value departs systematically (Mosteller 1978: 208–09). Some such departures are attributable to a consistent human motivation, and we shall call this phenomenon *motivated bias*. Thus, if we wish an indicator statistic to measure the well-being of a population, and a true value can be specified, we shall consider the statistic motivationally biased if it departs from the true value because it reflects the wishes of producers or affected groups to modify effects on themselves.

Some examples of possible sources of such bias in indicators are: (1) the modification by the French government in 1957 (under Premier Guy Mollet) of its index of the cost of living (*échelle mobile*) in order to prevent automatic increases in government expenditures; (2) pressures on the administrators of the 1980 United States Census to find more nonenumerated persons so as to increase the population counts of local areas and their corresponding eligibility for federal aid;[10] (3) the effort by students taking college admission tests to improve their scores by attending cramming schools or by cheating; (4) decreases in response rate by survey respondents when they believe that their responses may be used to harm them.

The first two of these examples involve potential collective effects of indicators working through public policies. The third involves direct effects on students through their test scores; although average test scores constitute an indicator statistic, the biases are caused by the prospect of individual decisions on college admissions. The last example may involve either collective or individual effects; but survey respondents are probably more often concerned

10. Many of the protesters claimed, however, that they wished only to produce a true count. In such situations, the true value is usually ill defined; the issue is consensus as well as truth (de Neufville 1984), and must be resolved through a process of argumentation (Mitroff et al. 1983).

with the potential effects of their responses on themselves than with effects on public policies. The last two examples, because they can be independent of policy use, illustrate a type of bias that can affect measures in basic research as well (Webb et al. 1966: 13–23).

Two contrasting viewpoints can be taken concerning motivated bias in policy indicators. On one hand it can be ignored, as is often done when we pursue pure science and do not intend policy applications. On the other, it can be assumed to be always present, as is often done by economists in connection with questioning respondents on their preferences for policy purposes.[11] We propose an intermediate viewpoint: that possible bias of this sort, or the presence of reactive measures, is an intrinsic feature of indicator variables used for policy purposes; that this bias is often related to the understandable desire of those affected to alter the effects of using the information; and that the conditions under which this occurs need to be studied as a guide to measurement of any indicator variable to be used for policy guidance.

However, we must recognize that many responses are not falsified. The conventional assurance of anonymity given by survey interviewers persuades many respondents that their responses will not be used to harm them. Those American taxpayers who are asked to report their income each spring do not grossly distort all their responses, being restrained by conscience or by the fear of audits and penalties. There *are* social mechanisms for reducing bias; our task in policy indicator measurement is to find and use them.[12]

In survey interviewing we usually try to obtain sincere, spontaneous expressions of a respondent's attitudes, feelings, or perceptions. We thus try to be sure that members of the respondent's family are not present during a personal interview; we try not to reveal particular policy purposes that the sponsor has in mind (though this effort is modified by our concern for respondents' rights); and we phrase questions and the subsequent probes in terms that do not imply favor for one side or the other of a controversy. Such procedures may be especially appropriate for measuring subjective well-being.

The fact that attitudes do not always successfully predict behavior suggests, however, that for some purposes we might prefer to observe a response con-

11. Subjective indicators may be especially liable to strategic manipulation (Clark 1973: 5). Economists thus tend to trust market data on valuation of goods more than survey data related to policies, since in the market the consumer and producer can suffer direct consequences for any falsification of preferences. Samuelson (1948) showed that in principle, indifference curves could be reconstructed ("revealed") from observations of market behavior responding to given prices.

12. We do not wish, however, to place the powers of government behind the gathering of all sorts of information, and especially not those sorts that might be construed as invasions of privacy (Duncan 1981).

In our discussion we tend to assume that the conventional procedures of social statistics lead to valid measurement and that the pressures of data producers and users bias the measures. In connection with the use of U.S. Census data in formula grant allocation programs, however, Seidman (1980) suggests that legal accountability through the courts is one means to ensure data quality.

strained by the conditions of real life rather than one that is spontaneous, private, and unconstrained. Market behavior is constrained and often visible to others. Similarly the behavioral expression of group prejudice may be more useful as a predictor if it occurs when the person in question is observed by others and knows it. Legislative roll call votes, though cast in a situation of public pressure, give information about the political system that legislators' private responses might not provide. Thus privacy or spontaneity does not necessarily correspond to an unbiased measurement.

The motivations of real life can also sometimes lead to changes in behavior that make indicator statistics better predictors of behavior than they would otherwise have been, as illustrated by a "self-fulfilling prophecy" that can occur in planning. Ordinarily, if our announcement of a prediction affects the predicted events, we consider this to bias our measurement of those events. When we wish to induce participants in an economic system to channel their activities into a desired plan, however, if the announcement of the plan leads them to comply with it, this result would seem desirable. Thus in indicative planning (Meade 1970) individuals and firms are asked their intentions; the aggregate of their intentions is fed back to participants as both a prediction and an encouragement of the next step in the action of the economy; and the result is a greater coordination of future actions than might have occurred otherwise. The information given out is not strictly a prophecy; but in the case of French economic planning after World War II, announcement of the goals (together with incentives) was expected to lead people to conform to them.

We must thus consider in empirical detail the possible sources of motivated bias in policy indicators. These include effects on our data that can occur from the initial gathering of our data, through their aggregation, to their dissemination. For simplicity we consider primarily the publication of descriptive data, even though similar biases can occur in analysis. We can then begin to classify the motivational sources of bias in terms of the actors who have opportunities to alter the data,[13] and of the prospective uses of the data that may lead them to do so.

We first set aside the biases that result from inattention or from giving low priority to the collection of accurate data. Examples are the falsification of survey data by interviewers, who wish merely to earn their pay easily; and the misrecording of organizational data because of competing commitments of personnel, and low priority to data recording (Garfinkel 1967: Chap. 6).

We thus consider those opportunities for motivated bias of statistics by persons who contribute to producing them and expect to be affected by their use.

13. Our distinction between bias in production of data and political pressures on their use is blurred, however, when producers are users. A further overlap derives from the values and ideologies of professionals and other employees in data-producing organizations (Henriot 1970: 245–46). In addition, the realm of political calculation merges into that of unconscious motivations that nevertheless result in behavior that furthers one's interests.

Our emphasis is similar to that of economists who consider lump sum payments not to distort the functioning of the economy, but expect repetitive payments that motivate actions of the recipient to lead to distortion. The collection of data at a single time might, however, be similarly biased if announced in advance as usable for policy purposes.

The data used in indicator statistics may be collected in various ways. Some, like those of the census, are gathered by governmental organizations whose principal mission is the provision of information, and which ask questions of citizens. Others, like those of the national economic accounts, depend on reports by private firms. Some statistics of crime or disease depend on the administrative records of organizations that deal with these conditions. Data on voting, or on college admissions tests, derive directly from decision processes. In all of these procedures, however, it is useful to distinguish between the *indicator-producing organizations* that collect, combine, and present the data and the individual or organizational *respondents*.

All indicator statistics pass through *collecting organizations*. This may be the principal stage in the generation of the data, if respondents are not used (e.g., when facilities are observed or the environment studied). It may also follow or accompany the gathering of data from respondents.

One of the most common sources of bias occurs when the collecting organization has major goals other than the collection of data. Local police departments, for example, contribute data to the Uniform Crime Reports, but also report to their communities; if they are concerned with showing either what they have accomplished or the size of the remaining task, they may bias their reports (Biderman 1970: 219–20). This bias can occur without deliberate falsification but by use of certain judgmental categories or classifications. But falsification can occur; some developing nations are said to have reported to international agencies that they had eradicated smallpox when they had not; foreign investment depended on the data.

In general, if the collecting organization expects its budget or functions to be affected by the results, it can be expected either to refuse access or to seek to influence the data (Weiss 1972: 100–01). This sort of motivation need not, however, be directed at data "quality" in a technical sense; it can lead to choice of variables, questions, and modes of presentation, or to suppression of variables and findings, which we discuss in the next section.

Organizations that collect and report indicator statistics are susceptible to motivated biases related more to the aggregate results than to individual responses; but particular respondents (organizations, test takers, survey interviewees) are more concerned with effects that could flow from the disclosure and use of their responses. A large firm providing economic data to the census may be concerned with possible use of the data by competitors. Persons eligible for public housing on the basis of income, as well as taxpayers generally, may bias their income reports downward. Ex-convicts whose responses are sought in a followup study may be especially difficult to locate (Rossi 1980: 899) because they wish to maintain their reputations and their chances of employment. Persons engaged in illegal or disapproved acts may conceal them if they fear that the organization gathering the data (including the interviewer) may exert social control or sanctions on them. We ask respondents about victimization but not about crimes they have committed, judging that victimization will be acknowledged more truthfully. Even then, there are crimes that the victim is often reluctant to report, including rape, bank fraud by computer crime, and confidence games in which the victim appears greedy or foolish.

Respondents may bias indicators when rewards or regulation are tied to a specific measure of one aspect of a general concept but not to the concept as a whole; Etzioni and Lehman (1967) refer to this as "fractional measurement." The respondents may then in effect change the meaning of the indicator so that it comes to measure little more than its operational definition and not the broader initial concept—especially when the fraction measured is small. Thus the use of students' test results to reward teachers and schools can lead to "teaching to the test" and thus to distortion of the meaning of measurements (Coltham 1972). The control of automobile emissions of carbon monoxide and unburned hydrocarbons led manufacturers to raise engine flame temperatures in such a way that more oxides of nitrogen were emitted (Gell-Mann 1971). Basing allocation of federal aid on indicators of economic need might restrain cities from annexing affluent suburbs (Nathan and Adams 1976: 61).[14] If people are affected by an information system centered about a "fractional measure" they are likely to try to respond to that measure rather than to the concept. The meaning of researchers' models as well as of their measures may thus be changed. This change is part of the larger problem of "opposed-systems design" (Wohlstetter 1968) in which an opponent can be expected to try to change the causal or technological relations underlying a policy or strategy.

The above examples of bias refer to respondents' concern not only with public policies but also with particular administrative or private decisions that might affect a respondent individually. Even a general practice of asking questions on topics considered to be sensitive or to be invasions of privacy may lead to reduced response rates, apart from direct effects on the respondents of whom they are asked.

The notion of a respondent may be enlarged somewhat when an indicator is based on one person's reporting on another, as in the generation of administra-

14. This is an example of a generic tactic for circumventing regulations, which needs to be considered routinely in policy analysis. If a particular status or condition is defined as measuring need and being a condition for policy benefits, potential recipients may seek to move into or remain in that condition—whether it be poverty (for public housing or for nursing home eligibility) or lack of facilities in an area. Means tests try to seek out this behavior but can be demeaning to those who are not engaging in it.

tive records on patients or clients. In an organization, the recorder of such data may well be motivated by organizational goals like those we have just considered. Institutions that have mistreated their patients or inmates may wish to suppress this information. Managers in productive organizations may wish to conceal inefficiency or internal conflict from their superiors.

A nonorganizational example of bias in one person's report on another occurs in the use of footnote citation counts to measure scientific productivity. The footnote citations that scientists use to refer to others' work have been organized (according to the works cited), published in the *Science Citation Index*, and studied in relation to possible conditions for scientific quality and productivity. As a result of this research—giving rise at first only to models of scientific behavior—it has been proposed that citation statistics be used for allocation of research support to persons or organizations. Such a policy use would increase the risk, however, that scientific footnotes would be used to bolster the citation counts of the authors' friends. This practice might contaminate not only future policy uses of the data, but also the flow of scientific information itself.

One might ask whether phenomena comparable to this possible distortion of the *Science Citation Index* could occur in the market. That index, like market statistics, constitutes a form of "revealed preference" for works cited (although there is no clear price associated with a citation), and each citer can contribute only a small amount to the overall statistics. Nevertheless, systematic biases may occur. Similarly, we may imagine that market prices could be used for important benefit-cost analyses, and that producers or consumers might wish to influence these analyses. If there is a systematic interest in the result of the analyses—for example, in having a higher price attributed to some required product—conceivably the producers will try to raise that price in order to influence the analysis. In the controversy over deposit-bottle bills, manufacturers have pointed to increases in beverage prices following such legislation as evidence against this policy, and opponents have accused them of raising the prices for this purpose.

Our listing of sources of motivated bias suggests general remedies for them, including either (1) reducing biasing motivations, or (2) introducing corrective motivations that counteract them. We shall classify these remedies in terms of individual motivations, but it is also useful to think of them in institutional terms. As de Neufville (1975: 226) expresses it, "Our goal should be to insulate, but not isolate, the statistics from the immediate vagaries of day-to-day politics and to set up institutions which will permit change in indicators to occur in an orderly fashion, with public scrutiny and public assent, and in a way relevant to changing concerns and perceptions."

Reducing motivated bias. We may reduce biasing motivations, on the part of either the data-producing organizations or the individual respondents, by reducing the extent to which the data threaten to affect these sources.[15] The production of data, for example, can often be placed in the hands of organizations whose budgets or functioning do not depend on the findings. An example is the design of social experiments so as to place data generation under the direction of persons who expect to be rewarded for the quality of the data rather than for the accomplishments of the program (D. Campbell 1969: 427).

One way to induce individual sources to report information honestly is to persuade them that the data will not be used to harm them. If the possible biasing effects bear on individuals and flow directly from their responses, we may give assurances of anonymity; or with questions that are particularly sensitive, we may randomly substitute other questions for them for further protection of respondents.[16]

If the biasing effects work through collective decisions or policies, we may choose the sample to be studied from outside the political system affected by the prospective policy. Suppose, for example, that a community wishes to decide on the allocation of funds for public recreation facilities. It can make this choice through voting and representation, but it may wish more precise data on the extent (intensity) to which its members prefer one or another type of recreation. To ask directly for monetary estimates of willingness to pay may lead to distortion of responses; but if these data can be gathered in another similar community, they may be useful and less liable to self-interested bias. On the basis of research in that other community we might discover an association between demand and variables such as income, race, and sex, and then return to the initial community and make less biased estimates of demand on the basis of these findings.

A similar separation of the collection of data from their use is involved in the use of policy models. Data may be collected at a time, or in a place, where no specific policy applications are contemplated. They may be analyzed to reveal the general conditions under which a policy might be effective. Knowledge of these general conditions may then be used later, without the collection of additional data, to guide policy choices.

A model may thus be based on data gathered in the past or prior to the proposal of new policies. This approach may be used to circumvent bias in educational policy, for example. If we measure pupils' improvement in scores on standardized tests and reward teachers or schools according to this indicator, we may produce a finely-tuned motivational system aimed at improving the test scores rather than the more general sort of learning that such scores represent. If, however, we use existing test score records, which were not

15. We have tacitly assumed that biasing motivations result from self-interest, but conceivably they can result from efforts to serve the general welfare.

16. Tracy and Fox (1981) argue, however, that the gains from this randomization procedure are limited.

generated for policy purposes, to discover and model the conditions for effective teaching, we may then try to promote those conditions through subsequent policies. Under these conditions, we can have more confidence that the test scores are indicators of learning in a more general sense. We can thus hope that in promoting these conditions we will encourage all the other sorts of learning that the tests were intended to sample. Bias may emerge, however, at the stage of monitoring.

We cannot, of course, go so far as to claim that policy models eliminate bias; they are selective representations of reality, and they fix our attention on certain variables rather than others. One of the main claims of advocates of social indicators has, of course, been that economic indicators and their associated models are themselves biased in the values they serve.

Further means of reducing these biasing motivations are (1) the avoidance of "policy overload" on indicators (Seidman 1980) and (2) the use of multiple or changing indicators. The first of these methods involves limiting the number of policy effects that depend on a particular indicator, in order to limit the political pressures that might bias it. Direct use of statistics for allocating valued things usually brings more pressure than merely using them to define problems. The second method, used by the Educational Testing Service, can combine continuity of indexes with change of items, by using statistical linkages that permit comparison of old items with new ones. Even this degree of variation in items among indicators of a given concept may be insufficient, however, to prevent students from concentrating their learning on a particular type of question; thus a still greater degree of diversity among indicators of a given concept may be desirable. At the same time, proliferation of measures can decrease their intelligibility to the public. Efforts to improve the Consumer Price Index have led to the presence of three competing indexes yielding different monthly inflation rates (*National Journal* 1982).

Introducing corrective motivations. We can also introduce corrective motivations into indicator-generating systems. Some ways of doing this are:

1. Special rules for collective decisions. When policies are to be chosen through aggregation of individuals' statements of their expected benefits, and if the individuals are to be taxed according to their statements, they are motivated to understate benefits; if they are not taxed, they are motivated to exaggerate benefits. Tideman and Tullock (1976) summarize a growing literature on "demand-revealing" procedures for validly eliciting consumer monetary preferences for collective goods. By taxing each "swing voter" according to the net decrement in the welfare of others that his vote produces, they propose to motivate the voter not to exaggerate his values; at the same time, strategic reduction in his valuations might risk the defeat of his preferred policy alternative.

An analogous approach proposed by Hylland and Zeckhauser (1979) assumes that each individual has a fixed budget of total influence. It involves combining voters' declarations of their amounts of incremental preference among a number of alternatives, by aggregating the square roots of the amounts they allocate to the alternatives. This procedure can be shown to motivate "sincere" allocation by voters of their declared valuations.

2. Professional training of persons involved in data collection, presentation, and interpretation. Although codes of professional ethics are not guarantees of virtue, they can influence behavior to some degree toward maintaining data quality. This motivation deriving from professional training resembles that of academic scientists, but faces stronger challenges when professionals deal with laymen and decision makers. Statisticians working in organizations whose purposes can conflict with unbiased reporting may have to be specially careful about professional standards. As Keyfitz (1978: 419–20) puts it: "In a hundred minor decisions of statistical compilation, an operating agency may be swayed by noting how the result will come out. . . . The political head of the department that includes the statistical office must accept the self-discipline of permitting decisions regarding statistics to be made on statistical considerations, not because such decisions are always correct but because, when they are wrong, they are wrong in a disinterested way."

When one's professional reputation depends on publicly available products that are evaluated by peers, career interest may strengthen professional standards and result in improved quality of data and inferences. Record keepers employed in hospitals may thus strive for data quality beyond the minimum necessary for service delivery (Moses and Mosteller 1972: 15). Biostatisticians may prefer evidence from randomized experiments to evidence based on nonrandom comparison groups, even when laymen would not be aware of the superiority of randomization (Meier 1972: 8).

3. Creation of professional constituencies for public data collection agencies. In addition to training individual professionals, associations of such professionals can exert important influences on data quality. De Neufville (1975: 82) shows how economists and statisticians played an important part in supporting an indicator of unemployment when it came under attack on the ground that it was an improper measure. The involvement of professionals in decisions about public data collection can not only guide the introduction of new variables and the release of information to the public, but can also protect data quality against irrelevant political attacks. One of the strongest such constituencies is an organization that is itself devoted to professional standards; in this respect the primarily statistical agencies of government differ from operating agencies (President's Commission 1971: I, 47–48).

4. Development of support from competing political constituencies. When indicator statistics affect competing interests, they can sometimes play a part in resolving conflict by supplying information that helps to define the situation and is not biased toward either side. The Consumer Price Index has this function. When it came under criticism during World War II, a committee includ-

ing business and labor members dealt with the dispute (de Neufville 1975: 186). Alternatively, an intragovernmental constituency might be created in the form of a Central Statistical Office that would make decisions on priorities (President's Reorganization Project 1981: 147).

The last three of these remedies are relevant not only to motivated bias in measurement of policy indicators, but also to the choice and use of such indicators; they are aimed at both these problems because the motivations of those affected by the use of information can be directed equally at bias or at changing the list of variables themselves. We now turn to these larger influences on policy indicators, which go beyond questions of technical quality and bias.

Political Pressures and the Use and Choice of Indicators

When individual respondents fear that their responses will be used against them, they may try to bias their responses. Affected groups and organizations that are powerful enough, however, may try to change the overall procedure for measurement by working through the political system. Even if this does not affect the technical quality of the data, and unbiased and valid[17] measurement is expected, those affected may still try to influence the use of the measures as well as the very choice of concepts to be measured. These problems of use and choice overlap with those of bias, but they are more political and less technical than the questions of bias we have discussed so far.

We shall analyze these problems largely as though the pressures derive from material and organizational interests; but equally important problems arise from clashes of basic values and ideologies. General orientations toward government, the state, the good life, morality, and inequality may divide groups in their choice of indicators. These divisions may be the subjects of political conflict, in which indicators are only one among many battlegrounds; or they may sometimes be subject to reconciliation of values (Chap. 2). When a value conflict is resolved by appeal to a more general value, an indicator may aid in this resolution; but when the conflict persists, indicators may remain among the subjects of conflict.

We may introduce these problems of use and choice of information by contrasting some types of accusations of "misuse" that are commonly made. The most frequent misuse of information, as suggested by the literature on policy

17. "Validity" may initially refer to the representativeness of the items in the index or indicator as a sample of a larger domain; those affected, however, can orient themselves to the sample rather than the domain. In a sense their behavior then becomes "invalid," i.e., is not motivated toward improvement in the domain in general.

implementation and on knowledge utilization, is nonuse. The rhetoric of implementation therefore favors increasing utilization through greater consideration of users' needs, motives, and constraints. Conceivably some information deserves not to be used; but if we have carefully designed an information system for a political community, we are inclined to try to see that it is used. In this effort we can encounter lack of interest, and associate it with our failure to communicate effectively;[18] or opposition, resulting from the fear of important actors that the information will harm them. These factors can hinder dissemination of policy models as well: the 1966 Coleman report encountered reinterpretation in its executive summary, as well as reduced publicity at the start; Martinson's report on prisoners' rehabilitation was not published for several years (Chap. 4). Stronger means of suppression can also be taken, equivalent to a public choice not to measure or report on a concept; and if information is publicized nevertheless, opponents may try to discredit the source.

The converse outcome is excessive use—though there will always be disagreement as to whether this has occurred, as the users will rarely regret their own use of information. Critics of the media interpretation of the various Coleman reports and of Ehrlich's work on capital punishment (Chap. 4) felt that this information was utilized too soon and overinterpreted. The question of excessive use is in part a political one; the possibility that new statistics will stimulate or legitimate new problem definitions requires judgment by the public or its representatives.[19] Experts can contend, however, that statistics are not sufficiently accurate, or that models are insufficiently tested, to be released. When expert communities make such judgments, they must seek to do so in terms of standards of quality that do not reflect their members' political preferences. The public may sometimes judge, however, that imperfect data are better than none.

Since we expect that policy indicator statistics will be used to define problems, it is normal that they will be used and publicized. It is also normal that the introduction of new issues in this way will face political opposition. The development of new indicator variables and series thus often depends on a crisis or unusual state of affairs in which there is wider demand than usual for information. Publicity given to crimes, or the discovery of inadequacy in military draftees' abilities, can lead to calls for accurate and continuing measurement. De Neufville (1975: 71) notes that measurement of unemployment

18. Factors such as lack of interest, or information overload, may limit information use without involving political resistance.

19. From a conservative perspective one might argue that public concern should not be aroused over problems that cannot be remedied. An activist reply would be that public concern is necessary if we are to devote resources to seeking remedies. Indicators can also be used and misused in individual economic actions, as by citizens responding to crime statistics by "deserting their cities, streets, and parks" (Biderman 1966a: 114).

received an impetus in 1921: "The country was in a depression, and unemployment was obviously high, though there were just guesses about the numbers." President Harding thus called a conference of business leaders on the subject—a step not likely to arouse suspicions of radical agitation.

This normal function of indicator statistics is also brought out by Nakamura and Smallwood (1980: 52–53, citing Munger) when they note that those policies, such as economic ones, for which outputs are measured by such statistics are visible, encounter continuous group pressure, attract pressures for rapid action, and can create self-fulfilling expectations (as in the case of inflation). Indicator statistics can raise public consciousness of conditions and issues, and we can expect that when they suggest problems they will be publicized not only by the media but also by those who wish to criticize incumbent officials and hold them responsible. Conversely, incumbents will publicize statistics when they show improvement.[20]

Excessive uses of statistics (in some critics' judgment) can also result from publicizing information that, though well measured, does not illuminate public debate. Hennis (1957) has contended that public opinion polls have an undesirable effect on representative government because they publicize unreflective mass judgments and give them more weight than they should have. This criticism does not apply, of course, to all uses of surveys. Surveys of citizens' reports of their personal reactions to policies and administration, as they affect specific life conditions, have been used in Britain at least since the 1950s (as in the Government Social Survey directed by Louis Moss), following in the tradition of surveys of the life of the poor in the 1890s; results of such surveys can be used as bases of public debate and can be more informative than opinion data.

Publicizing a statistic can also direct attention to one part of a problem at the expense of another. As Wildavsky (1979: 31) notes: "Movement on any indicator can be maximized provided society is prepared to ignore all other indicators." As we have seen for respondents' bias, fractional measurement in operational definitions or measures used can similarly divert public attention from the broader original sense of the concept in question. One result may be emphasis on those things that we can measure, in contrast to equally important values that we cannot; this was one of the early arguments of the advocates of social indicators as a means to counterbalance the influence of economic indicators, but the problem exists for noneconomic concepts and mea-

20. We might imagine that problems of underuse and overuse could be dealt with by market mechanisms; but in policy uses of information there are extensive externalities due to efforts to publicize conditions. There are also systemic effects of problem definition, not limited to particular policy outcomes, that must be considered in a larger perspective: the integration of the body politic, and its capacity for choice and action, can be increased by conflict resolution and decreased by an overload of anxiety and disaffection. For those interested in system change, of course, these latter problems may be seen as temporary, to be faced only under the next regime.

sures as well. A familiar academic example is the stress on professors' and departments' national research reputations, which can be quantified, at the expense of teaching accomplishment, which is harder to estimate; this emphasis can distort debates about higher education.

Even more important than the misuse of indicators is the choice of indicators and indicator concepts themselves, which is also subject to political influence. The omission, modification, or inclusion of variables through the pressure of affected groups must be considered carefully if we are to propose indicator concepts and expect them to be measured and used. Our choice of policy indicators is affected not only by the values we wish to further, but also by political interests and by costs (Alonso and Starr 1982). Federal tax expenditures, for example, have often been used by Congress in preference to the more visible expenditures made through the regular budget.

Organizational interests and concerns like those we noted in connection with bias can also affect the choice of concepts and measures. Choices about the form and mode of dissemination of indicator statistics are constrained by the standard operating procedures of producer organizations (Dunn 1974: 141–43). The continuation of an indicator series is supported by institutionalized requirements that it be used in decisions (de Neufville 1984). Variables that might cause embarrassment may be deemphasized; for example, in health policy, a displacement of goals may be reflected in a change from the measurement of health to measurement of "access to medical care." Although much effort may be devoted to measuring access to medical care, the measurement of health itself may be neglected. Indeed, the question may then be raised whether we should not really seek to increase access to medical care simply because the public demands it. More generally, "any objective that cannot be attained will be replaced by one that can be approximated" (Wildavsky 1979: 285). The converse is also possible, however: an indicator that risks showing that an organization is "working itself out of a job" is also likely to be resisted. As Benveniste (1977: 44) points out, output measures produced by an organization fit the official rhetoric of the organization.

These examples illustrate a general sort of effect that can bear on organizations that collect and report data when they are also potential objects of policies. If an organization delivering services (such as a local police department) or a profession (such as that of medicine) is a constituency of either the collector or reporter of the data, it can be expected to exert the same kind of influence as if it were handling the data itself. The stage at which the influence is exercised may depend only on the foresight with which the results are anticipated; expected results can be avoided by eliminating or changing the questions asked, while unexpected results obtained from a previously approved study can be suppressed or reinterpreted. In one extreme case, undesired printed estimates of the farm population were simply burned (Rosenbaum 1964).

Other constituencies that can limit the gathering of national indicator data are the states and localities. When the National Assessment of Educational Progress first proposed to gather information on the educational attainments of school pupils, it met opposition from state and local leaders concerned about federal domination and about possible unfavorable comparisons among states and schools.[21] For this reason the organization began by gathering data so as to deemphasize "the performance of individuals, schools, school districts, or states," but instead emphasizing "the performance of groups" (Martin 1979: 45–46). Even when the states were not literally the providers of the data, they could limit access to data and resources for data collection. Later, when measures of change over time in individual schools were desired, they were difficult to obtain from the data.

The same motivations that distort the meaning of indicators when they are used to allocate specific rewards and penalties (as in admissions tests) can also lead to suppression of variables if those affected are powerful enough. Thus an incentive pay system based on the success of individual employment counselors, devised by the Rand Corporation for manpower training in California, was defeated by the pressure of employees' groups (Greenberg et al. 1976). Salamon (1981) has observed that although policies involving precise performance measures and incentives are easier to administer than policies without such provisions, they are also less likely to be adopted in the first place.

So far we have characterized political pressures as deriving from the expected policy uses of information; the gainers or losers from those uses are thus expected to favor or oppose the information itself. But the creation of an information system can also aid the groups who are likely to control its use, regardless of who gains or loses from particular information outputs. Thus there can be power stakes in the use of policy indicators as well as substantive stakes in the policies that may result.

The most basic power stakes lie in the relation between government and citizens. Even though information can be efficiently supplied by government as a collective good, government is not necessarily guided by efficiency alone. Official statistics can be used to serve private interests and the interests of officials themselves—to support their actions or to conceal their shortcomings, as Henriot (1970) has pointed out.

Government failure has been a major theme of political philosophy, and not merely an afterthought appended to a theory of market failure. Governments may fail not only in the substantive outcomes they produce in the short run,

21. Indicators can sometimes be used for conflict resolution in connection with the equitable allocation of resources; such indicators include taximeters, the Consumer Price Index, the populations of geographical units, and the distribution of public services among neighborhoods. If, however, an indicator statistic risks placing blame on some units for conditions beyond their control (e.g. low educational achievement) it is likely to be resisted.

but in the working of their decision processes or structures of power, affecting many substantive outcomes in the longer run. Among the various ways in which we can protect ourselves against such failures, four features of information systems and their settings are particularly important: (1) expert control and monitoring to insure their quality; (2) their intelligibility to a wide public; (3) public scrutiny that can encourage the use of alternative information sources if these systems should be misused; and (4) avoidance of their use for monitoring political dissent.[22] As we move beyond the lesser types of failure that result from low technical quality or support for particular interests, freedom of speech and press can provide a counterweight to self-interested officials who wish to dominate government information systems. And for the major types of failure in which information on individuals or groups can be used to silence them politically, safeguards against such uses of information by government are vital.

Apart from the threat of direct manipulation, a number of critics have warned about the danger that indicator statistics may impose "official" definitions of problems. Green (1971: 25) urges that social scientists "oppose all efforts to institutionalize social reporting and data collection *at the Presidential level.*" Gates (1975b) contends that "government by indicator" would impose a static consistency even when social and economic reality changed, and that this fixed view of the world would sacrifice political processes and their attendant possibilities of reinterpreting the world.

Similar concerns have been expressed for the misuses of urban information systems. Downs (1967: 209), writing about urban data systems, has contended that they are likely to produce shifts in the power of city decision makers, and that the prospective beneficiaries (city planners, high ranking staff members, and some top politicians) would then favor relatively comprehensive data systems, while others would "drag their feet." A more recent study by Kraemer and Dutton (1979) of the use of computers in forty-two cities suggests that the differences in use of computers or data banks among persons occupying different positions in city government are not as great as might have been expected, but that they tend to reinforce preexisting power structures. To guard against such effects, indicator systems designed for use in public debate will have to be made open and intelligible, or will have to be used by countervailing power systems so that competition will bring information to the attention of the public.

The harm from possible bias or misuse lies not merely in its actual occurrence but in the public's suspicion that it might occur. The public's general confidence in public statistics can affect both the quality of responses and the potential for the use of these statistics. Lack of confidence may result partly

22. For this reason the measurement of citizens' alienation, or political views, by government agencies must be viewed critically.

from suspected bias but also from fear of violations of privacy or confidentiality. When such suspicions concern technical quality they can be dealt with by expert and public review. When they concern fear of breaches of privacy and confidentiality, however, public reactions may perhaps be expressed constructively through a prestigious citizen advisory board (President's Commission 1971: I, 222–25).

Dissemination and Mediating Roles

The motives of users of information, and of those affected by it, can not only modify a statistical system and its use, and affect power distributions, but also facilitate the use of policy indicators. The competition that takes place between political "ins" and "outs" can lead to selective emphases favoring some statistics as against others; but it can also foster the dissemination of better information than might otherwise become available. Political competition thus fulfills a necessary function in which other institutions and social roles can play a part.[23]

The publication of indicator information, or its inclusion in a data bank, is only the beginning of the potential use of that information. The choice of the audience, the mode of presentation, the need for accompanying information about the policy-relevant models, and the selection of the variables themselves, must all be considered in view of the decision process that they serve.

We must therefore examine more closely an assumption shared by many writers on social indicators—a "commitment to the idea that better social information will improve social policy" (Parke and Seidman 1978: 4). Although this idea may seem obvious, if unexamined it may lead us to accept uncritically that certain prevalent definitions of "better information"—information judged significant in a disciplinary, value-free perspective, gathered on national samples, and published in social reports or volumes such as *Social Indicators III* (U.S. Department of Commerce 1980)—will be the information that is most valuable. But some commentators on published volumes of social indicator statistics have in fact suggested that there should be more concern with modes of communication and with audiences. Brusegard (1978: 271–74) suggests that further explanation of the published statistics would increase their value. Caplan and Barton (1976) find only limited use of *Social Indicators, 1973* by upper-level executive branch policymakers. Johnston (1978: 294) suggests that the greatest use of such reports might well be in educational programs for future policymakers rather than by those now making pol-

23. This advantage will not be realized, however, if adversary information sources merely discredit one another rather than developing new sources of information with some consensus on their quality.

icy. To follow through on such specific suggestions requires more information on the actual uses of policy indicator statistics and models. One useful approach has been that of Statistics Sweden, which provides specially edited "current information to the general public in their role as citizens" as well as unabridged reports for researchers and community planners (Vogel 1982: 5).

The means of dissemination most often available for policy indicators require that users seek out information by interrogating a data source (e.g., telephoning or querying a computer), going to a library, or buying a periodical or book. Users may also obtain statistical information as part of a bundle of information-type goods (buying a newspaper, subscribing to a cable television service). But even if newspaper headlines or television news items alert readers to problems, someone, e.g. the media and their reporters, must have obtained the information first. We cannot rely on selective supply of information by political officials (Gandy 1982: 12) but must encourage the media to seek out this information actively. Similarly, if a public official uses indicator information to call the public's attention to a problem, the official or his or her staff must first have sought that information. In the utilization of evaluation research, Patton (1978: Chap. 4) has especially stressed the importance of having a strongly motivated client who needs and seeks the information.

To define problems for a political community, we need means for bringing signals of warning or of progress[24] to the attention of relevant publics; and we need to define special social roles such that persons in them may attend to such signals for the community (like forest rangers in observation towers), thereby enhancing the observation capacity of the public through a two-stage dissemination process. The use of statistical information for public decisions thus depends on the identity and motives of those who seek it out and interpret it.

These specially motivated users may be journalists, public officials, political candidates, party or group leaders, staff assistants of such leaders, scholars, or interested citizens. They thus include many of the active members of the political community. Their motives can range from personal advancement—for a news scoop or a telling campaign issue—to a concern for the public good, or a mixture of the two. They are often led, however, to interpret information in terms of the public good because of their need to gain the attention or support of uncommitted citizens. Conflicts between interested parties are often enlarged in scope by the involvement of the spectators (Schattschneider 1960: Chap. 1).

The dissemination of information to motivated users may be different from

24. Warning signals may be generated by current values of indicator variables, by extrapolation of current trends, or by more complex forecasting models. In addition to decreasing the sense of urgency of a problem, signals of progress may also encourage further efforts if it appears that previous efforts have borne fruit.

dissemination to the general public. It may require periodically updated data banks that can be accessed quickly for the user's purposes, disaggregated (within the limits that respondents' privacy permits), and shaped in different ways for different users (Johnston and Carley 1981: 252). In this respect, even apart from questions of time scale, information may sometimes be better disseminated from data banks than from books. Choldin (1980) favors the conversion of local social indicator publications from print to computer output; and although this risks limiting availability to trained users, conceivably it could be more widely available if there were widespread public education on computer use and if the designers of data sources also facilitated use by citizens.[25]

To a limited extent, analyses and interpretations of these data can also be made available in the same way as the data themselves (Mare 1981: 111–12). As in scholarly publication, however, the preparation of these analyses can cause delays in relation to the needs for use of the data. Moreover, various groups of users might require different sorts of screening of these analyses in terms of quality, intelligibility, and valuative bases; and official interpretations must not attempt to resolve controversial issues prematurely. Public employees, who deal with a wide variety of groups over time, must exercise special restraint.[26]

The appropriate channels and styles of dissemination of data and interpretations thus differ among user roles. We can recognize three types of motivated users in terms of the values they represent: (1) representatives of groups with material interests in policy outcomes; (2) persons who advocate policies in view of notions of the public interest—a category that partly overlaps the first; and (3) neutral experts who present information and interpretations with only minimal advocacy or espousal of controversial values. In the area of overlap between (1) and (2) are advocates of groups or of value-positions who argue in terms of various notions of the public interest. Their system of communication is more like that of weekly or monthly journals of opinion, published for limited and self-selected audiences, than that of newspapers or television. Among the neutral experts are journalists who, when governed by a

25. Improving information technology and capacities to use it have led to greater flexibility and diversity of use of public data. In earlier governmental usage, those agencies that were especially concerned with providing information to the public were "likely to concentrate on publication while discouraging requests for special tabulations or for provision of tapes or tape services" (President's Commission 1971: I, 47–48). Dunn (1974: 143) also observed that data adapted to "multiple use" in the precomputer sense provided a "static set of windows" on the world and were not adapted to the diverse current uses.

26. The problems of interpreting indicator statistics also overlap with those of disseminating information on causal models. Whittington and Grubb (1984) point out that cost-benefit analyses can aid public debate if they are conducted according to professional standards and properly disseminated.

code of objectivity, seek out information to report but not to recommend; somewhat similar are teachers who use indicator statistics. The data bank user who is an analyst and advocate for a group, and the reporter who allows the reader to become a citizen-analyst, thus represent alternative role types of central importance; they closely resemble alternative roles available for social scientists (Chap. 2).

Political Conditions and Indicator Development

Statisticians and other experts who wish to propose a new policy indicator variable must thus proceed with political conditions in mind. An initial impetus for a new indicator can come from the findings of research, when they demonstrate that a new concept (such as the national product was initially) is theoretically significant and measurable as well as practically important. Such an initial impetus need not lead directly to data collection and dissemination by a governmental agency, however. Sometimes trial data collection may be initiated to test users' reactions and an organization's capacity to gather the data. A state or local government may initiate such trial information. Private or semipublic organizations may initiate data collection in order to persuade government officials gradually of the usefulness of a new type of data; this has been the case with the development of national economic accounts by the National Bureau of Economic Research, with the National Assessment of Educational Progress (Bradburn 1973: 27–36) and with subjective indicators (Sheldon and Land 1972: 148). Initiatives may also come from statistical associations or from government officials, as in the history of the U.S. Census.

An assessment of benefits and costs, political if not economic, will necessarily be involved in the choice of indicator variables. Though a current acute problem may exist, the sustained interest of groups may also be necessary before public funds are devoted to data collection. The possibility that information can be used for multiple purposes or by multiple groups can reduce the effective cost of one particular use. Economic statistics, for example, serve the private decisions of firms as well as those of government. If private uses dominate, the market may even generate useful policy indicator statistics without government intervention.

Several examples of the gradual process of institutionalizing indicators have been presented by de Neufville (1975: Chap. 10), showing both successes and failures and their causes. Indicators of unemployment and of prices have been developed in the United States through a continual interplay among government officials, business and labor groups, and experts in statistics and economics. The process of institutionalization has typically taken place over a period of decades, with sporadic but important crises and periods when the indicators were attacked by citizen groups and required defense.

On the basis of United States experience, we suggest first that social scientists who wish to provide guidance for indicator systems recognize certain features of the existing process:

1. It is both political and technical at the same time (de Neufville 1978: 182). Social scientists in government respond both to their superiors and to their scientific communities. Testimony for a statistical system must include input from interested groups as well as from experts. Thus while a "pure science" perspective is unrealistic, politically sensitive social scientists can exert influence for the quality and continuity as well as for the usefulness of the data presented. At the same time, expert groups' research on policy models can be oriented to end-values and social goals of the political community (Chap. 10).

2. A period of decades may be required to institutionalize an indicator at the national level.[27] Short-run payoffs are likely to be unavailable for either politicians or academics. Immediate partial victories may be gained, but they should not be confused with the eventual goal of institutionalizing an indicator series. Trial and experimentation are essential for testing the usefulness of a data series. If this can be done locally, it will facilitate later national use. Potentially affected groups can participate in planning. The experiences of nations can also be compared. Demand will have to be developed as potential users learn what they can do with a new sort of information. Difficulties, biases, and potential misuses need also to be searched out through trials, even though they can be anticipated to some extent.

3. In government decision making about indicators, the valuative aspect of an indicator variable is often subordinated to consensus and continuity. The most general valuative questions, concerning the development of general end-value indicators, need to be examined first apart from government. The earliest measures of a value concept may be neither widely intelligible nor directly relevant to policy. Reliability and theoretical relevance may well have to be studied as preliminaries to the widespread use of such a concept. Only after some familiarity has been gained with its measurement as well as its meaning can it be proposed as a basis for government-sponsored indicator statistics.

4. Different political communities, such as states and cities, have different goals and means of realizing them, and thus different information needs.

Hand in hand with the problems of initiation of indicator series go those of termination. It has been recognized increasingly in recent years that public

27. The United States political system may, however, provide special opportunity for multiple sources of opposition. The development of a system of social indicators of inequality in Sweden, stimulated by the 1968 Level of Living Survey (Johansson 1973), has proceeded much more rapidly.

programs, once initiated, can continue beyond their usefulness; the same problem must inevitably arise for indicator series, however difficult it may seem to establish and maintain them. The need to use scarce resources efficiently and the problem of information overload for users require continuing review of our statistical priorities. The process of initiation cannot simply be reversed, however, by moving back from central governmental collection to private and local use, unless the need for information is locally concentrated. Termination of public programs seems to require rapid and forceful action (Behn 1975–1977).[28]

Indicator series can also need modification. A particular indicator statistic can become rigidly linked to a process of conflict resolution; an example is the regular use of a price index in wage bargaining between unions and management. If the statistic requires change for technical reasons, a temporary parallel use of the old and the revised indicators may reduce the bargainers' anxieties about the change.

For these reasons it is especially important that we focus attention on the processes by which indicators are chosen. We must be skeptical of rapid development of practical information systems, and of design by experts without continuing participation of users. We turn in the next chapter to the roles that scientific and technical communities can play in these processes.

28. This approach was used early in the Reagan administration to reduce what was judged to be an excessive level of federal domestic expenditure, including some statistical programs. The need for speed can conflict, however, with the need for careful estimates of priorities or of benefits and costs.

10 Technical Communities, Indicators, and Models

The successful use of policy indicators requires coordination between the political community that they serve and the communities of experts who contribute to their design. Experts engaged in this process must control the quality of one another's work (as all expert communities do), but must also participate in the political community through that work. This participation can involve communicating with citizens, advising government officials, or working for government. The development of policy indicators and models thus requires different sorts of expert communities from those of the existing basic sciences.[1] Practical needs may require special sorts of interdisciplinary cooperation, as between natural and social scientists (Chap. 3) or between economists and developers of indicators of subjective well-being (Chap. 5). But more important and more difficult is the need for the members of any such community to combine mutual quality control with participation in the political community; quality control in the basic sciences works best for scholarly publications and least well for public statements, testimony, or government employment. Not merely scientific accuracy, but relevance to policy, must be monitored. Because the work of these communities must be judged in part by its relevance to values, goals, and policies of the political community, we recognize this difference by referring to them as one type of *technical community* (MacRae 1976b)—an expert group that deals with laymen's practical problems, conducts related research, and monitors the quality of its members' work on both these tasks.

1. We have taken the political community largely as it is, suggesting that information systems should be adapted to its needs and capacities. We have refrained from suggesting changes in political structure because the needs of information use raise only a few of the many issues involved in structural reform. Governments with a given decision structure can, however, differ in their capacities to use information (Szanton 1981), and these capacities can be improved by training and staff selection.

Some technical communities are required for the development and application of causal models in social goal areas, such as education and crime reduction (Chap. 4), and for the development of models relating objective variables to subjective well-being (Chap. 6). Some will find themselves in an adversarial climate, especially those studying policy models related to compensatory equity (Chap. 7). Most closely involved in decisions about indicators will be those communities whose members propose and plan statistical operations, either in government or in nongovernmental organizations. Some such technical communities already exist, and their experiences can serve as guides for others that need to be created; others are developing, and we shall review some of these developments.

Scientific and Technical Communities

Expertise, in the modern world, is defined more by groups of experts than by individual experts (Ziman 1968); these groups influence their members' training, define their domains of competence, review their work, and certify their expertise. When presidential commissions are named to advise on public statistics or on the causes of change in policy indicators, they usually include outstanding statisticians, economists, and other persons with prestige in relevant scientific communities. When experts are employed by government, their training in such a community, as shown by a university degree, is usually a criterion of employment. Thus disciplines, professions, and similar groups of experts provide gateways to influence. Disciplines, through their members' work and communication, gradually choose and define accepted theories and models, identify variables in these models that are to be measured, and develop methods for measuring them. They allocate prestige in terms of accomplishment in these directions; and they supply much of the manpower, trained to emphasize these variables and methods, that investigates both measurement and causation. They are guided by empirical findings, but also by the selective attention that different scientific communities give to the same phenomena (Kuhn 1962: 50–51). The emphases that they choose, lasting longer than the interests of particular persons, are reflected in the choice of indicators for time series. Such emphases are expressed not only by the organizations that collect and present the information, but also by the expert groups that sustain and monitor its quality.

The type of collective decisions that are made, and the continuing interests chosen, are most clearly seen within a scientific community—a group of scientists reading one another's published research reports and building on them—in a discipline or specialty or communicating informally in an "invisible college." Such communities have been studied extensively by sociologists of science (e.g., Merton 1968: Chaps. 17, 18; Hagstrom 1965; Ravetz 1971;

Zuckerman 1977). The distinctively collective feature of a scientific community is the willingness of its members to join together in sharing contributions to knowledge that are judged by a special set of criteria: those of rigorous empirical inference and theoretical significance, unbiased by the personal identity of the researcher or the practical recommendations that would flow from the findings. The importance of these communities has grown during the past 150 years with the employment of most of their members in universities, where their research is primarily subject to peer scrutiny. One of the strongest motivations to enter and remain in them, developing in this period, has been the prospect of a calculable career (Ben-David 1971).

The collective choices of subject matter by scientific communities result largely, however, from a succession of judgments by individuals and small research groups. The value of each contribution is judged by the contributor's peers through refereeing, citation, criticism, channeling of monetary support, and choice of related topics in their own work. The result is effectively a decision by the scientific community, superficially like the allocation of resources by a market,[2] concerning variables and methods as well. The general control that these actions exert over other scientists works through the recognition of findings as important, and of lines of research or approaches as fruitful. Because this recognition is given only by one's fellow scientists, the scientific community forms a relatively closed system of communication.

A scientist who chooses to do practical research can move into roles that are less closely involved with the scientific community. He can become a part-time consultant while employed at a university, and in this role deal with clients more than with scientific colleagues. He can become a full-time employee of government or a consulting organization. He can also become affiliated with a pressure group or political party. In any of these roles he may provide expert judgments to laymen without new research; but even when he conducts research his findings are less likely to be reviewed by a community of fellow scientists.

The growth of scientific communities has required increasing material support from the outside society. University budgets have supported increasing numbers of professors who do research; governments and foundations have contributed increasingly to research support. The autonomy of scientific communities has depended on their controlling the use of these resources—disciplines exercising major influence on university appointments and research communities guiding peer review in competitions for research funds. They

2. These decisions have some stability because they are constrained by empirical findings. Market decisions, too, are constrained by consumers' needs; tastes cannot fluctuate without limit. Nevertheless, the judgments made by scientists on one another's work are not primarily matters of personal taste; they embody collective standards and norms of excellence and are themselves part of the scientific community's public discourse.

justify this control by pursuing the disinterested goals of scientific knowledge; their involvement with policy is limited by their need to justify support in terms of consensual values, and not merely by positivistic philosophy (Jennings and Callahan 1983: 5).

The scientist who works for government or a private firm is usually controlled less by his scientific community than is the academic scientist; his organizational responsibilities, rather than the theoretical significance of his work, then usually dictate the topics he studies. As Kornhauser (1962: 9) has pointed out: "The professional employee of an organization is subject to two often conflicting sets of expectations, demands, and identifications." On the one hand he is concerned with professional standards of work. But on the other, "*organizations limit professions*. For organizations strive to mobilize professional people to serve their own ends. . . ." In addition, much of the nonacademic scientist's work (like consulting work by professors) is likely to be unpublished and thus to escape the scrutiny of his scientific community.

In organizational employment, in consulting, and in activity as a citizen, the scientist is more often addressing not other scientists in his specialty, but laymen.[3] He is then likely to be subject to different standards from those of theoretical significance and methodological rigor that his scientific community requires. He may maintain that community's standards because of his training, but in doing so he is less likely to receive group support and control (Benveniste 1982).

The closed character of scientific communities, in contrast to the roles of scientists in organizations, reduces the pressure of demands from laymen and allows a central emphasis on questions of theory and method. However, this very closed character limits their members' concern, as members, with matters of practice or of public policy unless it affects the conditions or resources for research (Jennings and Callahan 1983: 5). It provides little motivation to seek out problems from "the world of action" (Coleman 1972b), to transmit findings that are intelligible and interesting to that world, or even to engage in disciplined criticism of those who do so. That motivation, when it exists, must usually come from sources outside the role of the scientist in a scientific community. At the same time those who move into the world of action lose some of the advantages of quality control that a scientific community provides.

Basic scientists who speak in the public arena can face a dilemma if they wish to provide expert information in such a way as to lead to reasoned policy judgment. The goal of providing expertise in the name of one's scientific community can conflict with that of viewing the policy decision as a whole. Those matters on which any basic scientific discipline claims expertise are likely to cover only a fraction of the considerations relevant to a given policy

3. The lay audience predominates in most bureaucracies; exceptions are those that, like universities, are guided by the standards of expert communities.

choice. This is true even if there is complete agreement on the values to be sought.

The introduction of expert information dealing with one aspect of a problem or decision is likely to bias the decision so as to amplify that aspect of it. Contributions from a single scientific community are likely to receive additional, perhaps excessive, weight because of the prestige of science. In the controversy over the supersonic transport, for example, expert information suggesting large economic benefits might weigh the scales in one direction; information on harmful effects on the upper atmosphere, in the other. An overall policy decision is also likely to require a combination of components, some precise and others imprecise, some having more grounding in the work of basic scientific communities and others less. For balanced assessment of policies we thus need ways to increase the likelihood that diverse basic sciences are related to particular policy problems, and that relevant information that none of them provides is sought out. This combination and balance require not merely the consultation of existing specialties, but the formation of more inclusive expert communities.

Scientists—especially social scientists—who deal with policy indicators and models need to make their work inclusive enough to contribute to reasoned policy judgments, while monitoring one another's work for quality. The reconciliation of these needs requires different types of social organization from that of either the basic scientific community (which encourages generality and rigor within its domain) or the political community (which requires a valuative and less specialized view of a policy problem). Even while dealing with laymen and facing controversies and political pressure, these scientists need to join together so as to monitor and reinforce their tendencies toward rigorous inference—as well as their concern with general models. As we have suggested (Chap. 4), the development of models relevant to particular social goal areas may be a useful focus for such groups; other such focuses are policies of continuing interest, and models linking particular social goal areas to one another.

The type of group we seek here is a special type of technical community, a body of scientists who engage in mutual quality control but who also deal with laymen.[4] At several points in the preceding chapters we have argued that

4. Dealing with laymen creates a problem for quality control only when they exercise pressure on the community to modify the substance or methods of its work. If the relationship is essentially one-way communication, like undergraduate teaching, this pressure will be small. Even if the communication from laymen involves submitting problems or data, as in the case of statisticians' consultation, it need not cause changes in the work of a technical community except that it may choose to devote more resources to specific applied problems. Some response to laymen can also increase the effectiveness of recommendations (Lehman 1980: 214–16). In the design of indicator statistics, laymen can suggest uses and anticipated criticisms. In the development of models, laymen can propose relevant values, represent the interests of affected groups, and formulate alternative models that command support and require testing.

one or another community of this sort needs to be created or strengthened so as to contribute to the development of policy indicators or models. Such communities must be encouraged among those who generate and present public statistics, both in particular social goal areas and in relation to more general end-values. They must also be encouraged among those who develop the causal models that aid the public and its leaders to interpret policy indicators and propose policies.

The members of a technical community, like those of a scientific community, exchange research findings and criticisms in view of shared criteria. But the distinctive activity of a technical community is its internal review of members' general recommendations made to laymen on the basis of scientific expertise. We have cited examples of such review in the areas of education and crime (Chap. 4), and it is important that the community reward such reviews. Many professions do not engage in this activity, including those that do not make scientific contributions and those that make only specific recommendations to individual patients or clients. Some physicians encounter general problems related to medical science, and those who publish and review recommendations about these problems constitute a technical community; the *New England Journal of Medicine* is one forum for it. General principles of engineering design are another basis for technical communities. Neither of these types of community is primarily concerned with public policy. Nor should we identify medical communities or engineering fields in their entirety with technical communities. Many physicians do not participate in this publication network, and their quality control over their peers is slight (Freidson 1970); they do not constitute a technical community. Similarly, Perrucci and Gerstl (1969) have referred to engineering as a "profession without community."

For the development and use of policy indicators, we need technical communities that deal with government and the public.[5] Their members' recommendations to governmental clients such as municipal government, or their general policy proposals of national interest, may be formulated and reviewed with the aid of scientific principles and findings. This sort of community encounters the same difficulties of organization and quality control as do the medical and engineering communities in private practice, but their activities are influenced (and perhaps distorted) less by the market and more by the political pressures inherent in public discourse and advocacy. One role that approximates participation in these communities has been that of the "statesman of science," assumed by prestigious members of a scientific community who are willing to serve part-time as governmental consultants, advisers, or members of commissions. This role can be useful when the persons occupying it have broad experience, and especially when the practical problems addressed

5. Such communities also seek nongovernmental sources of support in the initial trials of new indicators, and in continuing work on indicators that are not appropriate for government sponsorship.

correspond to the areas of competence of their communities. The prestige of the statesmen serves as a partial substitute for peer review. This role does not, however, fulfill the need for new technical communities with subject matter different from that of scientific communities, nor does it enlist the career motivations of younger scientists.

When the members of a technical community are full-time employees of nonacademic organizations, the community is likely to have serious problems in maintaining its internal quality control. Its members are then usually dispersed among employing organizations that determine their terms of work. Their communication with one another, either directly or through the literature, is limited unless they can work for an organization guided by standards of professional quality as well as organizational effectiveness (Lehman 1980: 214–16). When they become concerned with values, as we have suggested, their possibility of consensus and quality control is threatened from another direction. If, in addition, they become involved too deeply in politics, as the previous chapter suggested might occur, their consensus on methods and concepts may be further jeopardized.

The reconciliation of these two types of criteria of work—scientific rigor and significance, on the one hand, and valuation and policy relevance, on the other—is the problem of this chapter. It involves a type of boundary work (Gieryn 1983) that is more difficult than that of basic science. We shall address it by examining the activities of policy-relevant technical communities that now exist or are being developed. In doing so, we shall see that certain general procedures can be used by these communities to reconcile scientific quality with policy relevance, over and above the clarity of intellectual organization of the subject matter itself. One is the drawing of clear boundaries between the contributions of expert and layman, such as that between means and ends. Another is the assumption of a leadership role by organizations that are themselves largely devoted to research quality control, such as professional schools, government statistical agencies, and consulting research organizations. These organizations can play a role in sustaining technical communities that is similar to that of the basic-science parts of universities in sustaining scientific communities.

Our argument here is that some technical communities need to be created, and others need to be strengthened, if policy indicators and models are to be used successfully. In making such a proposal we face the counterargument that the inclusion of policy-related communication in the roles of experts will destroy internal quality control. In response we shall show in the next section that the field of statistics has succeeded in maintaining its internal standards while dealing with laymen, by setting limits on its political involvement. This field is especially important for the development of indicators. In the following section we shall compare scientific and technical communities as sources of policy models, showing that the major guidance for these models must come not from the basic sciences but from the decision sciences and practical fields—even though the policy contributions of these latter fields need to be strengthened.

The Community for Public Statistics

The field of statistics is central to the development of policy indicators. The corresponding expert community includes those experts who are concerned with the production and interpretation of statistics[6] generally: statisticians, together with economists, social scientists, and others who specialize in statistics. The organization of this community illustrates a combination of internal peer control and communication with laymen. Yet its organization is complex; it is neither a single scientific community nor entirely a technical community, and it has changed over time. It includes important parts of the profession and discipline of statistics, which differs from many other basic scientific disciplines in the importance of its applied component. In the United States, the American Statistical Association was founded in 1839; the publication series that became its present *Journal* began in 1888. It drew its membership from persons trained in other fields until the first university statistics departments were founded in the 1930s and 1940s. State universities in Iowa and North Carolina played a significant part in this development (Ben-David 1971: 147–52). Initially devoted to relatively descriptive applied work, the discipline became more concerned with theoretical questions in the years preceding and following World War II, with influence from Britain. Yet its *Journal* has continued to give an important place to applications, instituting a special part for them in 1970. The contributors to this part, insofar as they are reporting on practical consulting work, constitute a technical community of a very general character, not restricted to any one substantive area of application. The Association also has a number of sections devoted to areas of application, such as biometrics, business and economic statistics, physical and engineering sciences, and social statistics, which are links with particular applied fields and overlap with more substantively specialized technical communities. Economic statistics, particularly important for policy indicators, is also a subfield of economics.

The larger community of producers and interpreters of statistics, when it extends beyond the discipline of statistics as such, is less well defined and more diverse. Practicing applied statisticians and other contributors to public statistics are not only a somewhat different group from academic statisticians;

6. In connection with indicators we refer primarily to "statistics (plural) as numbers"; in contrast, "statistics (singular) as a body of methods . . . for obtaining and analyzing data in order to base decisions on them" (President's Commission 1971: I, 14) is more relevant to causal models.

they practice in disparate organizational settings and come to their work with various types and amounts of training. In the federal government, there has been a continuing concern with the recruitment and training of statisticians and with the monitoring of the quality of their work (President's Reorganization Project 1981: Chap. 3).[7]

This diversity has led to efforts to improve quality through the formation of a better-defined technical community. Kruskal (1973: 1256) has pointed out that even within the realm of federal statistics much of the wide range of applications "is not carried out by people called statisticians, or trained as statisticians. Much of it is not regarded as having important statistical components. Consequently, much of it is of poor quality." He may well have had in mind the unconcern of some agencies with amounts and sources of error, or their failure to use scientific sampling for economy and accuracy (President's Commission 1971: I, 32, 34).

Yet the formation of a technical community, especially if it is to act as a gatekeeper to influence, can lead to exclusiveness as well as to improved quality.[8] W. Edwards Deming, a distinguished statistician, is reported as having noted that "statistical work is not a sideline and is not a part-time job. One does not dabble in statistics. One is not engineer and statistician, nor economist and statistician, etc. He is one or the other, or neither. There may be exceptions, but it is poor management to hope to engage the exception" (President's Commission 1971: I, 142). He seems thus to be joining in Kruskal's concern for quality. But Kruskal's remark about "people called statisticians" provoked a letter to the editor of *Science* (Pollock 1973) and a rejoinder by Kruskal, reminding the readers that one can be trained in statistics without being called a statistician.

Within the federal government, statistical quality is generally better within specifically statistical agencies such as the Bureau of the Census than within operating agencies (Wallis 1971; President's Commission 1971: 34–35). This difference in quality illustrates a general problem in the work of technical communities. When an employing organization is seeking to prove its worth by delivering services, its statisticians will feel pressure to serve those ends. If, however, the organization is devoted to the production of statistics of high quality, its statisticians' goals will accord more with those of the discipline.[9]

7. These problems exist for other fields of expertise as well, and are much greater at present for cost-benefit analysis (Whittington and Grubb 1984).

8. An important aspect of maintaining quality is the recognition that "users have different demands for accuracy, timeliness, frequency of reporting, comparability of definitions, and detail. . ." (President's Commission 1971: I, 173). Indeed, quality includes the accuracy and adequacy of the descriptions given when statistics are presented.

9. A similar consistency between organizational and scientific goals exists in those organizations set up to administer experimental programs, where employees are seeking to produce valid and unbiased evidence (Chap. 9).

For this reason, among others, statisticians serving as consultants have repeatedly favored greater central coordination and quality control of government statistics.

The very existence of public statistics, and thus of policy indicators, depends on our being able to set some standards of quality for their production and use. If the producers of statistics are advocates or are unskilled and thus frame biased questions, draw biased samples, present results selectively, or fail to describe their procedures so that others may check them, we prefer not to call the results public statistics. In defining an indicator as a variable *justifiably* included in a public statistical system we have implied a number of criteria, some difficult to measure, for its inclusion; but certainly one of them is that statistical bias be sufficiently small, whether such bias arises from motivation or lack of care.

In justifying public statistics, we also face the question whether they are well used; as Dunn (1974: 133–35) points out, the use of statistics requires interpretation of meanings as well as statistical inferences. If statistics of impeccable quality (in a narrow sense) are provided to audiences who have no notion of their meaning, and thus either ignore them or interpret them in utterly different ways from those intended, we may well hesitate before recommending them. But this hesitation involves another value: the public has a right to reinterpret information and may sometimes be justified in doing so. The role of expert communities, in that case, may have to be restrained to questions of method—if these can be separated from substance. Kruskal (1973: 1256) points out that much use of statistics in public debate is conducted "with extraordinary naïveté, not to say tendentiousness"; yet the statistician cannot be a censor of the mass media, and can only try to educate journalists and the public in valid procedures of inference.

One way for a technical community, such as that on public statistics, to facilitate its consensus on quality is to define its expertise independently of value positions involved in public debate. Statistics, with its mathematical base, can apparently define its domain of competence clearly, set standards of quality and resist pressures to change them more easily than some of the social sciences. Statisticians who work in government are particularly bound to avoid the suspicion of serving partisan values in their work and to refrain from proposing controversial options.[10] They try to insulate the production of governmental statistics from political pressures (President's Reorganization Project 1981: 180), drawing a clear boundary between the contributions of expert and layman. Thus when the President's Commission on Federal Statistics (1971: I, 172) recommended the review by a statistical coordinator of "the use

10. They may, however, suggest and examine possible questions to be used in censuses. In one instance, however, a Swedish government statistician, Joachim Vogel, successfully argued his right to publish information on the effects of a public policy, doing so in his role as a private citizen after his superiors had declined to release the information.

of statistical evidence for decisions," it took pains to state that "professional statisticians are expert in statistics, not in making the decisions which may rely on statistics."[11] For similar reasons, "there is a strong inclination by statisticians to estimate only quantities for which there is a general professional consensus as to how the estimates should be made" (Roberts 1980: 217–18).[12]

Government statistical administrators, although they risk losing support for their agencies if they become involved in public debate, must nevertheless be sensitive to their organizations' environment. Budgetary and other support can depend on favorable attitudes of users and other agencies, as well as on avoidance of offending significant groups. Bureaucratic politics must be understood and practiced even by the leaders of scientific agencies that do largely value-free work.

Carroll D. Wright, who initially stressed the use of the best available statistical methods in the Bureau of Labor in the late nineteenth century, also took the lead in the "meticulous avoidance of [partisan] politics" (de Neufville 1975: 24) in place of a previous advocacy approach. This nonpartisanship has been maintained. But a governmental statistical system requires anticipation of important new data needs (President's Reorganization Project 1981: 185), and this demands political judgment—perhaps even anticipation of the problems that a newly elected party will choose to address.

Statisticians in government are thus concerned with priorities as well as techniques; but these priorities come largely from evaluation of outside requests. Internal initiatives do occur in all government agencies, but even then the responsible civil servants need not always take credit for them. The value-free stance can be maintained in a value-laden environment such as that of policy indicators by characterizing values as originating in an external constituency or clientele that one's agency serves.

Connections between government statistics and the academic research community have been made in several ways that are typical of technical communities. First, relevant academic training has increasingly become a criterion for employment in the Bureau of the Census and related agencies (Keyfitz 1978: 419–20). Second, a number of statisticians and specialists in related subject matter have moved back and forth between academic research and government employment during their careers. Third, academic statisticians

11. Such statements gloss over the fact that statistical expertise is invoked so as to stress the decision criterion of "statistical significance" at a conventional probability level. When practical significance is at stake, such levels are open to criticism and controversy, as professional statisticians know. See note 16 below.

12. A similar restraint has been urged on economists in benefit-cost analysis (Harberger 1971). These positions reflect the need for an expert group to have its expertise accepted by outsiders. By limiting the scope of their expert judgments they reduce the chances of public disagreements among their members, which might undermine outsiders' confidence in them.

have held consulting and advisory roles. Fourth, increasing use of census and other data by researchers has led to wider interest in their content and quality.

These relations date back to the mid-nineteenth century, but the creation of a permanent Bureau of the Census at the turn of the century (Alterman 1969: 219–20, 231–32) allowed them to expand through the employment of continuing personnel, an increasing proportion of whom were experts. In 1933 the American Statistical Association and the Social Science Research Council combined to form the Committee on Government Statistics and Information Services (COGSIS), which recommended establishment of a Central Statistical Board. This was established by executive order and subsequently became the Division of Statistical Standards in the Office of Management and Budget. Scientific sampling was introduced in the U.S. Census of 1940.

The developing conditions for outside consultation are illustrated by an incident concerning errors in the 1950 Census. Profs. Ansley J. Coale and Frederick F. Stephan of Princeton University discovered that an inordinate number of teenage widowers and an unusual variation among age groups in the Indian population had been reported. They brought this discrepancy to the attention of the Census Bureau by traveling to Washington, "paying their own fare" (Alterman 1969: 314–16). Eventually a punching error was discovered that had converted the records for a number of persons from one population category to another. Although much internal review and professional training had taken place by that time, the purely voluntary and unofficial nature of their visit suggests that better mechanisms for review and coordination were still needed. Even in 1962, when a *Reader's Digest* article attacked the unemployment index, President Kennedy had to appoint an ad hoc "committee of noted economists and statisticians" to review the question (de Neufville 1975: 82). Subsequently the procedures for obtaining outside advice were made more regular.

Additional support was given to the connection between government and expert statistical communities by President Nixon's appointment in 1970 of the President's Commission on National Statistics. This commission recommended in its report "continuous review of federal statistics . . . by a group of broadly representative professionals without direct relationships to the federal government." Following a more specific recommendation of this commission (President's Commission 1971: I, 175–76), the National Research Council created its Committee on National Statistics in 1972, which then became a part of its Assembly of Behavioral and Social Sciences, organized in 1973.[13]

13. A committee or commission of this sort, together with its expert audience, constitutes the central core of a technical community of a special kind. Its recommendations are based not only on research in its members' scientific area of expertise, but also on other knowledge such as that of the expected results of certain procedures and organizational forms. Its recommendations are published for an audience similar to that which would read more technical reports of applied

The establishment of the National Research Council (NRC) itself represents a contribution to the formation of a number of technical communities; we shall refer to the work of others below. Formed during World War I to contribute scientific advice to war mobilization, it was continued by an executive order of President Wilson in 1918. At first its special divisions centered about the natural sciences, but they were followed in 1919 by a division of anthropology and psychology (National Research Council 1933: 7, 39). The prestige of the National Academy of Sciences, with which the NRC was associated, was thus transferred to certain applied activities that might not otherwise have received so much attention from basic scientists.

The community for public statistics, as we have described it, can exert an important influence on improving and maintaining quality, but it pays little attention to two other goals we have set for policy indicators and models. It is not explicitly concerned with values (except for economic statisticians' interest in preference satisfaction and efficiency), nor does it develop the type of specific policy models we have recommended in Chapters 3 and 4. Its techniques of empirical inference and experiment are of great importance for the development of these models, especially in the less exact sciences, but we must look elsewhere for the communities that produce and refine these models for particular social goal areas.

Communities Dealing with Policy Models

The technical community for public descriptive statistics, though heterogeneous, is closely related to a single scientific field—that of applied statistics. The development of policy models, however, can involve numerous scientific and technical communities that differ in their audiences, the substantive content with which they deal, their value orientations, and their ways of contributing to policy models. We must now distinguish them from one another so as to clarify the relations between the communities that now exist and those that we advocate. We shall deal in this section with their audience and content, and in the next with their value orientations. Both sections will refer to the classification shown in Table 10–1, in which the lower right part will be related most directly to policy indicators.

The primary distinction between a scientific community and a technical community is that members of the former take as their audience only their peers, while members of the latter have a dual audience: peers and laymen. The column headings at the top of Table 10–1 distinguish these two types of

research. The *American Statistician* has served as a forum for some proposals for the development and improvement of federal statistical systems (Malkiel 1978; Moses 1979).

Table 10–1. Classification of Scientific and Technical Communities

	Audience and Value Orientation			
Audience	Peers only (scientific community)	Peers and laymen (technical community)		
Value orientation	Value-free	Value-free	End-values	Specific values
Content-free	Mathematics Theoretical statistics, methodology	Applied statistics (general) Operations research		
			Public policy analysis, generic planning	
Content related to existing basic sciences	Basic disciplines in natural and social science, interdisciplinary research areas	Applied sciences: engineering (some fields); subfields of applied statistics		
	Economics		Applied economics, benefit-cost, distribution	
Content				
	Psychology		Models of subjective well-being	
	Positive political science		Normative political theory	
Content related to social goal areas or values	?	Engineering (some fields)		Health sciences Environmental sciences Education Crime reduction Urban planning Social work
			General studies of equity	Models of domain satisfactions

audiences.[14] The effect of the lay audience on the problems, concepts, and values of a technical community varies somewhat among these communities, as we shall see.

Our classification of technical communities in the right-hand part of the diagram as speaking to peers and laymen, however, reveals only a small part of the possible variation in their relations with laymen.[15] Models vary from general to specific, and their relevance varies from continuous to sporadic. Table 10–1 classifies expert communities that can focus on questions that are of long-term interest and thus are of some generality; but not all policy models are of this kind. Each technical community concerned with persisting questions may, therefore, be linked with practitioners who can give advice in the short term, if their expertise is relevant. It may also be linked with capacities for mounting research operations in the moderately short term. Because of time constraints it must also have some concern with anticipating problems rather than merely refining answers to old ones. Moreover, because problems change, the practical orientation of these communities can lead them to be less concerned than basic scientific communities with developing theories for the longer run.

The rows of Table 10–1 also distinguish among scientific and technical communities in terms of their substantive content. We have emphasized (Chaps. 3, 4) the distinction between the substantive content of the basic scientific disciplines, defined by their respective bodies of theory (center row), and that of many policy models, especially those guided by particular social goals (bottom row). The top row deals, however, with communities whose subject matter is content-free; statistics, for example, is relevant to problems that range across all the disciplines and social goal areas shown. The row labels in Table 10–1 thus deal with three aspects of content. At any given time these types of content are represented by existing scientific and technical communities; but over time the scientific communities evolve in response to research findings and paradigm changes, while technical communities evolve in response to practical problems and technological and institutional possibilities (as do our lists of relevant contributory variables). This process is not entirely determinate, however, and we shall make recommendations for it.

The basic scientific disciplines, in spite of their prestige in the universities, cannot be the primary source of policy models. The center row of Table 10–1 includes those expert communities whose subject matter is defined in relation to the basic sciences. Only in a few subfields of social science, as we shall see, is there a central concern with values that can guide our choice of policies and of relevant subject matter. The basic disciplines are thus quite likely

14. They also distinguish value-orientations, which we discuss in the next section.

15. Different lay audiences vary greatly in their education, their use of language, and the types of arguments that they find to be persuasive.

to contribute components of policy models, but not to guide the development of these models in their entirety.

The discipline of statistics, more nearly a technical community than a basic discipline, appears in three distinct cells in Table 10–1. In the upper left cell is theoretical statistics, classified as having an audience of peers only (like other basic scientific communities) and as content-free, together with mathematics and methodological studies more generally. Its contribution to policy models lies in such areas as general decision theory (Chernoff 1978) and the study of probability distributions appropriate to particular types of statistical tests.

Applied statistics, in its general aspects, appears in the upper right cell, being practiced by a technical community having laymen as well as peers as its audience; here it joins operations research as a "decision science." These fields provide general procedures such as those of regression and linear programming that are applicable to policy model construction in a wide variety of substantive areas. These procedures become components of particular policy models when made specific by combination with data, causal theories, parameter estimates, and valuative criteria.[16]

The particular subfields of applied statistics (such as biostatistics, psychometrics, or business and economic statistics) appear in the center row and right-hand cell, among the applied sciences that define their subject matter similarly to corresponding basic sciences. The multiple locations of statistics in the table, and the apparent success of the discipline in maintaining them all, remind us that the organization of this discipline is worthy of study and perhaps emulation by other decision sciences and technical communities.

The decision sciences in the upper right-hand cell include public policy analysis and generic planning; these fields include some content-free contributions of decision science, but also require specific bodies of knowledge from substantive areas (Alterman and MacRae 1983). Unlike many of the other technical communities to be discussed they can claim to produce actual policy recommendations and thus require a relatively balanced combination of expert contributions to policy choice. The technical community for public policy analysis, organized as the Association for Public Policy Analysis and Management, combines graduate professional schools in which practitioners and researchers are trained, with major independent research contracting organizations whose reputation depends on the quality of their work.

In the left-hand cell in the center row of Table 10–1 we find the basic empirical scientific disciplines and those interdisciplinary fields that arise from their interactions in research and theory (such as biochemistry or social psy-

16. Tests of statistical significance using conventional probability levels may seem to beginning students to permit decisions without consideration of values—a procedure that is impossible in the realm of policy. Kruskal (1978: 952) recognizes that instead of conventional levels such as probabilities of .05 and .01, "special circumstances may dictate tighter or looser levels," reflecting the values attached to various possible events and decisions.

chology). The corresponding scientific communities, centered in universities, are organized for the pursuit of truth in a general form, that is, the development of verified general theories. They are governed by canons of inference similar to those of technical communities, but they are subject more exclusively to peer control because of their closer communication in university departments and through the literature, leading to their closed character. Because of this closure (screening out of communications with laymen) they are less concerned with practical questions than are technical communities; and when their members venture into the public sphere, their pronouncements are not subjected to the usual processes of peer evaluation and quality control.

The right-hand cell in the center row contains the applied sciences, areas of application whose subject matter lies largely within the domain of particular basic sciences or other research-defined fields.[17] Because the corresponding communities deal with laymen, they are technical communities; but they are atypical in drawing only on single disciplines. The foundations of chemical engineering, for example, lie largely within chemistry; those of mechanical and electrical engineering relate largely to physics. These engineering fields nevertheless select and adapt certain parts of the corresponding discipline in view of their practical tasks. Some technical communities that utilize the basic sciences, however, require task-related (and not merely research-related) combinations of contributions that extend beyond a single basic science; those that do will fall in the lower right cell.

In the center row of the table three disciplines—economics, psychology, and political science—are set off from the others in the left center cell because they encompass corresponding value-related subfields. At the bottom of the right center cell we show these three special valuative fields: applied economics, embodying the value of efficient preference satisfaction; the study of models of subjective well-being, related to psychology; and normative political theory.[18] We shall consider these fields in the next section.

The bottom row in Table 10–1 shows fields whose content is defined in relation to the tasks of society rather than to the theoretical organization of basic scientific communities. We classify these tasks according to the goals or values they seek to further; some may be the social goals we have considered, others more general values. These fields are not mere applications of basic

17. The reader should recall that we are referring to applied sciences in a narrow sense, related to utilization of knowledge that originates in particular existing basic sciences.

The communities for those applied fields that are closely associated with basic scientific communities but are concerned with public policy (and thus with values) have been called "fringe associations" (MacRae 1976a: 75). They often express latent values characteristic of the disciplines in question. A fringe association or intradisciplinary applied field can be organized as a subfield of a discipline.

18. These fields have only limited interaction with laymen. Business administration and applied psychology (not shown) can involve work for lay clients, guided by market values.

science because their guiding values help to define what parts of the basic sciences are relevant, how they should be combined, and what additional knowledge not provided by the basic sciences is needed.[19]

The values that guide them may be derived from the market (e.g., providing transportation, as a value guiding highway or automotive engineers) or from a professional notion of a social value (e.g., health as a value guiding the health professions).[20] Market values that guide technical communities call forth certain combinations of knowledge but are not themselves shaped by practitioners; we refer to such a community as value-free. Professional values are defined in considerable measure by the profession itself. Physicians tell patients what their state of health is and what treatment it requires, but the designers of automobiles do not try to persuade the public in the same way as to what sort of transportation they really need. Those communities that are guided by their own values are better prepared to extend their concerns to the justification of public policies.

The lower left cell in Table 10–1 is empty. We have defined the left column in the table by limiting the experts' audience to peers alone; but experts with practical goals must have a wider audience. Those experts responding to market values must deal with customers, and those serving professional values must deal with clients.

The lower right cell shows three types of technical communities in its three columns. In the left column of this cell are interdisciplinary fields of engineering; even when an engineering community is not organized about a single corresponding basic science, it is still largely oriented to the market. An example is aeronautical engineering, which draws together aerodynamics and principles of material construction and propulsion—thus utilizing aspects of various basic sciences—but makes no effort to prescribe goals for transportation or military strategy. In the rightmost column of the cell (except for its bottom entry) are a number of professions in the stricter sense of that term: communities of experts whose members are trained to serve lay clients, claiming collective autonomy from society in return for service to social goals (Wilensky 1964; MacRae 1977: 4–10). These include the health sciences, the environmental sciences, education, crime reduction, urban planning, and social work. Finally, in the bottom row of the cell are two more general value-guided fields without close attachment to corresponding disciplines, concerned with equity and with models of domain satisfactions. Although all the technical communities in this cell are guided by social goals, the engineering fields are

19. Even for the applied fields listed in the center-right cell of Table 10–1, practical goals guide their choice of aspects of the corresponding basic disciplines that are relevant to their tasks, just as the health sciences use the goal of health to guide their choice of relevant aspects of biological science.

20. These orientations correspond to Ravetz's (1971: Chaps. 12, 13) distinction between technical and practical problems.

more oriented to the market; the professions, to goals that they themselves shape outside the market; and the last two research fields, to more general values with less direct connection to particular clienteles.

In the engineering and professional fields, the corresponding technical communities are closely tied to the education of practitioners. The faculty and advanced graduate students in these schools conduct research related to the goals of their fields. This research can contribute to policy models if it is not too narrowly concerned with the delivery of services by individual professionals. At present, the technical communities based on professional schools do not always stress the widest range of policy models; their first priority tends to be given to questions that face their graduates. Physicians in their private practice, for example, will not be inclined to deal with patients' problems through legislation or behavioral change and will tend to emphasize the adequate delivery of health services more than the improvement of health through policies for environmental change, exercise, or nutrition.

We have suggested (Chap. 4) that the technical communities for education and crime reduction need strengthening. There are actually two corresponding professions, teaching and the administration of justice, for which professional schools and programs exist; but the training of teachers and justice professionals centers more on the administration of given policies than on models for policy choice. The research literature relevant to educational policy is found in periodicals related to education, sociology, and psychology; but some of it also relates to economics and demography. Researchers in the applied subfields of the relevant basic disciplines are apparently not sufficiently motivated to provide the policy-related synthesis needed, and schools of education are often handicapped by the lower average qualifications of their students. Similarly, criminology is studied sometimes as a subfield of sociology or political science and sometimes in independent schools, but the teachers in these schools do not enjoy the support and prestige necessary to take the lead in a corresponding synthesis.[21] The academic prestige system and the needs for training professional personnel have impeded the strengthening of the technical communities needed for developing policy models. Independent research organizations and "think tanks" play an important part in research on crime policy, but their research is often unpublished; thus quality control is less explicit and careers are less attractive than in basic science.

The other professional fields listed in the lower right cell, though we have not discussed their policy models in detail, combine concepts from various basic sciences, drawing them together on grounds other than scientific theory. The health sciences combine biological science, material technology, and hu-

21. In criminology in the 1960s there may well have been a consensus in a technical community in favor of nonpunitive policies, but this community may have also been somewhat self-contained rather than in contact with a broad range of policymakers, administrators, and the public.

man relations; the environmental sciences, production technology, dispersion by air and water, and health effects. Urban planning covers an especially wide range of subject matters, since all the social goals mentioned so far can be sought in urban areas. Social work also combines findings from several social sciences with specific information about institutions and government programs.

In classification of scientific and technical communities, we have not simply been dealing with communities that exist, but have asked which ones required support or improvement. We have dealt briefly with the ways in which their content, internal criticism, and organization could be improved; but improving their external support is of equal importance.

We see in the four vertical columns of Table 10–1 four types of activities, supported by society in different ways: those motivated by the search for knowledge (first column), those carried out in response to the market (second column), those guided by general social values shared within a technical community (third column), and those guided by more specific social goals (last column). The research activities of the basic sciences, motivated by the search for knowledge, are not directly supported by the market. Rather, they are supported as a collective good or social overhead activity in view of their expected contribution to education and general enlightenment and in turn to production, citizenship, and consumption.[22]

Statistics and decision science have more direct and marketable applications; their use in business administration, or in consulting for private firms, provides them with sources of support that the purely basic disciplines do not have. The application of these fields to public policy can, however, create collective goods that are not produced sufficiently in response to market demand. In particular, their desirable level of contribution to the design of policy indicator systems and development of policy models goes beyond what is demanded by the market.

Among the fields with more specific content (in the middle and bottom rows of the table) are some that respond primarily to market demand, especially the fields of engineering. The types of goods and services demanded by the market may be producible with knowledge derived largely from a single basic discipline, or may require combination of two or more, as we have noted. The needs of production tend to call for practical knowledge that transcends the scope of the basic disciplines, and this seems especially true when two or more basic disciplines must be combined. When the required practical knowledge extends beyond individual professional services to public policy judgments, an additional combination of types of knowledge is needed. Not

22. In this perspective, we might expect the beneficiaries of the applications to be taxed for an appropriate degree of support of basic science; but in fact this "tax" is levied on higher education, which is organized so that new knowledge is produced by persons employed as educators.

all the engineering fields are directly relevant to policy indicators, as they do not all provide this combination; perhaps most relevant are those that deal with the environment and transportation, in addition to economic production generally.

This additional need for policy-relevant content and organization arises in the professional schools as well. It is here that a second type of social overhead must be recognized by the universities: in addition to basic research, they contribute to policy choice, citizenship, and the use of policy indicators. The values embodied in the professions resemble the social goals that figure in indicator systems. Insufficiently demanded by the market, these topics related to the professions can provide a collective good for society through their contribution to policy indicators and models, and to undergraduate education for citizenship as well.

We need, therefore, to develop new units in universities, or to strengthen those that exist, in social goal areas related to policy, as well as in public policy analysis and decision sciences. University research in these goal areas is concerned only to a limited extent with proposing policy indicators, as this function is presently centered in government. University research concerns policy models more directly; here our need is for a more practically oriented combination of scientific disciplines and professions, dealing with policy problems that are general, continuing, and outside the battleground of partisan politics if possible. The concern of technical communities with values, to which we now turn, illustrates why these communities need to avoid political controversy.

Technical Communities and Values

Policy indicators measure values—end-values, social goals, and other contributory variables. These values must be consonant with values of the public, in the sense either of a broad consensus or of a coalition to measure a combination of more specific values. They must be lasting if they are to be measured in longer time series; end-values, and to a lesser extent social goals, have this stability. Even though the means to these values may change, and though some policy indicators need to be measured only for a short period, many policy indicators can serve continuing needs by their incorporation in longer-term statistical time series.

We must now ask whether scientific and technical communities can be sufficiently concerned with values of this kind while still avoiding partisan controversy. In Table 10–1 we have classified the closed basic scientific communities (left-hand column) as value-free. This means that the criteria of relevance and excellence for contributions to the literature of such a community shall have no connection with values that favor or oppose particular policies or groups outside that community. To be value-free does not mean ignoring the internal scientific values of empirical verification, logical rigor, and generality, or the norms of the scientific community concerning modes of publication and criticism. It does require that scientists ignore the effects of their discoveries outside the realm of science (such as the changes in worldview occasioned by Copernicus, Newton, or Darwin), and the effects of application of basic science (such as the development of new goods or weapons). To be value-free in natural science thus means to seek knowledge for its own sake without regard to the purposes for which it might be used. The fact that it might be misused then leads to the problem of the scientist's social responsibility (Haberer 1969).

In basic social science, to be value-free means to avoid explicit espousal of social or political values in one's contributions to the scientific community. Implicit incorporation of values in "secondarily evaluative terms," however, is common. Such values include those of stable democracy, development, per capita welfare expenditures, real national income, social mobility, personal adjustment, mental health, reduction of prejudice, and equity (MacRae 1976a: 78). They include end-values, social goals, and valued contributory variables such as we discussed in Chapter 1. To varying degrees, they may be made explicit—more by technical than by scientific communities—and used as dependent variables in policy models.

Only economics has developed an applied field based on values consistent with those of the American public and evolved from it a technical literature applicable to policy choice. Benefit-cost analysis (Chap. 5) has this character; we thus show it in the middle row, right-hand cell of Table 10–1 as reflecting an end-value. Psychology includes a concern with well-being, expressed in individual therapy but also having policy relevance. The literature of political theory is valuative, dealing with the foundations of regimes and their contributions to the good life, but does not incorporate technical means of calculation. With these exceptions, we classify all the basic scientific disciplines, as regards that part of their members' communication whose audience consists only of their peers, as value-free. The official stance of these disciplines is that valuative statements are irrelevant to the judgment of the quality of work done in them. By this stance they largely avoid partisan controversy, but also sharply limit their usefulness for policy indicators and models.

The value-related aspects of economics, psychology, and political science are expressed, however, in certain areas of work that are guided by end-values but largely unrelated to lay audiences. One of these is theoretical welfare economics, not shown in Table 10–1, which deals with the foundations of preference satisfaction through public policy. A second is the fundamental study of global subjective well-being and its causes, which has so far been pursued with little policy application or lay involvement. Thus the literature we discussed in the first two sections of Chapter 6, on subjective well-being, came from a scientific (not technical) community centered in social psychology. In order to avoid complicating the table, we have placed this topic in the center row and third column of Table 10–1, on the basis of its value orientation but not its audience; it may eventually gain a lay audience as it acquires policy relevance. We have suggested that it overlaps the health sciences in connection with the study of quality-adjusted life years; this part of the field would belong in the lower right cell of the table. Neither of these fields, involving the study of end-values, has a specific clientele like that for a social goal area, unless it be a broad philosophically oriented group. The impetus for further study of subjective well-being must thus apparently come from disciplines linked with psychology, or possibly from a recognition of general needs of the decision sciences or policy analysis. Links between the study of subjective well-being and economics are also desirable (Scitovsky 1976) but they have not been made extensively. A third value-related subfield, normative political theory, could potentially be involved in major political controversies, but is somewhat protected from this sort of involvement by its central concern with the history of political thought. The discipline of philosophy itself, though not a science, provides another expert community that must interact with these fields.

Among the technical communities shown in Table 10–1 we also classify engineering, applied science, operations research, and applied statistics as value-free. By this we mean that they serve values brought to them by their clients. Engineers serve the market as consultants or employees, and rarely place general professional restrictions on the purposes they will serve in terms of formal or informal codes of ethics. Applied statisticians are concerned with data and problems of inference or decision brought to them from a wide variety of sources.[23]

The professional groups we have listed in the rightmost column in the lower-right cell, however, serve social goals with considerable consensus internally and with support from the outside society; these goals are health, environmental quality, learning and opportunity, crime reduction, planned urban development, and the welfare of the less fortunate.[24] These last two fields are of interest because of their valuative commitments to the interests of ultimate clients (Alterman and MacRae 1983). Each of these goals can be sought ex-

23. Applied sociology, even though permeated by a concern for equity and for the disadvantaged that is shared by many sociologists (MacRae 1976a: Chap. 8), also includes sociologists who work for private businesses and serve the market.

24. A challenge to such a consensual goal from within the community can, however, call for valuative or political discourse that is not always part of such a community's literature; an example is radical criminology (e.g., Quinney 1974), which might be acceptable to an academic audience but not to the practitioner clientele of criminology. Further questions then arise as to which lay clienteles or constituencies should enter into the discourse in question, for example, whether prisoners or school children, or parents, should enter along with prison personnel or school teachers.

plicitly within a technical community, within constraints defined by the outside society, and connection with such a goal may be a criterion of relevance for contributions to the community's literature. At times one such goal can clash with another, especially when scarce resources must be allocated between them. Social goals may change, as is evidenced by the increasing concern for the environment in the United States in recent decades and by the decline of such concern in the policies of the Reagan administration. The association of social goals with patients, clienteles, or institutional constituencies can, however, contribute to their continuity.

The values expressed in existing technical communities can be bases for organizing the development of policy models. Some of these communities require greater continuity in model development than they now provide (Chap. 4); others can benefit from attention to questions of policy in addition to those of professional practice. Some technical communities shown in Table 10–1, however, are relatively new, may be organized only informally, and have special potential: public policy analysis (in the upper right-hand cell), model development for subjective well-being (center right), and general studies of equity and models of domain satisfactions (lower right). We have suggested that policy indicators would benefit from further development of these value-guided technical communities and we must thus assess their prospects.

The field of public policy analysis, as a decision science, is placed between the second and third columns of Table 10–1. This is because there is some division of opinion within this field as to whether it should be value-free, taking its values from clients; center about economic values; or deal with a wider range of values. The last position is characteristic of undergraduate programs, whose graduates are not immediately seeking professional jobs and may express their values as citizens (MacRae and Wilde 1979; Alterman and MacRae 1983). This field is directly concerned with policy models and has taken values from applied economics and benefit-cost analysis. Consensual values other than economic benefit may also be used in policy analyses; Chapters 6 and 7 have been arguments not only for scientific development of indicators related to subjective well-being and equity, but also for the closer connection of that development with public policy analysis.

Models for subjective well-being, which require further development, are referred to in the center right cell. These models require connection with those of domain satisfactions (lower right corner of table); but the latter variables need also to be connected to objective policy-manipulable contributory variables that lie outside the concerns of psychology.

In the lower right-hand cell, we have suggested the need for a technical community concerned with general studies of equity. The beginning of such a community seems to have been made in the meetings leading up to the work of Alvarez et al. (1979). Consensus will be especially difficult to maintain in this field, however, because of the controversial nature of public (and scientists') valuations concerning equity. Generic planning, closely related to public policy analysis, is more concerned with equity and with service to the urban residents who are its ultimate clients; but these values may be closely connected with the specific field of urban planning and not easily extended to many other professional fields.

The development of policy indicators therefore requires a careful balance between the values deriving from relevance and the dangers of discontinuity and politicization. On the one hand, because policies revolve about values, so must the relevant indicators. On the other, some values are controversial and changing and thus potentially incompatible with stable data-gathering organizations and technical communities. The recognition of this problem played a part in the substantial conversion of the social indicators movement from a concern with policy to an emphasis on the scientific study of social change (Introduction).

Much of the realm of value related to public policy is linked to controversy and partisan politics. Some of the broadly supported values for which we now have policy indicators and models, including those of environmental quality and equity, owe their present acceptance to the past struggles of political groups and social movements to legitimize them. Some support for these groups and movements came from the analysis of statistics that were not expressly gathered for this purpose; but these statistics were available to both sides of the conflicts and have also been used to limit the claims of the environmental, feminist, and minority movements. Proponents of one side can make a contribution to public statistics by publicizing statistics as part of their argument. The other side can try to do likewise. Expertise in making these inferences would be desirable, but cannot be required by government regulation. Some experts may wish to join in this process as advocates; others must remain as the restrained suppliers of information and models, sustaining standards of statistical quality and retaining the confidence of both sides if possible. We must help the public to distinguish these roles clearly.

A major requirement for the further development of policy indicators is thus the organization of technical communities. Neither the internal development of basic science nor the demands of the market for professionals will adequately fulfill this need. From this perspective—analogous to that of market failure—we may be able to identify the technical communities needed for policy indicators and work through private as well as public channels to create or improve them.

The viability of technical communities depends on reducing the risks of external political attacks and internal political dissension. The members of a technical community concerned with policy indicators and models must be continually concerned with the possibility of political controversy, because they cannot avoid values as basic scientific communities do. A technical community can, however, reduce the risk of political controversy in several ways:

1. By centering its activities about a recognized profession that is the bearer of one or more values associated with its practice. Even these values may become controversial, however, when moved from the realm of professional practice to that of policy.

2. By serving external constituencies and their values without overly espousing one such value over another.

3. By serving other relatively noncontroversial values, such as that of economic efficiency. Such values might include the other end-values that we have proposed—subjective well-being and equity—though we must guard against possible controversies centered about them as well. Subjective well-being, though a counterweight to net economic benefit, has manipulative implications. Equity involves concern with redistributive policies and is difficult to study without taking the side of one group against another.

Because of the pervasiveness of values in the technical communities we need, the development of these communities will be facilitated if the part to be played by these values is widely understood. This requires an understanding on the part of academics in the basic sciences of the ways in which they can cooperate with such value-guided communities. It also requires understanding on the part of citizens and public leaders of the possible role of end-values in reconciliations and tradeoffs among more specific values (Chap.2). The development of these sorts of understanding will require time and effort; but if it can be achieved, it will contribute to the quality of our public deliberation.

11 The Choice of Policy Indicators

The choice of policy indicators raises many questions that have been largely neglected in the social indicators literature. These questions concern the anticipated use, and the best possible use, of public statistics within a political community. They include the choice of those values that are to be expressed by indicator statistics—a task in which indicator designers must interact with that community. They concern the development of causal models that will aid that community to choose courses of action when problems or opportunities are perceived. The effective development of these models requires better organization of practical research. If, in addition, we wish to reconcile or trade off particular social goals in terms of end-values, we must compare the uses of economic and social indicators for this purpose, seeking noneconomic indicators that permit tradeoffs and reconciliation. For this purpose, we are led to give particular emphasis to subjective well-being and equity. The integration of social systems, which has been discussed but not explored systematically in the social indicators literature, can also be given a place as a type of contributory variable. Once we make general proposals for indicators, we must also anticipate bias that can result from policy applications.

I have argued for a more practical concern on the part of expert communities, and a concern with more general ends in public debate, bringing the discourse of experts and the public closer together and improving the quality of public judgments. The policy indicators that we choose must thus link expert and political communities with one another. This task has led me to trace a complex web of interrelated factors that we must consider in our choice. Many of these factors are not easily measurable, including the anticipated benefits of using policy indicator statistics. Nevertheless, a wise choice of policy indicators requires that we keep such factors in mind.

Interrelated Factors in Indicator Choice

We may review the factors involved in indicator choice with the aid of Figure 11–1, which shows their (not strictly causal) interrelations.

The four blocks heavily outlined in the center row, from left to right, emphasize the potential effects of choosing the particular variables that are to be measured by policy indicators. These four blocks show a major sequence from choice of indicators, through the preparation and dissemination of statistics, to the public definition of problems, to policy choices and their benefits and costs (valued or disvalued results). This four block sequence was described in the Introduction and Chapter 1, where we stressed that the choice of policy indicators should depend on the expected effects of the resulting public policies.[1] Measurement of the incidence of a disease in time series, for example, might be expected to guide action to control that disease; measurement of unemployment might similarly lead to better calculated action to reduce it. If we are looking to the future, judgments about uses of indicator statistics are important and can be informed by analysis, even if such judgments are only qualitative conjectures. If we are reviewing our experience with an indicator series, we can judge more concretely that certain actions might have been neglected had the statistic not been available. In either case, such consequences should be a major concern in choosing of policy indicators.

A dashed reverse (leftward) arrow between the first two blocks indicates that the feasibility of preparing statistics should also affect the choice of variables. This arrow indicates not a simple effect of the choice of variables on later consequences, but the need for considering those consequences when we choose variables. If space permitted, we should include such reverse dashed arrows from "public definition of problems" and from "policy choices and their benefits and costs" back to "choice of policy indicator variables" as well.

The two blocks in the left-hand column, to the left of the heavily outlined blocks, show additional factors that can enter into our choice of variables. We can make a preliminary judgment concerning the benefits to be expected from indicators by assessing the information needs of the political community in question (Introduction, Chaps. 1, 9). Thus in planning information use for a city we can draw on the experience of other cities, which may be more relevant than the experience of nations. A nation stressing the private sector over the public may require statistics to guide private action more than policy indicators.

We must also consider the costs and other (by-product) benefits of introducing a variable. Costly information, such as measures of organizational integration based on extensive surveys (Chap. 8), may be low in our priorities even if it could be put to good use. Information that aids private as well as public action, such as statistics on business conditions (Chap. 1), may have a

1. The consequences of making indicator statistics available are not limited to direct effects on policies, however. These statistics (and the interpretations made of them) may arouse the public's hopes or assuage their concerns; they may enhance conflict or consensus, affecting not only what problems are dealt with and who gets what, but also the capacity of the political community to make decisions in the future.

Figure 11-1. Interrelated Factors in Choice of Policy Indicators (relevant chapter numbers are indicated in parentheses)

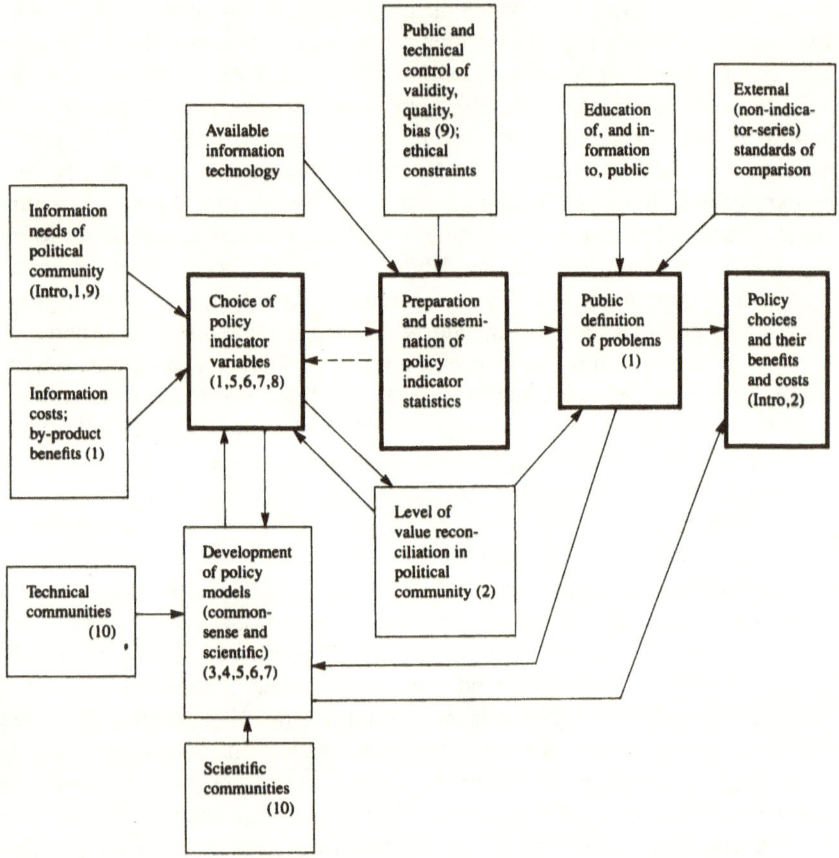

lesser incremental cost as an aid to policy, and may be produced to some extent in response to private demand through the market.

The top row of blocks, above the heavily outlined blocks, introduces further variables that affect the preparation of statistics and the public definition of problems. In preparing and disseminating public statistics we must consider the available information technology, including the public's ability to use it. In addition to printing aggregate reports and special tabulations, we can consider placing data on tape for analysis by interested groups; placing tabulations in computer storage available for access, together with explanatory text that the user can request; and (perhaps by private rather than governmental sources) making more vivid video presentations as well as text supplements available to explain statistics. Statistics based on technically designed indicator series thus overlap with those used by the mass media and can serve as sources for them.

The large top center block refers to various external controls that bear on the persons who prepare and disseminate indicator statistics. The validity of the measures is tested by both the public and relevant technical communities. So, too, is the statistical quality of sampling, data gathering, and reporting, as well as possible bias; these are largely questions for expert scrutiny, but the results of this scrutiny must be available to the public. In addition, there are ethical constraints concerning the protection of privacy and confidentiality, and the proper role of government in collecting data; these may be proposed by experts but are the concern of citizens generally.

Moving to the right of the large top center block, we see that the public's definition of problems is influenced by the education and information it receives apart from indicator statistics: news, background articles, literature, the arts, conversation, and personal experience form a context within which statistics are interpreted. In addition, problem definition depends on the standards of comparison against which present or anticipated conditions, based on statistics, are judged to be matters of concern. The rightmost block reminds us that these standards need not be derived solely from last year's statistics or from trends, but can come from comparison with other nations (e.g., infant mortality rates in Sweden) or from groups' efforts to redefine problems (e.g., equality for minorities and women).

The blocks below the main row of heavily outlined blocks recall some more complex conditions for indicator choice. Below the second of the heavily outlined blocks ("Preparation and dissemination") we have included the "level of value reconciliation" in the political community in question (Chap. 2). This level affects the choice of policy indicators (diagonal arrow upward to left). If there is a willingness in a given community to consider tradeoffs in terms of general end-values, then indicators of variables such as net economic benefit, subjective well-being, and general equity (Chaps. 5–7) may be of special value in that community.

When general end-values play a major role in shaping policy indicators, detailed study of the measurement and causes of these end-values is required. Expert communities and the public must consider their ethical meanings and seek their causes, relating them to more specific social goals. The values, models, and tradeoffs involved will not be the same for subjective well-being and equity as for economic benefit, but the interrelations among these end-values need to be studied. The expert communities who study them must both reach across disciplinary boundaries, and deal eventually with the public.

In Chapter 2 I stressed the possible advantages to a political community of debating public issues in such a way that particular values are reconciled in terms of more general values. By stressing general end-values in Chapters 5 through 7, I have suggested ways in which that debate can be raised above the level of social goal areas such as health, education, crime, and the environment, by comparing and reconciling them in terms of more general values: net economic benefit, subjective well-being, and equity. In addition, I have suggested (Chap. 8) that the integration of social systems can provide another general focus of public discussion—not as an end-value but as a contributory value that can affect many ends that we seek.

Our public debates do not at present center about ends of this degree of generality. Insofar as the political debates of the community center about more specific goals or administrative outcomes, then indicators of general end-values will be less useful. In such circumstances, we shall continue to reconcile particular values through ad hoc, casuistic arguments and through the political strength of the groups that espouse them. My hope in stressing general indicators, however, is to encourage a type of politics and public argument that involves a greater degree of reasoned reconciliation among particular values.

Some of our public leaders are concerned with general end-values, and the discipline of economics has led us to stress economic benefit; but if broader debates are to deal with a variety of end-values, a gradual process of education of the public must take place. Indicators of general end-values, and the interaction of the public with the technical communities that study them, can play a part in this education. There is thus a diagonal arrow from "choice of policy indicator variables" (toward the lower right) to "level of value reconciliation." Although the level of value reconciliation affects our choice of indicators, choice of indicators may also be expected to affect value reconciliation; the availability of economic statistics on the GNP, for example, may have encouraged the public to think more in terms of tradeoffs based on this more general value. Similarly, statistics on subjective well-being or general aspects of equity might encourage more consideration of tradeoffs based on them.

The level of value reconciliation in the community also affects the public's definition of problems (diagonal arrow to right and upward); in fact, it is a manifestation of the public's willingness to deal with general problems and tradeoffs.

The remaining three boxes, at the lower left of the figure, reflect the role of causal models in the choice and use of indicators. Directly below the first heavily outlined block is "development of policy models (commonsense and scientific)." These models of causation should affect our choice of indicators (upward arrow) because we should give priority to measuring those conditions on which we can act so as to improve them (Introduction). But sometimes we choose to measure conditions on which we can act only to a limited degree (e.g., unemployment or the diseases of old age), and this measurement increases our efforts to devise or discover better causal models to deal with them (downward arrow).

Our causal models are interrelated not only with our choice of policy indicators, but also with the public's definition of problems and the political community's choice of policies (last two heavily outlined blocks). There is thus an arrow downward and to the left from "public definition of problems" to "development of policy models," reflecting the fact that public concern can heighten the priority of developing models relevant to particular problems. If public concern is temporary and fluctuates, expert analyses cannot always supply corresponding models in time. Scientific communities (lowest block in diagram) are unlikely to respond to this concern, and technical communities (lower left block) can do so only if they have the capacity to prepare for it—by anticipating trends that lead to public concern, by maintaining a continuing interest in problems that arise repeatedly, or by developing adaptable skills and resources for model development.

Our models are also of major importance in influencing our policy choices (last heavily outlined block), and are thus connected with policy choices by an arrow to the right and upward. I have of course drastically abbreviated the complex process of democratic and representative policy formation by this single arrow; but I wish simply to show that the same models that affect our choice of policy indicators, and the same valued variables measured by those indicators and entering into the models, should be relevant to our choice of policies. Both the values measured (to a large extent), and models that tell us how we may influence them (to a considerable extent), can be expected to be chosen by the public and its representatives.

Policy Indicators and Technical Communities: Recommendations

The two boxes at the lower left, referring to scientific communities (bottom box) and technical communities (left, bottom), summarize a major argument of this book: that expert communities as presently constituted are inadequate for guiding the design and use of policy indicators. Basic scientific communities have important contributions to make to policy models; but technical communities should be developed on an equal footing with them. Institutional innovations are thus required within the universities, emphasizing the decision sciences and professional schools together with systematic ethics, to provide both better education and more career opportunities for members of these technical communities. Practical questions relevant to public debate need to be given equal importance with the theoretical concerns of basic disciplines. More importance must be given to values, especially general end-values, and to knowledge that aids decisions affecting them. This knowledge must be cultivated within organized technical communities, inside or outside existing disciplines, and must be part of a citizen's general education.

Our analysis of the choice of policy indicators has led to an extensive study of practical social science, including recommendations as to how it can be strengthened. This study is necessary for the choice of policy indicators, but it has a much wider relevance. The technical communities that need to be created and strengthened to support policy indicators are also needed to contribute to policy-relevant social science, including its combinations with natural science, for other purposes. Not all of our practical problems that call for public action are defined by indicator statistics; but the values and policy models that we need to supplement indicator statistics are likely to be relevant for many other policy choices as well. The creation of technical communities thus not only aids the use of policy indicators, but contributes as a joint product to the enlightenment of other policy choices.[2] When Sheldon and other critics of policy uses of social indicators judged that our related causal knowledge was inadequate to guide policy (Introduction), they were partly correct. The knowledge supplied by basic science is inadequate because it can at best provide only components of our policy models, not the entire models or even the combination of the components; technical communities are needed to organize these models more completely.[3] Our knowledge of social system models is inadequate, but we may not need general social system models to make policy. The public's knowledge of the expected effects of policies seems inadequate, as it is not usually subjected to the requirements of method that soundly based knowledge requires; but it is and will be a central ingredient of policy formation, and our task as practical scientists and educators is to improve it rather than to try to supplant it.

The critics of policy indicators were also correct in warning against the usurpation by experts of the democratic function of choosing priorities and values. It is doubtful that this sort of usurpation has actually occurred, either in the choice of social indicator concepts in *Toward a Social Report* (U.S. HEW 1969), in the OECD (1976) choice of indicator topics, or in other national or local social indicator systems. Some would argue, as the early proponents of social indicators did, that economists have assumed too large a place in forming our public values. The relative acceptance of their values in the United States suggests, however, that they have been in tune with dominant values of the society such as materialism and individualistic competition. I

2. Another joint product relationship that similarly complicates our choice of policy indicators is the use of policy indicators for private purposes.

3. Even when a technical community draws on only a single discipline, it still selects and organizes aspects of that discipline for its own purposes.

recommend that expert groups who advocate other values for policy indicators depart no further from the public's values than economists have done. It is unlikely that they can, because their proposals and their resources are limited by democratic processes; but even if they could, they should be cautious about doing so.

The design of policy indicator systems is thus not solely a question of analyzing the systems (such as the economy or the society) to which indicators refer; rather, it centrally involves what is to be done with this information. If we pay attention to these questions of use—the special needs of public policy debate, the reconciliation of values, the types of models needed, the avoidance of bias, and the kind of support needed from expert communities—we may become better able to choose and use policy indicators.

References

Acton, Jan Paul
1973 *Evaluating Public Programs to Save Lives: The Case of Heart Attacks*. Report R-950-RC. Santa Monica, Calif.: Rand Corporation.

Adams, Arvil V.
1981 "The American Work Force in the Eighties." In *America Enters the Eighties: Some Social Indicators*, edited by Conrad Taeuber. *Annals of the American Academy of Political and Social Science* 453 (January): 123–29.

Adams, F. Gerard, and Duggal, Vijaya G.
1976 "Anticipations Variables in an Econometric Model: Performance of the Anticipations Version of the Wharton Mark III." In *Econometric Model Performance*, edited by Lawrence R. Klein and Edwin Burmeister, pp. 9–26. Philadelphia: University of Pennsylvania Press.

Alexander, Karl L., Pallas, Aaron M., and Cook, Martha A.
1981 "Measure for Measure: On the Use of Endogenous Ability Data in School-Process Research." *American Sociological Review* 46, no. 5 (October): 619–31.

Allardt, Erik
1976 "Dimensions of Welfare in a Comparative Scandinavian Study." *Acta Sociologica* 19, no. 3: 227–39.
1977 "On the Relation Between Objective and Subjective Predicaments." Research Report No. 16, Research Group for Comparative Sociology, University of Helsinki.
1979 *Implications of the Ethnic Revival in Modern, Industrialized Society*. Helsinki: Societas Scientarum Fennica.

Allison, Graham T.
1971 *Essence of Decision*. Boston: Little, Brown.

Almond, Gabriel A., and Coleman, James Smoot, eds.
1960 *The Politics of the Developing Areas*. Princeton: Princeton University Press.

Alonso, William, and Starr, Paul
1982 "The Political Economy of National Statistics." *Items* 36, no. 3 (September): 29–35. [Social Science Research Council, New York].

Alterman, Hyman
1969 *Counting People: The Census In History*. New York: Harcourt, Brace and World.

Alterman, Rachelle, and MacRae, Duncan, Jr.
1983 "Planning and Policy Analysis: Converging or Diverging Trends?" *Journal of the American Planning Association* 49, no. 2 (March): 200–15.

Altshuler, Alan A.
1965 *The City Planning Process*. Ithaca: Cornell University Press.

Alvarez, Rodolfo, Lutterman, Kenneth G., and Associates
1979 *Discrimination in Organizations: Using Social Indicators to Manage Social Change*. San Francisco: Jossey-Bass.

Alves, Wayne M., and Rossi, Peter H.
1978 "Who Should Get What? Fairness Judgments in the Distribution of Earnings." *American Journal of Sociology* 84, no. 3 (November): 541–64.

Andersen, Kristi
1981 Review of Jonathan Cole, *Fair Science*. *American Political Science Review* 75, no. 3 (September): 765–66.

Andorka, Rudolf
1980 "Long-Term Social Development of Hungary, Measured by Social Indicators." *Social Indicators Research* 8, no. 1 (March): 1–13.

Andrews, Frank M.
1981 "Subjective Social Indicators, Objective Social Indicators, and Social Accounting Systems." In *Social Accounting Systems*, edited by F. Thomas Juster and Kenneth C. Land, pp. 377–419. New York: Academic Press.

Andrews, Frank M., and McKennell, Aubrey C.
1980 "Measures of Self-Reported Well-Being: Their Affective, Cognitive, and Other Components." *Social Indicators Research* 8, no. 2 (June): 127–55.

Andrews, Frank M., and Withey, Stephen B.
1976 *Social Indicators of Well-Being*. New York: Plenum.

Angell, Robert Cooley
1951 "The Moral Integration of American Cities." *American Journal of Sociology* 57, no. 1, part 2 (July): 1–140.
1974 "The Moral Integration of American Cities, II." *American Journal of Sociology* 80, no. 3 (November): 607–29.

Angrist, Shirley S., Belkin, Jacob, and Wallace, William A.
1976 "Social Indicators and Urban Policy Analysis." *Socio-Economic Planning Sciences* 10, no. 5: 193–98.

Argyris, Chris
1980 "Some Inner Contradictions in Management Information Systems." In *The Information Systems Environment*, edited by Henry C. Lucas, Jr., Frank F. Land, Timothy J. Lincoln, and Konrad Supper, pp. 15–25. Amsterdam: North-Holland.

Aristotle
1942 *Ethics: Ethica Nichomachea*. The Student's Oxford Aristotle, edited by W. D. Ross, vol. 5. London: Oxford University Press.

Armor, David J.
1980 "White Flight and the Future of School Desegregation." In *School Desegregation: Past, Present and Future*, edited by Walter G. Stephan and Joe R. Feagin, pp. 187–226. New York: Plenum.

Arrow, Kenneth J.
1963 *Social Choice and Individual Values*. 2d ed. New Haven: Yale University Press.

Ascher, William
1981 "The Forecasting Potential of Complex Models." *Policy Sciences* 13, no. 3 (June): 248–67.

Austin, J. L.
1961 "A Plea for Excuses." In *J. L. Austin, Philosophical Papers*, edited by J. O. Urmson and G. J. Warnock, pp. 123–52. London: Oxford University Press.

Averch, Harvey A., Carroll, Stephen J., Donaldson, Theodore S., Kiesling, Herbert J., and Pincus, John
1971 *How Effective Is Schooling? A Critical Review and Synthesis of Research Findings*. Santa Monica, Calif.: Rand Corporation.

Azar, Edward E., and Ben-Dak, Joseph D.
1975 *Theory and Practice of Events Research: Studies in Inter-Nation Actions and Interactions*. New York: Gordon and Breach.

Bailyn, Lotte, and Schein, Edgar H.
1976 "Life/Career Considerations as Indicators of Quality of Employment." In *Measuring Work Quality for Social Reporting*, edited by Albert D. Biderman and Thomas F. Drury, pp. 151–68. New York: Wiley.

Balkin, Steven
1979 "Victimization Rates, Safety and Fear of Crime." *Social Problems* 26, no. 3 (February): 343–58.

Barraclough, Robert E.
1971 "Geographic Coding." In President's Commission on Federal Statistics, *Federal Statistics*, vol. 2, pp. 221–95. Washington, D.C.: U.S. Government Printing Office.

Barry, Brian
1965 *Political Argument*. London: Routledge & Kegan Paul.
1979 "Justice as Reciprocity." In *Justice*, edited by Eugene Kamenka and Alice Erh-Soon Tay, pp. 50–78. London: Edward Arnold.
1983 "Intergenerational Justice in Energy Policy." In *Energy and the Future*, edited by Douglas MacLean and Peter G. Brown, pp. 15–30. Totowa, N.J.: Rowman and Littlefield.

Bator, Francis M.
1958 "The Anatomy of Market Failure." *Quarterly Journal of Economics* 72, no. 3 (August): 351–79.

Bauer, Raymond A.
1966 "Detection and Anticipation of Impact: The Nature of the Task." In *Social Indicators*, edited by Raymond A. Bauer, pp. 1–67. Cambridge, Mass.: M.I.T. Press.
1967 "Societal Feedback." *Annals of the American Academy of Political and Social Science* 373 (September), 180–92.

Baumgardt, David
1952 *Bentham and the Ethics of Today*. Princeton: Princeton University Press.

Baumol, William J.
1982 "Applied Fairness Theory and Rationing Policy." *American Economic Review* 72, no. 4 (September): 639–51.

Becker, Gary S.
1975 *Human Capital*. 2d ed. New York: National Bureau of Economic Research, Columbia University Press.
1976 *The Economic Approach to Human Behavior*. Chicago: University of Chicago Press.
1977 "A Theory of the Production and Allocation of Effort." Working Paper No. 184, Center for Economic Analysis of Human Behavior and Social Institutions, National Bureau of Economic Research, Stanford, California.

Behn, Robert D.
1975 "Termination: How the Massachusetts Department of Youth Services Closed the Public Training Schools" (1975); "Termination II" (1976); "Termination III" (1977). Working papers, Institute of Policy Sciences and Public Affairs, Duke University.

Behn, Robert D., and Vaupel, James W.
1982 *Quick Analysis for Busy Decision Makers*. New York: Basic Books.

Ben-David, Joseph
1971 *The Scientist's Role in Society: A Comparative Study*. Englewood Cliffs, N. J.: Prentice-Hall.

Bentham, Jeremy
1948 *The Principles of Morals and Legislation*. New York: Hafner. First published 1789.

Benveniste, Guy
1977 *Bureaucracy*. San Francisco: Boyd and Fraser.
1982 "Professionalization of Policy Experts." Paper presented at the Tenth World Congress of Sociology, Mexico City, 16–21 Aug. 1982.

Beresford, J., Billot, P., Hansen, H., Ollivier, Y., Willis, J., Staack, G., Gremion, P., and Jamous, H.
1976 *Information Systems in Government* 3. Marseille: Data for Development.

Berkowitz, S. D.
1982 *An Introduction to Structural Analysis*. Toronto: Butterworth.

Bernard, Jessie
1973 *The Sociology of Community*. Glenview, Ill.: Scott, Foresman.

Bharadwaj, Lakshmi, and Wilkening, E. A.
1977 "The Prediction of Perceived Well-Being." *Social Indicators Research* 4, no. 4: 421–39.

Biderman, Albert D.
1966a "Social Indicators and Goals." In *Social Indicators*, edited by Raymond A. Bauer, pp. 68–153. Cambridge, Mass.: M.I.T. Press.
1966b "Anticipatory Studies and Stand-by Research Capabilities." In *Social Indicators*, edited by Raymond A. Bauer, pp. 272–301. Cambridge, Mass.: M.I.T. Press.
1970 "Information, Intelligence, Enlightened Public Policy: Functions and Organization of Societal Feedback." *Policy Sciences* 1, no. 2 (Summer): 217–30.
1979 "Aversions to Social Concepts." *Proceedings of the Social Statistics Section*, pp. 328–33. [American Statistical Association, Washington, D.C.].

Biderman, Albert D., and Drury, Thomas F.
1976 "Social and Moral Qualities of Work as Social Indicators." In *Measuring Work Quality for Social Reporting*, edited by Albert D. Biderman and Thomas F. Drury, pp. 223–40. New York: Wiley.

Bidwell, Charles E., and Kasarda, John D.
1975 "School District Organization and Student Achievement." *American Sociological Review* 40, no. 1 (February): 55–70.
1980 "Conceptualizing and Measuring the Effects of School and Schooling." *American Journal of Education* 88, no. 4 (August): 401–30.

Black, Donald
1983 "Crime as Social Control." *American Sociological Review* 48, no. 1 (February): 34–45.

Blau, Peter M.
1964 *Exchange and Power in Social Life*. New York: Wiley.

Blau, Peter M., and Scott, W. Richard
1962 *Formal Organizations*. San Francisco: Chandler.

Blechman, Elaine A.
1982 "Are Children with One Parent at Psychological Risk? A Methodological Overview." *Journal of Marriage and the Family* 44, no. 1 (February): 179–95.

Bordua, David J., and Reiss, Albert J., Jr.
1967 "Law Enforcement." In *The Uses of Sociology*, edited by Paul F. Lazarsfeld, William H. Sewell, and Harold L. Wilensky, pp. 275–303. New York: Basic Books.

Bowen, Howard R.
1943 "The Interpretation of Voting in the Allocation of Economic Resources." *Quarterly Journal of Economics* 58, no. 1 (November): 27–48.

Bowles, Samuel, and Levin, Henry M.
1968 "The Determinants of Scholastic Achievement—An Appraisal of Some Recent Evidence." *Journal of Human Resources* 3, no. 1 (Winter): 3–24.

Bradburn, Norman M.
1969 *The Structure of Psychological Well-Being*. Chicago: Aldine.
1973 "The Generation and Utilization of Social Data." *Ethics* 84, no. 1 (October): 22–37.

Bradburn, Norman M., and Caplovitz, David
1965 *Reports on Happiness*. Chicago: Aldine.

Braybrooke, David, and Lindblom, Charles E.
1963 *A Strategy of Decision*. New York: Free Press.

Brecher, Charles, and Horton, Raymond D., eds.
1982 *Setting Municipal Priorities, 1982*. New York: Russell Sage Foundation.

Brenner, Harvey
1976 "Estimating the Social Costs of National Economic Policy: Implications for Mental and Physical Health and Criminal Aggression." Paper no. 5, Joint Economic Committee, Congress of the United States. Washington, D.C.: U.S. Government Printing Office.

Brewer, Garry D., and Brunner, Ronald D., eds.
1975 *Political Development and Change: A Policy Approach*. New York: Free Press.

Brewer, Garry D., and deLeon, Peter
1983 *The Foundations of Policy Analysis*. Homewood, Ill.: Dorsey.

Brickman, Philip, and Campbell, Donald T.
1971 "Hedonic Relativism and Planning the Good Society." In *Adaptation Level Theory: A Symposium*, edited by M. H. Appley, pp. 287–302. New York: Academic Press.

Brooks, Harvey
1982 "Science Indicators and Science Priorities." *Science, Technology, & Human Values* 7, no. 38 (Winter): 14–31.

Brookshire, David S., d'Arge, Ralph C., and Schulze, William D.
1980 "Experiments in Valuing Public Goods." In *Advances in Applied Micro-Economics*, vol. I, edited by V. Kerry Smith, pp. 123–72. Greenwich, Conn.: JAI Press.

Brown, B. W., Jr.
1972 "Statistics, Scientific Method, and Smoking." In *Statistics: A Guide to the Unknown*, edited by Judith M. Tanur et al., pp. 40–51. San Francisco: Holden-Day.

Brusegard, David A.
1978 "*Social Indicators, 1976* and *Perspective Canada II*: Elixirs of Reason or of Sleep?" In *America in the Seventies: Some Social Indicators*, edited by Richard D. Lambert. *Annals of the American Academy of Political and Social Science* 435 (January): 268–76.

Buchanan, James M., and Tullock, Gordon
1962 *The Calculus of Consent*. Ann Arbor: University of Michigan Press.

Bulmer, Martin
1982 *The Uses of Social Research*. London: George Allen & Unwin.
1983 "Science, Theory, and Values in Social Science Research on Poverty: The United States and Britain." In *Comparative Social Research*, vol. 6, edited by Richard F. Tomasson, pp. 353–69. Greenwich, Conn.: JAI Press.

Bumpass, Larry, and Rindfuss, Ronald R.
1979 "Children's Experience of Marital Disruption." *American Journal of Sociology* 85, no. 1 (July): 49–65.

Burns, Arthur F.
1966 *The Management of Prosperity*. New York: Columbia University Press.

Bursik, Robert J., and Meade, Anthony C.
1982 "A New Look at the Concept of Delinquency Seriousness." Paper presented at the World Congress of Sociology, Mexico City, August 1982.

Burt, Ronald S.
1978 "Cohesion Versus Structural Equivalence as a Basis for Network Subgroups." *Sociological Methods and Research* 7, no. 2 (November): 189–212.
1980 "Models of Network Structure." *Annual Review of Sociology* 6: 79–141.

Burt, Ronald S., Wiley, James A., Minor, Michael J., and Murray, James R.
1978 "Structure of Well-Being: Form, Content, and Stability Over Time." *Sociological Methods and Research* 6, no. 3 (February): 365–407.

Burton, Nancy W., and Jones, Lyle V.
1982 "Recent Trends in Achievement Levels of Black and White Youth." *Educational Researcher* 11, no. 4 (April): 10–14.

Buttel, Frederick H., Wilkening, E. A., and Martinson, Oscar B.
1977 "Ideology and Social Indicators of Quality of Life." *Social Indicators Research* 4, no. 3 (July): 353–69.

Cain, Glen G., and Watts, Harold W.
1970 "Problems in Making Policy Inferences from the Coleman Report." *American Sociological Review* 35, no. 2 (April): 228–42.

Calabresi, Guido, and Bobbitt, Philip
1978 *Tragic Choices*. New York: Norton.

Campbell, Angus
1981 *The Sense of Well-Being in America*. New York: McGraw-Hill.

Campbell, Angus, Converse, Philip E., and Rodgers, William L.
1976 *The Quality of American Life*. New York: Russell Sage Foundation.

Campbell, Donald T.
1969 "Reforms as Experiments." *American Psychologist* 24, no. 4 (April): 409–29.
1971 "Methods for the Experimenting Society." Paper presented to the American Psychological Association.
1976 "Focal Local Indicators for Social Program Evaluation." *Social Indicators Research* 3, no. 2 (July): 237–56.

Caplan, Nathan, and Barton, Eugenia
1976 "Social Indicators, 1973: A Study of the Relationship Between the Power of Information and Utilization by Federal Executives." Program in Social Research and Public Policy, University of Michigan.

Caplow, Theodore
1976 *How to Run Any Organization*. Hinsdale, Ill.: Dryden.
1982 *Middletown Families*. Minneapolis: University of Minnesota Press.

Carley, Michael
1980 *Rational Techniques in Policy Analysis*. London: Heinemann.
1981 *Social Measurement and Social Indicators*. London: George Allen and Unwin.

Carlisle, Elaine
1972 "The Conceptual Structure of Social Indicators." In *Social Indicators and Social Policy*, edited by Andrew Shonfield and Stella Shaw, pp. 23–32. London: Heinemann.

Carter, Genevieve W.
1966 "Measurement of Need." In *Social Work Research*, edited by Norman A. Polansky, pp. 201–21. Chicago: University of Chicago Press.

Chen, Huey-Tsyh, and Rossi, Peter H.
1980 "The Multi-Goal, Theory-Driven Approach to Evaluation: A Model Linking Basic and Applied Social Science." *Social Forces* 59, no. 1 (September): 106–22.

Chen, Milton M., Bush, J.W., and Patrick, Donald L.
1975 "Social Indicators for Health Planning and Policy Analysis." *Policy Sciences* 6, no. 1 (March): 71–89.

Cherlin, Andrew J.
1981 *Marriage, Divorce, Remarriage*. Cambridge: Harvard University Press.

Chernoff, Herman
1978 "Decision Theory." In *International Encyclopedia of Statistics*, vol. I, edited by William H. Kruskal and Judith M. Tanur, pp. 131–35. New York: Free Press.

Choldin, Harvey M.
1980 "Electronic Community Fact Books." *Urban Affairs Quarterly* 15, no. 3 (March): 269–89.

Christ, Carl F.
1976 "Judging the Performance of Econometric Models of the U. S. Economy." In *Econometric Model Performance*, edited by Lawrence R. Klein and Edwin Burmeister, pp. 322–42. Philadelphia: University of Pennsylvania Press.

Cicchetti, Charles J., Fisher, Anthony C., and Smith, V. Kerry
1976 "An Econometric Evaluation of a Generalized Consumer Surplus Measure: The Mineral King Controversy." *Econometrica* 44, no. 6 (November): 1259–76.

Clark, Terry N.
1973 "Community Social Indicators: From Analytical Models to Policy Applications." *Urban Affairs Quarterly* 9, no. 1 (September): 3–36.

Clawson, Marion
1959 "Methods of Measuring the Demand for and Value of Outdoor Recreation." Resources for the Future, Washington, D.C.

Coase, R. H.
1960 "The Problem of Social Cost." *Journal of Law and Economics* 3: 1–44.

Cohen, David K., and Garet, Michael S.
1975 "Reforming Educational Policy with Applied Social Research." *Harvard Educational Review* 45, no. 1 (February): 17–43.

Cohen, Jacqueline
1978 "The Incapacitation Effect of Imprisonment: A Critical Review of the Literature." In *Deterrence and Incapacitation: Estimating the Effects of Criminal Sanctions on Crime Rates*, edited by Alfred Blumstein, Jacqueline Cohen, and Daniel Nagin, pp. 187–243. Washington, D.C.: National Academy of Sciences.

Cohen, Lawrence E., and Felson, Marcus
1979 "On Estimating the Social Costs of National Economic Policy: A Critical Examination of the Brenner Study." *Social Indicators Research* 6, no. 2 (April): 251–59.

Cohon, Jared L.
1978 *Multiobjective Programming and Planning*. New York: Academic Press.

Cole, Jonathan R.
1979 *Fair Science: Women in the Scientific Community*. New York: Free Press.

Coleman, James Samuel

1961 *The Adolescent Society*. New York: Free Press.

1970 "Political Money." *American Political Science Review* 64, no. 4 (December): 1074–87.

1972a "Integration of Sociology and the Other Social Sciences Through Policy Analysis." In *Integration of the Social Sciences through Policy Analysis*, edited by James C. Charlesworth, pp. 162–74. Philadelphia: American Academy of Political and Social Science.

1972b *Policy Research in the Social Sciences*. Morristown, N.J.: General Learning Press.

1974 *Power and the Structure of Society*. New York: Norton.

1975 "Racial Segregation in the Schools: New Research with New Policy Implications." *Phi Delta Kappan* 57, no. 2 (October): 75–78.

1980 "The Structure of Society and the Nature of Social Resources." *Knowledge* 1, no. 3 (March): 333–50.

Coleman, James S., Campbell, Ernest Q., Hobson, Carol J., McPartland, James, Mood, Alexander M., Weinfeld, Frederic D., and York, Robert L.

1966 *Equality of Educational Opportunity*. Washington, D.C.: U.S. Government Printing Office.

Coleman, James S., Hoffer, Thomas, and Kilgore, Sally

1981 *Public and Private Schools*. Draft report released by the National Center for Education Statistics, Washington, D.C., 7 April 1981.

1982 *High School Achievement: Public, Catholic, and Private Schools Compared*. New York: Basic Books.

Coleman, James S., Kelly, Sara D., and Moore, John A.

1975 *Trends in School Segregation, 1968–73*. Washington, D.C.: The Urban Institute.

Coleman, James S., and MacRae, Duncan, Jr.

1960 "Electronic Processing of Sociometric Data for Groups up to 1,000 in Size." *American Sociological Review* 25, no. 5 (October): 722–27.

Coltham, Jeannette B.

1972 "Educational Accountability: An English Experiment and Its Outcome." *University of Chicago School Review* 81, no. 1 (November): 15–34.

Cook, Philip J.

1980 "Research in Criminal Deterrence: Laying the Groundwork for the Second Decade." In *Crime and Justice: An Annual Review of Research*, vol. 2, edited by Norval Morris and Michael Tonry, pp. 211–68. Chicago: University of Chicago Press.

Cook, Thomas D., and Campbell, Donald T.

1979 *Quasi-Experimentation: Design and Analysis Issues for Field Settings*. Boston: Houghton Mifflin.

Coser, Lewis A.

1974 *Greedy Institutions*. New York: Free Press.

Costa, Paul T., Jr., and McCrae, Robert R.
1980 "Influence of Extraversion and Neuroticism on Subjective Well-Being: Happy and Unhappy People." *Journal of Personality and Social Psychology* 38, no. 4 (April): 668–78.

Crain, Robert L., and Weisman, Carol Sachs
1972 *Discrimination, Personality, and Achievement.* New York: Seminar.

Cronbach, Lee J.
1982 *Designing Evaluations of Educational and Social Programs.* San Francisco: Jossey-Bass.

Culyer, A. J., Lavers, R. J., and Williams, Alan
1972 "Health Indicators." In *Social Indicators and Social Policy*, edited by Andrew Shonfield and Stella Shaw, pp. 94–118. London: Heinemann.

Danziger, James N.
1977 "Computers, Local Governments, and the Litany to EDP." *Public Administration Review* 37, no. 1 (January/February): 28–37.

Darling, A. H.
1973 "Measuring Benefits Generated by Urban Water Parks." *Land Economics* 49, no. 1 (February): 122–34.

Dasgupta, Partha, Sen, Amartya, and Marglin, Stephen
1972 *Guidelines for Project Evaluation.* New York: United Nations.

Davis, Robert C.
1972 "The Beginnings of American Social Research." In *Nineteenth-Century American Science: A Reappraisal*, edited by George H. Daniels, pp. 152–78. Evanston, Ill.: Northwestern University Press.

Deacon, Robert and Shapiro, Perry
1975 "Private Preference for Collective Goods Revealed Through Voting on Referenda." *American Economic Review* 65, no. 5 (December): 943–55.

de Neufville, Judith Innes
1975 *Social Indicators and Public Policy.* New York: Elsevier.
1978 "Validating Policy Indicators." *Policy Sciences* 10, no. 2/3 (December): 171–88.
1984 "Functions of Statistics in State and Local Decision Making." Paper prepared for a monograph on "The Political Economy of Statistics," edited by William Alonso and Paul Starr, sponsored by The Social Science Research Council (New York).

Dentler, Robert A.
1980 "The Boston School Desegregation Case: Implications for Policy Research and Development." In *Problems in American Social Policy Research*, edited by Clark C. Abt, pp. 205–11. Cambridge: Abt.

Deutsch, Karl W.

1953 *Nationalism and Social Communication*. Cambridge: Technology Press, and New York: Wiley.

1956 "Shifts in the Balance of Communication Flows: A Problem of Measurement in International Relations." *Public Opinion Quarterly* 20, no. 1 (Spring): 143–60.

1963 *The Nerves of Government*. New York: Free Press.

Dibble, Vernon K.

1962 "Occupations and Ideologies." *American Journal of Sociology* 68, no. 2 (September): 229–41.

Dickson, Paul

1971 *Think Tanks*. New York: Ballantine.

Diemer, Joel A., and McKean, John R.

1978 "The Assessment of Community Preference: A Methodology and a Case Study." *Land Economics* 54, no. 2 (May): 244–52.

Dorfman, Robert

1977 "Incidence of the Benefits and Costs of Environmental Programs." *American Economic Review* 67, no. 1 (February): 333–40.

Dornbusch, Rudiger, and Fischer, Stanley

1978 *Macroeconomics*. New York: McGraw-Hill.

Downs, Anthony

1967 "A Realistic Look at the Final Payoffs from Urban Data Systems." *Public Administration Review* 27, no. 3 (September): 204–10.

1972 "Up and Down with Ecology—the 'Issue-Attention Cycle.'" *The Public Interest* 28 (Summer): 38–51.

Dreeben, Robert, and Thomas, J. Alan

1980 "Introduction." In *The Analysis of Educational Productivity, Vol. I: Issues in Microanalysis*, edited by Robert Dreeben and J. Alan Thomas, pp. 1–12. Cambridge: Ballinger.

Duncan, Otis Dudley

1967 "Discrimination Against Negroes." In *Social Goals and Indicators for American Society*, edited by Bertram M. Gross. *Annals of the American Academy of Political and Social Science* 371 (May): 85–103.

1969 *Toward Social Reporting: Next Steps*. New York: Russell Sage Foundation.

1975 *Introduction to Structural Equation Models*. New York: Academic Press.

1981 "Statement." In Panel on Survey Measurement of Subjective Phenomena, National Research Council, *Surveys of Subjective Phenomena: Summary Report*, pp. 93–95. Washington, D.C.: National Academy Press.

Dunn, Edgar S., Jr.

1974 *Social Information Processing and Statistical Systems—Change and Reform*. New York: Wiley.

Durkheim, Emile
1947 *The Division of Labor in Society*. Translated by George Simpson. Glencoe, Ill.: Free Press. Original French edition, 1893.
1951 *Suicide: A Study in Sociology*. Translated by John A. Spaulding and George Simpson. Glencoe, Ill.: Free Press. Original French edition, 1897.

Dye, Thomas R.
1981 *Understanding Public Policy*. 4th ed. Englewood Cliffs, N.J.: Prentice-Hall.

Easterlin, Richard A.
1974 "Does Economic Growth Improve the Human Lot? Some Empirical Evidence." In *Nations and Households in Economic Growth*, edited by Paul A. David and Melvin W. Reder, pp. 89–125. New York: Academic Press.

Easton, David
1965 *A Systems Analysis of Political Life*. New York: Wiley.

Eckland, Bruce K.
1980 "The Efficient Use of Educational Data: A Proposal for the National Longitudinal Study of the High School Class of 2002." In *The Analysis of Educational Productivity, Vol. II: Issues in Macroanalysis*, edited by Charles E. Bidwell and Douglas M. Windham, pp. 93–151. Cambridge: Ballinger.

Ehrlich, Isaac
1975 "The Deterrent Effect of Capital Punishment: A Question of Life and Death." *American Economic Review* 65, no. 3 (June): 397–417.
1981 "On the Usefulness of Controlling Individuals: An Economic Analysis of Rehabilitation, Incapacitation, and Deterrence." *American Economic Review* 71, no. 3 (June): 307–22.

Elder, Glen H., Jr.
1974 *Children of the Great Depression*. Chicago: University of Chicago Press.

Emery, James C.
1975 "Integrated Information Systems and Their Effects on Organizational Structure." In *Information Systems and Organizational Structure*, edited by Erwin Grochla and Norbert Szyperski, pp. 95–103. Berlin: Walter de Gruyter.

Erlandson, Robert F.
1981 "A Community Developed Knowledge Base System and Its Impact on a School Closing Decision." *IEEE Transactions on Systems, Man, and Cybernetics* SMC-11, no. 4 (April): 253–61.

Etzioni, Amitai
1968 *The Active Society*. New York: Free Press.
1975 *A Comparative Analysis of Complex Organizations*. Rev. and enlarged ed. New York: Free Press.

Etzioni, Amitai, and Lehman, Edward W.
1967 "Some Dangers in 'Valid' Social Measurement." *Annals of the American Academy of Political and Social Science* 373 (September): 1–15.

Fairweather, George W., ed.
1980 *The Fairweather Lodge, a Twenty-Five Year Retrospective*. San Francisco: Jossey-Bass.

Fairweather, George W., and Tornatzky, Louis G.
1977 *Experimental Methods for Social Policy Research*. New York: Pergamon.

Felson, Marcus, and Kenneth C. Land
1980 "Linking Education to the Larger Society with Social Indicator Models." In *The Analysis of Educational Productivity, Vol. II: Issues in Macroanalysis*, edited by Charles E. Bidwell and Douglas M. Windham, pp. 65–91. Cambridge: Ballinger.

Ferejohn, John, and Page, Talbot
1978 "On the Foundations of Intertemporal Choice." *American Journal of Agricultural Economics* 60, no. 2 (May): 269–75.

Fergus, Esther O.
1980 "Maintaining and Advancing the Lodge Effort." In *The Fairweather Lodge, a Twenty-Five Year Retrospective*, edited by George W. Fairweather, pp. 43–56. San Francisco: Jossey-Bass.

Fernandez, Roberto M., and Kulik, Jane C.
1981 "A Multilevel Model of Life Satisfaction: Effects of Individual Characteristics and Neighborhood Composition." *American Sociological Review* 46, no. 6 (December): 840–50.

Feuer, Lewis S.
1959 *Marx and Engels: Basic Writings on Politics and Philosophy*. Garden City, N.Y.: Doubleday Anchor.

Fienberg, Stephen E., and Goodman, Leo A.
1974 "*Social Indicators, 1973*: Statistical Considerations." In *Social Indicators, 1973: A Review Symposium*, edited by Roxann A. Van Dusen, pp. 63–82. Washington, D.C.: Social Science Research Council.

Fisher, Anthony C., and Peterson, Frederick M.
1976 "The Environment in Economics: A Survey." *Journal of Economic Literature* 14, no. 1 (March): 8–11.

Fisher, Franklin M., Griliches, Zvi, and Kaysen, Carl
1962 "The Costs of Automobile Model Changes Since 1949." *Journal of Political Economy* 70, no. 5 (October): 433–51.

Fisher, Franklin M., and Nagin, Daniel
1978 "On the Feasibility of Identifying a Crime Function in a Simultaneous Model of Crime Rates and Sanction Levels." In *Deterrence and Incapacitation: Estimating the Effects of Criminal Sanctions on Crime Rates*, edited by Alfred Blumstein, Jacqueline Cohen, and Daniel Nagin, pp. 361–99. Washington, D.C.: National Academy of Sciences.

Flemming, J. S.
1976 *Inflation*. London: Oxford University Press.

Fox, John W.
1980 "Gove's Specific Sex-Role Theory of Mental Illness: A Research Note." *Journal of Health and Social Behavior* 21, no. 2 (September): 260–67.

Fox, Karl A.
1974 *Social Indicators and Social Theory: Elements of an Operational System*. New York: Wiley.

Frankena, William K.
1963 *Ethics*. Englewood Cliffs, N.J.: Prentice-Hall.

Freedman, Jonathan
1978 *Happy People*. New York: Ballantine.

Freeman, A. Myrick, III.
1977 "Project Design and Evaluation with Multiple Objectives." In *Public Expenditure and Policy Analysis*, 2d ed., edited by Robert H. Haveman and Julius Margolis, pp. 239–56. Chicago: Rand McNally.
1980 "On Measuring Public Goods Demand from Market Data." In *Advances in Applied Micro-Economics*, vol. 1, edited by V. Kerry Smith, pp. 13–29. Greenwich, Conn.: JAI Press.

Freidson, Eliot
1970 *Profession of Medicine*. New York: Dodd, Mead.

Fromm, Gary, and Taubman, Paul
1968 *Policy Simulations with an Econometric Model*. Washington, D.C.: Brookings.

Furstenburg, Frank F., Jr., Nord, Christine Winquist, Peterson, James L., and Zill, Nicholas
1983 "The Life Course of Children of Divorce: Marital Disruption and Parental Contact." *American Sociological Review* 48, no. 5 (October): 656–68.

Gallup, George H.
1977 "Human Needs and Satisfactions: A Global Survey." *Public Opinion Quarterly* 41 (Winter): 459–67.

Gandy, Oscar H., Jr.
1982 *Beyond Agenda Setting: Information Subsidies and Public Policy*. Norwood, N.J.: Ablex.

Garfinkel, Harold
1967 *Studies in Ethnomethodology*. Englewood Cliffs, N.J.: Prentice-Hall.

Garn, Harvey A., Flax, Michael J., Springer, Michael, and Taylor, Jeremy B.
1976 *Models for Indicator Development: A Framework for Policy Analysis*. Washington, D.C.: Urban Institute.

Gates, Bruce L.
1975a "Needs-Based Budgeting: Considerations of Effectiveness, Efficiency and Justice in the Delivery of Human Services." Paper presented at the annual meeting of the American Political Science Association, San Francisco (September).

1975b "Knowledge Management in the Technological Society: Government by Indicator." *Public Administration Review* 35, no. 6 (November/ December): 589–93.

Gauron, A., Guillaume, H., Maurice, J., and Milleron, J. C.
1982 "Use of Macro-economic Studies in Preparation of the French VIIIth Plan." In *Evaluating the Reliability of Macro-economic Models*, edited by Gregory C. Chow and Paolo Corsi, pp. 281–301. New York: Wiley.

Gell-Mann, Murray
1971 "How Scientists Can Really Help." *Physics Today* 24, no. 5 (May): 23–25.

Gibbs, Jack P.
1975 *Crime, Punishment, and Deterrence*. New York: Elsevier.

Gieryn, Thomas F.
1983 "Boundary-Work and the Demarcation of Science from Non-Science: Strains and Interests in Professional Ideologies of Scientists." *American Sociological Review* 48, no. 6 (December): 781–95.

Gilbert, John P., Light, Richard J., and Mosteller, Frederick
1975 "Assessing Social Innovations: An Empirical Base for Policy." In *Evaluation and Experiment*, edited by Carl A. Bennett and Arthur A. Lumsdaine, pp. 39–193. New York: Academic Press.

Gintis, Herbert M.
1970 "Neo-Classical Welfare Economics and Individual Development." *Occasional Papers of the Union of Radical Political Economists*, no. 3.

Goldberger, Arthur S., and Cain, Glen G.
1982 "The Causal Analysis of Cognitive Outcomes in the Coleman, Hoffer, and Kilgore Report." *Sociology of Education* 55, no. 2/3 (April/ July): 103–22.

Goode, William J.
1982 *The Family*. 2d. ed. Englewood Cliffs, N.J.: Prentice-Hall.

Goodin, Robert E.
1982a "Discounting Discounting." *Journal of Public Policy* 2, no. 1 (February): 53–72.
1982b *Political Theory and Public Policy*. Chicago: University of Chicago Press.

Gordon, C. Wayne, and Babchuck, Nicholas
1959 "A Typology of Voluntary Associations." *American Sociological Review* 24, no. 2 (February): 22–29.

Gove, Walter R., and Tudor, Jeannette F.
1973 "Adult Sex Roles and Mental Illness." *American Journal of Sociology* 78, no. 4 (January): 812–35.

Gramlich, Frederick W.
1977 "The Demand for Clean Water: The Case of the Charles River." *National Tax Journal* 30, no. 2 (June): 183–94.

Granovetter, Mark
1982 "The Strength of Weak Ties: A Network Theory Revisited." In *Social Structure and Network Analysis*, edited by Peter V. Marsden and Nan Lin, pp. 105–30. Beverly Hills, Calif.: Sage Publications.

Green, Philip
1971 "The Obligations of American Social Scientists." *Annals of the American Academy of Political and Social Science* 394 (March): 13–27.

Greenberg, David, Lipson, Al, and Rostker, Bernard
1976 "Technical Success, Political Failure: The Incentive Pay Plan for California Job Agents." *Policy Analysis* 2, no. 4 (Fall): 545–75.

Greene, Kenneth V., Neenan, William B., and Scott, Claudia D.
1974 *Fiscal Interactions in a Metropolitan Area*. Lexington, Mass.: Lexington.

Greenwood, Michael J.
1975 "Research on Internal Migration in the United States: A Survey." *Journal of Economic Literature* 13, no. 2 (June): 397–433.

Greenwood, Peter W.
1983 "Controlling the Crime Rate Through Imprisonment." In *Crime and Public Policy*, edited by James Q. Wilson, pp. 251–69. San Francisco: ICS Press.

Griliches, Zvi
1977 "Estimating the Returns to Schooling: Some Econometric Problems." *Econometrica* 45, no. 1 (January): 1–22.
1978 "Economic Problems of Measuring Returns on Research." In *Toward a Metric of Science*, edited by Yehuda Elkana et al., pp. 171–77. New York: Wiley.

Grodzins, Morton
1956 *The Loyal and the Disloyal*. Chicago: University of Chicago Press.

Gross, Bertram M.
1966a "Preface: A Historical Note on Social Indicators." In *Social Indicators*, edited by Raymond A. Bauer, pp. ix-xvii. Cambridge: M.I.T. Press.
1966b "The State of the Nation: Social Systems Accounting." In *Social Indicators*, edited by Raymond A. Bauer, pp. 154–271. Cambridge: M.I.T. Press.

Gusfield, Joseph R.
1975 *Community: A Critical Response*. New York: Harper & Row.

Gustafsson, Gunnel, and Richardson, J. J.
1979 "Concepts of Rationality and the Policy Process." *European Journal of Political Research* 7, no. 4 (December): 415–36.

Haberer, Joseph
1969 *Politics and the Community of Science*. New York: Van Nostrand Reinhold.

Hagger, A. J.
1977 *Inflation: Theory and Policy*. London: Macmillan.
Hagstrom, Warren
1965 *The Scientific Community*. New York: Basic Books.
Hall, John
1975 "The Relationship between Subjective and Objective Indicators of Individual Well-Being—A Linear Modelling Approach." Unpublished paper, Survey Unit, Social Science Research Council, London (July).
Hampshire, Stuart
1973 "Morality and Pessimism." *The New York Review of Books* 19, no. 11–12 (25 January): 26–33.
Hanke, Steve H., and Walker, Richard A.
1977 "Benefit-Cost Analysis Reconsidered: An Evaluation of the Mid-State Project." In *Public Expenditure and Policy Analysis*, 2d ed., edited by Robert H. Haveman and Julius Margolis, pp. 329–54. Chicago: Rand McNally.
Hansen, W. Lee, and Weisbrod, Burton A.
1969 *Benefits, Costs and Finance of Public Higher Education*. Chicago: Markham.
Hanushek, Eric A.
1979 "Conceptual and Empirical Issues in the Estimation of Educational Production Functions." *Journal of Human Resources* 14, no. 3 (Summer): 351–88.
1981 "Throwing Money at Schools." *Journal of Policy Analysis and Management* 1, no. 1 (Fall): 19–41.
Hanushek, Eric A., and Kain, John F.
1972 "On the Value of *Equality of Educational Opportunity* as a Guide to Public Policy." In *On Equality of Educational Opportunity*, edited by Frederick Mosteller and Daniel P. Moynihan, pp. 116–45. New York: Vintage.
Harberger, Arnold C.
1971 "Three Basic Postulates for Applied Welfare Economics: An Interpretive Essay." *Journal of Economic Literature* 9, no. 3 (September): 785–97.
Hargrove, Erwin C., and Morley, Samuel
1984 "Introduction." In *The President and the Council of Economic Advisers, Interviews with CEA Chairmen*, pp. 1–44. Boulder, Colo.: Westview Press.
Harnischfeger, Annegret, and Wiley, David E.
1980 "Determinants of Pupil Opportunity." In *The Analysis of Educational Productivity, Vol. I: Issues in Microanalysis*, edited by Robert Dreeben and J. Alan Thomas, pp. 223–66. Cambridge: Ballinger.
Harris, Anthony H.
1980 "The Hedonic Technique and the Valuation of Environmental Quality." In *Advances in Applied Micro-Economics*, vol. 1,

edited by V. Kerry Smith, pp. 31–49. Greenwich, Conn.: JAI Press.

Harrison, Jeffery, and Grant, Peter
1976 *The Thames Transformed: London's River and its Waterfowl.* London: André Deutsch.

Harvard Educational Review
1981 *Harvard Educational Review* 51, no. 4 (November).

Haskins, Ron, Burnett, Charles K., and Dobelstein, Andrew W.
1983 "Single Parent Families: Policy Recommendations for Child Support." Unpublished paper, Bush Institute for Child and Family Policy, University of North Carolina at Chapel Hill.

Hatry, Harry P.
1978 "The Status of Productivity Measurement in the Public Sector." *Public Administration Review* 38, no. 1 (January/February): 28–33.

Hatry, Harry P., and Fisk, Donald M.
1972 *Improving Productivity and Productivity Measurement in Local Governments*. Washington, D.C.: Prepared for the National Commission on Productivity, 1971.

Hauser, Philip M.
1975 *Social Statistics in Use*. New York: Russell Sage Foundation.

Haveman, Robert H., and Margolis, Julius, eds.
1977 *Public Expenditure and Policy Analysis*. 2d ed. Chicago: Rand McNally.

Haveman, Robert H., and Weisbrod, Burton A.
1975 "Defining Benefits for Public Programs: Some Guidance for Policy Analysts." *Policy Analysis* 1, no. 1 (Winter): 169–96.

Headey, Bruce, Holmström, Elsie, and Wearing, Alexander
1984a "Well-Being and Ill-Being: Different Dimensions?" *Social Indicators Research* 14, no. 2 (February): 115–39.
1984b "The Impact of Life Events and Changes in Domain Satisfactions on Well-Being." *Social Indicators Research* (forthcoming).

Helfgot, Joseph
1974 "Professional Reform Organizations and the Symbolic Representation of the Poor." *American Sociological Review* 39, no. 4 (August): 475–91.

Henderson, Lawrence J.
1913 *The Fitness of the Environment*. New York: Macmillan.

Henderson, Scott, with Byrne, Donald G., and Duncan-Jones, Paul
1981 *Neurosis and the Social Environment*. New York: Academic Press.

Hennis, Wilhelm
1957 *Meinungsforschung und Repräsentative Demokratie*. Tübingen: Mohr.

Henriot, Peter J.
1970 "Political Questions about Social Indicators." *Western Political Quarterly* 23, no. 2 (June): 235–55.

Hensher, David A.
1976 "Review of Studies Leading to Existing Values of Travel Time." In *Value of Travel Time. Transportation Research Record*, no. 587: 30–41.

Heyns, Barbara
1978 *Summer Learning and the Effects of Schooling*. New York: Academic Press.
1980 "Models and Measurement for the Study of Cognitive Growth." In *The Analysis of Educational Productivity, Vol. I: Issues in Microanalysis*, edited by Robert Dreeben and J. Alan Thomas, pp. 13–52. Cambridge: Ballinger.

Hillery, George A., Jr.
1982 *A Research Odyssey: Developing and Testing a Community Theory*. New Brunswick, N.J.: Transaction.

Hirsch, Fred
1976 *Social Limits to Growth*. Cambridge: Harvard University Press.

Hoffman, Carl, and Reed, John Shelton
1981 "Sex Discrimination?—The XYZ Affair." *The Public Interest* 62 (Winter): 21–39.

Hoffmann, Stanley H.
1984 "Universities and Human Rights." *Human Rights Quarterly* 6, no. 1 (February): 5–20.

Holloway, Don C.
1973 "Evaluating Health Status for Utilization Review." In *Health Status Indices*, edited by Robert L. Berg, pp. 89–98. Chicago: Hospital Research and Educational Trust.

Horn, R. V.
1980 "Social Indicators: Meaning, Methods, and Applications." *International Journal of Social Economics* 7, no. 8: 421–60.

House, James S.
1981 "Social Indicators, Social Change, and Social Accounting: Toward More Integrated and Dynamic Models." In *Social Accounting Systems*, edited by F. Thomas Juster and Kenneth C. Land, pp. 421–52. New York: Academic Press.

Hughes, G. David
1973 *Demand Analysis for Marketing Decisions*. Homewood, Ill.: Irwin.

Hughes, G. David, and Ray, Michael L.
1974 *Buyer/Consumer Information Processing*. Chapel Hill: University of North Carolina Press.

Hunt, Leon G.
1971 "Management Information Systems." In *Federal Statistics*, vol. 2, President's Commission on Federal Statistics, pp. 437–54. Washington, D.C.: U.S. Government Printing Office.

Huxley, Aldous L.
1932 *Brave New World*. Garden City, N.Y.: Sun Dial.

Hylland, Aanund, and Zeckhauser, Richard
1979 "A Mechanism for Selecting Public Goods When Preferences Must Be Elicited." Discussion Paper Series. Cambridge, Mass.: John F. Kennedy School of Government, Harvard University.

Inglehart, Ronald
1977 "Values, Objective Needs, and Subjective Satisfactions among Western Publics." *Comparative Political Studies* 9, no. 4 (January): 429–58.

Inkeles, Alex, and Diamond, Larry
1980 "Personal Development and National Development: A Cross-National Perspective." In *The Quality of Life: Comparative Studies*, edited by Alexander Szalai and Frank M. Andrews, pp. 73–109. Beverly Hills, Calif.: Sage.

Jacob, Philip E., and Toscano, James V., eds.
1964 *The Integration of Political Communities*. Philadelphia: Lippincott.

Janowitz, Morris
1967 *The Community Press in an Urban Setting*. 2d ed. Chicago: University of Chicago Press.
1970 *Political Conflict*. Chicago: Quadrangle Books.
1978 *The Last Half-Century*. Chicago: University of Chicago Press.
1980 "Observations on the Sociology of Citizenship: Obligations and Rights." *Social Forces* 59, no. 1 (September): 1–24.

Jennings, Bruce, and Callahan, Daniel
1983 "Social Science and the Policy-Making Process." In *Hastings Center Report*, Special Supplement (February), "Ethics and Social Inquiry": 3–8.

Johansson, Sten
1973 "The Level of Living Survey: A Presentation." *Acta Sociologica* 16, no. 3: 211–19.
1976 "Towards a Theory of Social Reporting." Paper presented at the III. Nordic Symposium on Social Policy, 26–29 September 1976, at Hanasaari, Finland. (Trans. by Phyllis Arora.)

Johnson, Ronald W., and Lewin, Arie Y.
1984 "Management and Accountability Models." In *Public Sector Performance: A Conceptual Turning Point*, edited by Trudi C. Miller, pp. 224–50. Baltimore: Johns Hopkins University Press.

Johnston, Denis F.
1978 "Postlude: Past, Present, and Future." In *America in the Seventies: Some Social Indicators*, edited by Richard D. Lambert. *Annals of the American Academy of Political and Social Science* 435 (January): 286–94.

Johnston, Denis F., and Carley, Michael J.
1981 "Social Measurement and Social Indicators." In *America Enters the Eighties: Some Social Indicators*, edited by Conrad Taeuber. *Annals of the American Academy of Political and Social Science* 453 (January): 237–53.

Juster, F. Thomas, Courant, Paul N., and Dow, Greg K.
1981 "The Theory and Measurement of Well-Being: A Suggested Framework for Accounting and Analysis." In *Social Accounting Systems*, edited by F. Thomas Juster and Kenneth C. Land, pp. 23–94. New York: Academic Press.

Kadushin, Charles
1982 "Social Density and Mental Health." In *Social Structure and Network Analysis*, edited by Peter V. Marsden and Nan Lin, pp. 147–58. Beverly Hills, Calif.: Sage.

Kahneman, Daniel, and Tversky, Amos
1979 "Prospect Theory: An Analysis of Decision Under Risk." *Econometrica* 47, no. 2 (March): 263–91.

Kaldor, Nicholas
1939 "Welfare Propositions of Economics and Interpersonal Comparisons of Utility." *Economic Journal* 49, no. 3 (September): 549–52.

Kammann, Richard, Farry, Marcelle, and Herbison, Peter
1984 "The Analysis and Measurement of Happiness as a Sense of Well-Being." *Social Indicators Research* 15, no. 2 (August): 91–115.

Kasarda, John D., and Janowitz, Morris
1974 "Community Attachment in Mass Society." *American Sociological Review* 39, no. 3 (June): 328–39.

Kaufman, Herbert
1973 *Administrative Feedback: Monitoring Subordinates' Behavior*. Washington, D.C.: Brookings.

Keech, William R., and Matthews, Donald R.
1976 *The Party's Choice*. Washington, D.C.: Brookings.

Keeler, Emmett B., and Cretin, Shan
1982 "Discounting of Nonmonetary Effects." *Rand Notes* N-1875-HHS (June). Santa Monica, Calif.: Rand Corporation.

Keeney, Ralph L., and Raiffa, Howard
1976 *Decisions with Multiple Objectives: Preferences and Value Tradeoffs*. New York: Wiley.

Kelling, George L., Pate, Tony, Dieckman, Duane, and Brown, Charles E.
1974 *The Kansas City Preventive Patrol Experiment: A Technical Report*. Washington, D.C.: Police Foundation.

Kerr, Richard A.
1983 "A Chance to Predict Next Month's Weather?" *Science* 220 (6 May): 590–91.

Keyfitz, Nathan
1978 "Government Statistics." In *International Encyclopedia of Statistics*, edited by William H. Kruskal and Judith M. Tanur, pp. 413–25. New York: Free Press.

Kiepke, Clare
1980 "The Lodge as a Community Volunteer Activity." In *The Fairweather Lodge, a Twenty-Five Year Retrospective*, edited by George W. Fairweather, pp. 67–73. San Francisco: Jossey-Bass.

Klages, Helmut
1973 "Assessment of an Attempt at a System of Social Indicators." *Policy Sciences* 4, no. 3 (September): 249–61.

Klein, Lawrence R.
1969 "Econometric Analysis of the Tax Cut of 1964." In *The Brookings Model: Some Further Results*, edited by James S. Duesenberry et al., pp. 458–72. Chicago: Rand McNally.
1975 "Research Contributions of the SSRC-Brookings Econometric Model Project—A Decade in Review." In *The Brookings Model: Perspective and Recent Developments*, edited by Gary Fromm and Lawrence R. Klein, pp. 11–29. Amsterdam: North-Holland.

Klein, Lawrence R., Forst, Brian, and Filatov, Victor
1978 "The Deterrent Effect of Capital Punishment: An Assessment of the Estimates." In *Deterrence and Incapacitation: Estimating the Effects of Criminal Sanctions on Crime Rates*, edited by Alfred Blumstein, Jacqueline Cohen, and Daniel Nagin, pp. 336–60. Washington, D.C.: National Academy of Sciences.

Knetsch, Jack L., and Davis, Robert K.
1966 "Comparisons of Methods for Recreation Evaluation." In *Water Research*, edited by Allen V. Kneese and Stephen C. Smith. Baltimore: Johns Hopkins Press for Resources for the Future, 125–42.

Koopmans, Tjalling C.
1960 "Stationary Ordinal Utility and Impatience." *Econometrica* 28, no. 2 (April): 287–309.
1979 "Economics Among the Sciences." *American Economic Review* 69, no. 1 (March): 1–13.

Kornhauser, William
1959 *The Politics of Mass Society*. New York: Free Press.
1962 *Scientists in Industry: Conflict and Accommodation*. Berkeley: University of California Press.

Kraemer, Kenneth L.
1977 "Local Government, Information Systems, and Technology Transfer: Evaluating Some Common Assertions about Computer Application Transfer." *Public Administration Review* 37, no. 4 (July/August): 368–82.

Kraemer, Kenneth L., and Dutton, William H.
1979 "The Interests Served by Technological Reform: The Case of Computing." *Administration and Society* 11, no. 1 (May): 80–106.

Kraemer, Kenneth L., and King, John Leslie
1977 *Computers and Local Government: Vol. 1, A Manager's Guide*. New York: Praeger.

Kraemer, Kenneth L., and Perry, James L.
1979 "The Federal Push to Bring Computer Applications to Local Governments." *Public Administration Review* 39, no. 3 (May/June): 260–70.

Kress, Paul F.
1969 "The Web and the Tree: Metaphors of Reason and Value." *Midwest Journal of Political Science* 13, no. 3 (August): 395–414.

Krieger, Martin H.
1981a *Advice and Planning*. Philadelphia: Temple University Press.
1981b "The End of the World as a Policy Problem." Lecture at the University of North Carolina at Chapel Hill, 26 February 1981.

Kruskal, William H.
1973 "The Committee on National Statistics." *Science* 180 (22 June): 1256–58.
1978 "Significance, Tests of." In *International Encyclopedia of Statistics*, edited by Kruskal and Judith M. Tanur, pp. 944–56. New York: Free Press.

Krutilla, John V., and Fisher, Anthony C.
1975 *The Economics of Natural Environments*. Baltimore, Md.: Johns Hopkins Press for Resources for the Future.

Kuh, Edwin
1969 "A Progress Report." In *The Brookings Model: Some Further Results*, edited by James S. Duesenberry et al., pp. 3–16. Chicago: Rand McNally.

Kuhn, Thomas S.
1962 *The Structure of Scientific Revolutions*. Chicago: University of Chicago Press.

Lancaster, Kelvin J.
1966 "A New Approach to Consumer Theory." *Journal of Political Economy* 74, no. 2 (April): 132–57.

Land, Kenneth C.
1975a "Social Indicator Models: An Overview." In *Social Indicator Models*, edited by K. C. Land and Seymour Spilerman, pp. 5–36. New York: Russell Sage Foundation.
1975b "Theories, Models, and Indicators of Social Change." *International Social Science Journal* 27, no. 1: 7–37.
1979 "Modeling Macro Social Change." In *Sociological Methodology 1980*, edited by Karl F. Schuessler, pp. 219–78. San Francisco: Jossey-Bass.
1982 "Social Indicators: Past Developments and Prospects for the Future." Texas Population Center Research Papers, no. 4.007. Austin: The University of Texas at Austin.
1983 "Social Indicators." *Annual Review of Sociology* 9: 1–26. Revised version of Land (1982).

Land, Kenneth C., and Felson, Marcus
1976 "A General Framework for Building Dynamic Macro Social Indicator Models: Including an Analysis of Changes in Crime Rates and Police Expenditures." *American Journal of Sociology* 82, no. 3 (November): 565–604.

Land, Kenneth C., and Juster, F. Thomas
1981 "Social Accounting Systems: An Overview." In *Social Accounting Systems: Essays on the State of the Art*, edited by F. Thomas Juster and Kenneth C. Land, pp. 1–21. New York: Academic Press.

Land, Kenneth C., and McMillen, Marilyn M.
1980 "A Macrodynamic Analysis of Changes in Mortality Indexes in the United States, 1946–75: Some Preliminary Results." *Social Indicators Research* 7 (January): 1–46.
1981 "Demographic Accounts and the Study of Social Change with Applications to the Post-World War II United States." In *Social Accounting Systems*, edited by F. Thomas Juster and Kenneth C. Land, pp. 241–306. New York: Academic Press.

Landecker, Werner S.
1951 "Types of Integration and Their Measurement." *American Journal of Sociology* 66, no. 4 (January): 332–40.

Langbein, Laura I.
1980 *Discovering Whether Programs Work*. Santa Monica, Calif.: Goodyear.

Larch, Billie B.
1980 "The Lodge as an Extension of the Hospital." In *The Fairweather Lodge, a Twenty-Five Year Retrospective*, edited by George W. Fairweather, pp. 59–65. San Francisco: Jossey-Bass.

Larkey, Patrick D., and Sproull, Lee S.
1981 "Models in Theory and Practice: Some Examples, Problems, and Prospects." *Policy Sciences* 13, no. 3 (June): 233–46.

Lasch, Christopher
1977 *Haven in a Heartless World: The Family Besieged*. New York: Basic Books.

Laumann, Edward O., Marsden, Peter V., and Prensky, David
1983 "The Boundary Specification Problem in Network Analysis." In *Applied Network Analysis*, edited by Ronald S. Burt and Michael J. Minor, pp. 18–34. Beverly Hills, Calif.: Sage.

Laumann, Edward O., and Pappi, Franz U.
1976 *Networks of Collective Action: A Perspective on Community Influence Systems*. New York: Academic Press.

Lazarsfeld, Paul F.
1959 "Problems in Methodology." In *Sociology Today*, edited by Robert K. Merton, Leonard Broom, and Leonard S. Cottrell, Jr., pp. 39–78. New York: Basic Books.

Lazarsfeld, Paul F., and Menzel, Herbert
1965 "On the Relations between Individual and Collective Properties." In *Complex Organizations*, edited by Amitai Etzioni, pp. 422–40. New York: Holt, Rinehart, and Winston.

Lazarsfeld, Paul F., and Reitz, Jeffrey G.
1975 *An Introduction to Applied Sociology*. New York: Elsevier.

Lazarsfeld, Paul F., Sewell, William H., and Wilensky, Harold L., eds.
1967 *The Uses of Sociology*. New York: Basic Books.

Lazear, Edward
1978 "Intergenerational Externalities." Report 7811, Center for Mathematical Studies in Business and Economics, University of Chicago.

Lederer, Katrin
1980 "Introduction." In *Human Needs: A Contribution to the Current Debate*, edited by Katrin Lederer, pp. 1–14. Cambridge, Mass.: Oelgeschlager, Gunn, and Hain.

Lehman, Edward W.
1980 "Policy Research: Industry or Social Movement?" In *Improving Policy Analysis*, edited by Stuart S. Nagel, pp. 201–18. Beverly Hills, Calif.: Sage Publications.

Leibenstein, Harvey
1976 *Beyond Economic Man*. Cambridge: Harvard University Press.

Levin, Martin A.
1981 "Conditions Contributing to Effective Implementation and Their Limits." In *Research in Public Policy Analysis and Management*, vol. 1, edited by John P. Crecine, pp. 65–111. Greenwich, Conn.: JAI Press.

Levins, Richard
1973 "Fundamental and Applied Research in Agriculture." *Science* 181 (August 10, 1973): 523–24.

Lewis, Clarence Irving
1946 *An Analysis of Knowledge and Valuation*. La Salle, Ill.: Open Court.

Lijphart, Arend
1975 *The Politics of Accommodation: Pluralism and Democracy in the Netherlands*. 2d ed. Berkeley: University of California Press.
1977 *Democracy in Plural Societies: A Comparative Exploration*. New Haven, Conn.: Yale University Press.

Lincoln, James R., and Miller, Jan
1979 "Work and Friendship Ties in Organizations: A Comparative Analysis of Relational Networks." *Administrative Science Quarterly* 24, no. 2 (June): 181–99.

Lincoln, James R., and Zeitz, Gerald
1980 "Organizational Properties from Aggregate Data." *American Sociological Review* 45, no. 3 (June): 391–408.

Lindahl, Erik
1967 "Just Taxation—A Positive Solution." In *Classics in the Theory of Public Finance*, edited by Richard A. Musgrave and Alan T. Peacock, pp. 168–76. New York: St. Martin's Press. Original German version published in 1919.

Lindberg, Leon N.
1982 "Economists as Policy Intellectuals and Economics as a Policy Profession: Reflections on the Retreat from Keynesianism and

from the Interventionist State." Paper presented at the Twelfth World Congress of the International Political Science Association, Rio de Janeiro, 9–14 August 1982.

Lindblom, Charles E.

1965 *The Intelligence of Democracy*. New York: Free Press.

1979 "Still Muddling, Not Yet Through." *Public Administration Review* 39, no. 6 (November/December): 517–26.

Lindblom, Charles E., and Cohen, David K.

1979 *Usable Knowledge*. New Haven: Yale University Press.

Lipset, Seymour Martin

1960 *Political Man*. Garden City, N.Y.: Doubleday.

Lipset, Seymour Martin, and Bendix, Reinhard

1960 *Social Mobility in Industrial Society*. Berkeley: University of California Press.

Lipset, Seymour Martin, and Schneider, William

1983 *The Confidence Gap: Business, Labor, and Government in the Public Mind*. New York: Free Press.

Lipsky, Michael

1980 *Street-Level Bureaucracy*. New York: Russell Sage Foundation.

Lipton, Douglas, Martinson, Robert, and Wilks, Judith

1975 *The Effectiveness of Correctional Treatment: A Survey of Treatment Evaluation Studies*. New York: Praeger.

Lowry, Ira S.

1966 *Migration and Metropolitan Growth: Two Analytical Models*. San Francisco: Chandler.

Lucas, J. R.

1980 *On Justice*. Oxford: Clarendon Press.

Lutgens, Frederick

1979 *The Atmosphere: An Introduction to Meteorology*. Englewood Cliffs, N.J.: Prentice-Hall.

Machlup, Fritz

1961 "Are the Social Sciences Really Inferior?" *Southern Economic Journal* 27, no. 3 (January): 173–84.

1962 *The Production and Distribution of Knowledge in the United States*. Princeton: Princeton University Press.

1980 *Knowledge: Its Creation, Distribution, and Economic Significance*, Knowledge and Knowledge Production, vol. 1. Princeton: Princeton University Press.

MacLean, Douglas

1980 "Benefit-Cost Analysis, Future Generations and Energy Policy, A Survey of the Moral Issues." *Science, Technology and Human Values* 31 (Spring): 3–10.

MacRae, Duncan, Jr.

1960 "Direct Factor Analysis of Sociometric Data." *Sociometry* 23, no. 4 (December): 360–71.

1976a *The Social Function of Social Science*. New Haven: Yale University Press.

1976b "Technical Communities and Political Choice." *Minerva* 14, no. 2 (Summer): 169–90.

1977 "Professions and Social Sciences as Sources of Public Values." *Soundings* 60, no. 1 (Spring): 1–21.

1980a "Policy Analysis Methods and Governmental Functions." In *Improving Policy Analysis*, edited by Stuart S. Nagel, pp. 129–51. Beverly Hills, Calif.: Sage.

1980b "Present and Future in the Valuation of Life." Paper presented at annual meeting of the American Political Science Association, Washington D.C., August.

1980c "Progressive Taxation, the 'Marriage Tax', and Ethical Consistency." In *Public Policy and the Family*, edited by Zee I. Giraldo, pp. 103–07. Lexington, Mass.: Heath.

1981 "Combining the Roles of Scholar and Citizen." In *Models for Analysis of Social Policy*, edited by Ron Haskins and James J. Gallagher, pp. 103–52. Norwood, N.J.: Ablex.

1983 "The Science of Politics and Its Limits." Paper presented at the Lasswell Symposium, annual meeting of the American Political Science Association, Chicago, Ill., September.

MacRae, Duncan, Jr., and Wilde, James A.

1979 *Policy Analysis for Public Decisions*. North Scituate, Mass.: Duxbury.

Malkiel, Burton G.

1978 "Problems with the Federal Economic Statistical System and Some Alternatives for Improvement." *American Statistician* 32, no. 3 (August): 81–88.

Mann, Leon

1969 "Queue Culture: The Waiting Line as a Social System." *American Journal of Sociology* 75, no. 3 (November): 340–54.

Mannheim, Karl

1949 *Ideology and Utopia*. New York: Harcourt, Brace. First published 1936.

March, James G., and Simon, Herbert A.

1958 *Organizations*. New York: Wiley.

Mare, Robert D.

1981 "Trends in Schooling: Demography, Performance, and Organization." In *America Enters the Eighties: Some Social Indicators*, edited by Conrad Taeuber. *Annals of the American Academy of Political and Social Science* 453 (January): 96–122

Margolis, Julius

1977 "Shadow Prices for Incorrect or Nonexistent Market Values." In *Public Expenditure and Policy Analysis*, 2d ed., edited by Robert H. Haveman and J. Margolis, pp. 204–20. Chicago: Rand McNally.

Marsden, Peter V., and Campbell, Karen E.

1984 "Measuring Tie Strength." *Social Forces* 63, no. 2 (December): 482–501.

Martin, Wayne H.
1979 "National Assessment of Educational Progress." In *Insights from Large Scale Surveys: New Directions for Testing and Measurement* 2, edited by John E. Milholland, pp. 45–67. San Francisco: Jossey-Bass.

Martinson, Robert
1974 "What Works?—Questions and Answers about Prison Reform." *The Public Interest* 35 (Spring): 22–54.

Mason, Karen Oppenheim
1980 "Sex and Status in Science." A review of Jonathan R. Cole, *Fair Science*. *Science* 208 (18 April): 277–78.

Mayeske, George W., Okoda, T., Beaton, A. E., Jr., Cohen, W. M., and Wisler, C. E.
1973 *A Study of the Achievement of Our Nation's Students*. Washington, D. C.: U. S. Department of Health, Education, and Welfare.

McConnell, Grant
1966 *Private Power and American Democracy*. New York: Knopf.

McKennell, Aubrey C.
1978 "Cognition and Affect in Perceptions of Well-Being." *Social Indicators Research* 5, no. 4 (October): 389–426.

McKennell, Aubrey C., Atkinson, Tom, and Andrews, Frank M.
1980 "Structural Constancies in Surveys of Perceived Well-Being." In *The Quality of Life: Comparative Studies*, edited by Alexander Szalai and Frank M. Andrews, pp. 111–28. Beverly Hills, Calif.: Sage.

McMillan, Melville L., Reid, Bradford G., and Gillen, David W.
1980 "An Extension of the Hedonic Approach for Estimating the Value of Quiet." *Land Economics* 56, no. 3 (August): 315–28.

Meade, James E.
1970 *The Theory of Indicative Planning*. Manchester, England: Manchester University Press.

Meehan, Eugene J.
1975 "Social Indicators and Policy Analysis." In *Methodologies for Analyzing Public Policies*, edited by Frank P. Scioli, Jr., and Thomas J. Cook, pp. 33–46. Lexington, Mass.: Heath.
1982 *Economics and Policymaking: The Tragic Illusion*. Westport, Conn.: Greenwood.

Meier, Paul
1972 "The Biggest Public Health Experiment Ever: The 1954 Field Trial of the Salk Poliomyelitis Vaccine." In *Statistics: A Guide to the Unknown*, edited by Judith M. Tanur et al., pp. 2–13. San Francisco: Holden-Day.

Merelman, Richard M.
1980 "Democratic Politics and the Culture of American Education." *American Political Science Review* 74, no. 2 (June): 319–32.

Merewitz, Leonard, and Sosnick, Stephen H.
1971 *The Budget's New Clothes*. Chicago: Rand McNally.

Merriam, Ida C.
1978 "Social Security and Social Welfare Indicators." In *America in the Seventies: Some Social Indicators*, edited by Richard D. Lambert. *Annals of the American Academy of Political and Social Science* 435 (January): 117–39.

Merton, Robert K.
1968 *Social Theory and Social Structure*. Enlarged edition. New York: Free Press.

Merton, Robert K., and Rossi, Alice S.
1968 "Contributions to the Theory of Reference Group Behavior." In *Social Theory and Social Structure*, enlarged edition, edited by Robert K. Merton, pp. 279–334. New York: Free Press.

Michael, Robert T., and Becker, Gary S.
1973 "On the New Theory of Consumer Behavior." *Swedish Journal of Economics* 75, no. 4: 378–95.

Michalos, Alex C.
1978 "Social Indicators Research." *Policy Studies Journal* 6, no. 3 (Spring): 393–404.

Mill, John Stuart
1910 "Utilitarianism." In his *Utilitarianism, Liberty, and Representative Government*. London: J.M. Dent. First published 1861.

Miller, Harold D.
1981 "Projecting the Impact of New Sentencing Laws on Prison Populations." *Policy Sciences* 13, no. 1 (February): 51–73.

Miller, Trudi C.
1981 "Political and Mathematical Perspectives on Educational Equity." *American Political Science Review* 75, no. 2 (June): 319–33.

Mindlin, Albert, and Levy, Nathan
1980 "A Local Government Computerized Statistical Information System." In *Computers in Local Government: Urban and Regional Planning*, edited by Auerbach Publishers, Module 2.5.1. Pennsauken, N.J.: Auerbach.

Mishan, Ezra J.
1982 *Cost-Benefit Analysis*. 3d ed. London: George Allen & Unwin.

Mitroff, Ian I., Mason, Richard O., and Barabba, Vincent P.
1983 *The 1980 Census, Policymaking amid Turbulence*. Lexington, Mass.: Heath.

Moore, Wilbert E., and Sheldon, Eleanor Bernert
1965 "Monitoring Social Change: A Conceptual and Programmatic Statement." *Proceedings of the Social Statistics Section*, pp. 144–49. [American Statistical Association, Washington, D.C.]

Morgan, James N., and Smith, James D.
1969 "Measures of Economic Well-Offness and their Correlates." *American Economic Review* 59, no. 2 (May): 450–62.

Morgan, M. Granger, and McMichael, Francis C.
1981 "A Characterization and Critical Discussion of Models and Their Use in Environmental Policy." *Policy Sciences* 13, no. 3 (June): 345–70.

Morss, Elliott R., and Rich, Robert F.
1980 *Government Information Management: A Counter-Report of the Commission on Federal Paperwork*. Boulder, Colo.: Westview Press.

Moses, Lincoln E.
1979 "Early Steps Toward a National Energy Information System." *American Statistician* 33, no. 3 (August): 97–101.

Moses, Lincoln E., and Mosteller, Frederick
1972 "Safety of Anesthetics." In *Statistics: A Guide to the Unknown*, edited by Judith M. Tanur et al., pp. 14–22. San Francisco: Holden-Day.

Mosteller, Frederick
1978 "Errors: I. Nonsampling Errors." In *International Encyclopedia of Statistics*, edited by William H. Kruskal and Judith M. Tanur, pp. 208–29. New York: Free Press.

Mosteller, Frederick, and Moynihan, Daniel P., eds.
1972 *On Equality of Educational Opportunity*. New York: Vintage.

Mosteller, Frederick, and Nogee, Philip
1951 "An Experimental Measurement of Utility." *Journal of Political Economy* 59, no. 5 (October): 371–404.

Moynihan, Daniel P.
1968 "Sources of Resistance to the Coleman Report." *Harvard Educational Review* 38, no. 1 (Winter): 23–36.
1973 *The Politics of a Guaranteed Income*. New York: Vintage.

Murnane, Richard J.
1983 "How Clients' Characteristics Affect Organization Performance: Lessons from Education." *Journal of Policy Analysis and Management* 2, no. 3 (Spring): pp. 403–17.
1984 "A Review Essay—Comparisons of Public and Private Schools: Lessons from the Uproar." *Journal of Human Resources* 19, no. 2 (Spring): 263–77.

Musgrave, Richard A., and Musgrave, Peggy B.
1976 *Public Finance in Theory and Practice*. 2d ed. New York: McGraw-Hill.

Nagar, A. L.
1969 "Stochastic Simulation of the Brookings Econometric Model." In *The Brookings Model: Some Further Results*, edited by James S. Duesenberry et al., pp. 425–56. Chicago: Rand McNally. McNally.

Nakamura, Robert T., and Smallwood, Frank
1980 *The Politics of Policy Implementation*. New York: St. Martin's.

Naroll, Raoul
1983 *The Moral Order*. Beverly Hills, Calif.: Sage.

Nathan, Richard P., and Adams, Charles
1976 "Understanding Central City Hardship." *Political Science Quarterly* 91, no. 1 (Spring): 47–62.

National Journal
1982 *National Journal* (May 1): "Playing Around with the Consumer Price Index," p. 782.

National Research Council
1933 *A History of the National Research Council, 1919–1933*. Washington, D.C.: U.S. Government Printing Office.

National Science Board
1979 *Science Indicators 1978*. Washington, D.C.: U.S. Government Printing Office.

Nelkin, Dorothy
1979 "Scientific Knowledge, Public Policy, and Democracy: A Review Essay." *Knowledge* 1, no. 1 (September): 106–22.

Nelson, Jon P.
1980 "Measuring Benefits of Environmental Improvements, Aircraft Noise and Hedonic Prices." In *Advances to Applied Micro-Economics*, edited by V. Kerry Smith, pp. 51–73. Greenwich, Conn.: JAI Press.

Niemi, Donald R.
1966 "Sweden's Municipal Consolidation Reforms." Ph. D. dissertation, University of Chicago.

Noam, Eli M.
1979 "The Optimal Distribution of Government Expenditures." Research working paper No. 180A, Graduate School of Business, Columbia University.

Nordhaus, William, and Tobin, James
1972 "Is Growth Obsolete?" In *Economic Growth*, Economic Research: Retrospect and Prospect, 50th Anniversary Colloquium, vol. 5, National Bureau of Economic Research, pp. 1–80. New York: Columbia University Press.

Nozick, Robert
1974 *Anarchy, State, and Utopia*. New York: Basic Books.

Oakeshott, Michael
1962 *Rationalism in Politics and Other Essays*. New York: Basic Books.

Odum, Eugene P., et al.
1976 "Totality Indices for Evaluating Environmental Impact: A Test Case—Relative Impact of Highway Alternates." In *Environmental Impact Assessment*, edited by Marlan Blissett, pp. 153–88. New York: Engineering Foundation.

OECD (Organisation for Economic Co-operation and Development)
1976 *Measuring Social Well-Being: A Progress Report on the Development of Social Indicators*. Paris: OECD.

Okun, Arthur M.
1971 "Should GNP Measure Social Welfare?" *Brookings Bulletin* 8, no. 3 (Summer): 4–7.
1975 *Equality and Efficiency: The Big Tradeoff.* Washington, D.C.: Brookings.

Olson, Mancur
1970 "An Analytic Framework for Social Reporting and Policy Analysis." *Annals of the American Academy of Political and Social Science* 388 (March): 112–26.

Olson, Mancur, and Bailey, Martin J.
1981 "Positive Time Preference." *Journal of Political Economy* 89, no. 1 (February): 1–25.

Orcutt, Guy, Caldwell, Steven, and Wertheimer, Richard, II, et al.
1976 *Policy Exploration through Microanalytic Simulation.* Washington, D.C.: Urban Institute.

Osipov, G. V., and Andreenkov, V. G., eds.
1979 *Issledovanie Postroeniia Pokazetelei Sotsialnogo Razvitiia i Planirovania* (Investigation of the Construction of Indicators of Social Development and Planning). Moscow: Nauka.

Ouchi, William G., and Johnson, Jerry B.
1978 "Types of Organizational Control and Their Relationship to Emotional Well Being." *Administrative Science Quarterly* 23, no. 2 (June): 293–317.

Page, Ellis B., and Keith, Timothy Z.
1981 "Effects of U.S. Private Schools: A Technical Analysis of Two Recent Claims." *Educational Researcher* 10, no. 7 (August/September): 7–17.

Palmer, John L., and Barth, Michael C.
1977 "The Distributional Effects of Inflation and Higher Unemployment." In *Improving Measures of Economic Well-Being*, edited by Marilyn Moon and Eugene Smolensky, pp. 201–39. New York: Academic Press.

Parfit, Derek
1983 "Energy Policy and the Further Future: The Social Discount Rate." In *Energy and the Future*, edited by Douglas MacLean and Peter G. Brown, pp. 31–37. Totowa, N. J.: Rowman and Littlefield.

Parke, Robert, and Seidman, David
1978 "Social Indicators and Social Reporting." In *America in the Seventies: Some Social Indicators*, edited by Conrad Taeuber. *Annals of the American Academy of Political and Social Science* 435 (January): 1–22.

Parke, Robert, and Sheldon, Eleanor B.
1973 "Social Statistics for Public Policy." *Proceedings of the Social Statistics Section*, pp. 105–12. [American Statistical Association, Washington, D.C.]

Parsons, Talcott
1937 *The Structure of Social Action*. New York: McGraw-Hill.
1951 *The Social System*. Glencoe, Ill.: Free Press.
1954 *Essays in Sociological Theory*. Glencoe, Ill.: Free Press.
1960 *Structure and Process in Modern Societies*. Glencoe, Ill.: Free Press.

Patton, Michael Q.
1978 *Utilization-Focused Evaluation*. Beverly Hills, Calif.: Sage.

Pearce, Diana
1980 "Breaking Down Barriers: New Evidence on the Impact of Metropolitan School Desegregation on Housing Patterns." Center for National Policy Review, School of Law, Catholic University of America, Washington, D.C.

Perelman, Chaim
1967 *Justice*. New York: Random House.

Perrucci, Robert, and Gerstl, Joel E.
1969 *Profession without Community: Engineers in American Society*. New York: Random House.

Peterson, Paul E.
1981 *City Limits*. Chicago: University of Chicago Press.

Pettigrew, Thomas F., and Back, Kurt W.
1967 "Sociology in the Desegregation Process: Its Use and Disuse." In *The Uses of Sociology*, edited by Paul F. Lazarsfeld, William H. Sewell, and Harold L. Wilensky, pp. 692–722. New York: Basic Books.

Pettigrew, Thomas F., and Green, Robert L.
1976 "School Desegregation in Large Cities: A Critique of the Coleman 'White Flight' Thesis." *Harvard Educational Review* 46, no. 1 (February): 1–53.

Pfeffer, Jeffrey, and Moore, William L.
1980 "Average Tenure of Academic Department Heads: The Effects of Paradigm, Size, and Departmental Demography." *Administrative Science Quarterly* 25, no. 3 (September): 387–406.

Pfeffer, Jeffrey, and Ross, Jerry
1982 "The Effects of Marriage and a Working Wife on Occupational and Wage Attainment." *Administrative Science Quarterly* 27, no. 1 (March): 66–80.

Phillips, David P.
1980 "The Deterrent Effect of Capital Punishment: New Evidence on an Old Controversy." *American Journal of Sociology* 86, no. 1 (July): 139–48.

Polanyi, Michael
1958 *Personal Knowledge*. Chicago: University of Chicago Press.

Pollock, Stephen M.
1973 "Quality of Statistics." *Science* 182 (12 October): 113.

President's Commission on Federal Statistics
1971 *Federal Statistics, Report of the President's Commission*, 2 vols. Washington, D.C.: U.S. Government Printing Office.

President's Reorganization Project for the Federal Statistical System
1981 "Improving the Federal Statistical System: Issues and Options." *Statistical Reporter* 81–5 (February): 133–221.

Price, Don K.
1965 *The Scientific Estate*. Cambridge: Harvard University Press.

Psychology Today
1982 "Job Quality Slipping for Management." Press release 8 August in *The Chapel Hill Newspaper*.

Quandt, Richard E., ed.
1970 *The Demand for Travel: Theory and Measurement*. Lexington, Mass.: Heath.

Quarmby, D. A.
1967 "Choice of Travel Mode for the Journey to Work: Some Findings." *Journal of Transport Economics and Policy* 1, no. 3 (September): 1–42.

Quinn, Robert P.
1976 "Strategy Issues in the Development of Quality of Employment Indicators." In *Measuring Work Quality for Social Reporting*, edited by Albert D. Biderman and Thomas F. Drury, pp. 32–43. New York: Wiley.

Quinney, Richard
1974 *Critique of Legal Order*. Boston: Little, Brown.

Raiffa, Howard
1968 *Decision Analysis*. Reading, Mass.: Addison-Wesley.

Rajković, Vladislav
1980 "Development of an Information System in a Self-Management Environment." In *The Information Systems Environment*, edited by Henry C. Lucas, Frank F. Land, Timothy J. Lincoln, and Konrad Supper, pp. 123–27. Amsterdam: North-Holland.

Ramsey, Frank P.
1928 "A Mathematical Theory of Saving." *Economic Journal* 38, no. 4 (December): 543–59.

Ramsøy, Natalie Rogoff
1974 "Social Indicators in the United States and Europe: Comments on Five Country Reports." In *Social Indicators, 1973: A Review Symposium*, edited by Roxann A. Van Dusen, pp. 41–62. Washington, D.C.: Social Science Research Council.

Randall, Alan, and Stoll, John R.
1980 "Consumer's Surplus in Commodity Space." *American Economic Review* 70, no. 3 (June): 449–55.

Ravetz, Jerome R.
1971 *Scientific Knowledge and Its Social Problems*. Oxford: Clarendon.

Rawls, John
1955 "Two Concepts of Rules." *Philosophical Review* 44, no. 1 (January): 3–32.
1971 *A Theory of Justice*. Cambridge: Harvard University Press.

Rein, Martin
1976 *Social Science and Public Policy*. Harmondsworth, England: Penguin.

Rescher, Nicholas
1966 *Distributive Justice*. Indianapolis: Bobbs-Merrill.

Rich, Robert F.
1981 *Social Science Information and Public Policy Making*. San Francisco: Jossey-Bass.

Riecken, Henry W., and Boruch, Robert F., eds.
1974 *Social Experimentation*. New York: Academic Press.

Robbins, Lionel
1937 *An Essay on the Nature and Significance of Economic Science*. London: Macmillan.

Roberts, Harry V.
1980 "Comments." *Procedures of the Conference on Census Undercount*, 25–26 February 1980, Arlington, Va. Washington, D.C.: Bureau of the Census.

Robinson, John P.
1977 *How Americans Use Time*. New York: Praeger.

Rockwell, Richard C.
1983 "Social Indicators of Organizations and Relationships." Paper presented at the annual meeting of the Southern Sociological Society, Atlanta, Georgia, 8 April 1983.

Rodgers, Willard L.
1982a "Density, Crowding and Satisfaction with the Residential Environment." *Social Indicators Research* 10, no. 1 (January): 75–102.
1982b "Trends in Reported Happiness within Demographically Defined Subgroups." *Social Forces* 60, no. 3 (March): 826–42.

Roethlisberger, F. J., and Dickson, William J.
1946 *Management and the Worker*. Cambridge: Harvard University Press.

Rose, Richard
1972 "The Market for Policy Indicators." In *Social Indicators and Social Policy*, edited by Andrew Shonfield and Stella Shaw, pp. 119–41. London: Heinemann.

Rosenbaum, Walter A.
1964 *The Burning of the Farm Population Estimates*. Syracuse, N.Y.: Inter-University Case Program, no. 83.

Rosenberg, Alexander
1980 "A Skeptical History of Microeconomic Theory." *Theory and Decision* 12, no. 1 (March): 79–93.

Rosenthal, Robert A., and Weiss, Robert S.
1966 "Problems of Organizational Feedback Processes." In *Social Indicators*, edited by Raymond A. Bauer, pp. 302–40. Cambridge: M.I.T. Press.

Ross, H. Laurence.
1973 "Law, Science, and Accidents: The British Road Safety Act of 1967." *Journal of Legal Studies* 2, no. 1 (January): 1–78.

Rossi, Peter H.
1972 "Community Social Indicators." In *The Human Meaning of Social Change*, edited by Angus Campbell and Philip E. Converse, pp. 87–126. New York: Russell Sage Foundation.
1980 "The Presidential Address: The Challenge and Opportunities of Applied Social Research." *American Sociological Review* 45, no. 6 (December): 889–904.

Rossi, Peter H., and Berk, Richard A.
1981 "An Overview of Evaluation Strategies and Procedures." *Human Organization* 40, no. 4 (Winter): 287–99.

Rossi, Peter H., Berk, Richard A., and Lenihan, Kenneth J.
1980 *Money, Work, and Crime*. New York: Academic Press.

Rossi, Peter H., and Freeman, Howard E.
1982 *Evaluation: A Systematic Approach*. 2d ed. Beverly Hills, Calif.: Sage.

Rossi, Peter H., and Henry, J. Patrick
1980 "Seriousness: A Measure for All Purposes?" In *Handbook of Criminal Justice Evaluation*, edited by Malcolm W. Klein and Katherine S. Teilman, pp. 489–505. Beverly Hills, Calif.: Sage.

Rossi, Peter H., Waite, Emily, Base, Christine E., and Berk, Richard E.
1974 "The Seriousness of Crimes: Normative Structure and Individual Differences." *American Sociological Review* 39, no. 2 (April): 224–37.

Rule, James B.
1978 *Insight and Social Betterment*. New York: Oxford University Press.

Runyan, William M.
1980 "The Life Satisfaction Chart: Perceptions of the Course of Subjective Experience." *International Journal of Aging and Human Development* 11, no. 1: 45–64.

Ryan, William
1971 *Blaming the Victim*. New York: Pantheon Books.

Sabatier, Paul A., and Mazmanian, Daniel A.
1981 "The Implementation of Public Policy: A Framework of Analysis." In *Effective Policy Implementation*, edited by Daniel A. Mazmanian and Paul A. Sabatier, pp. 3–35. Lexington, Mass.: D.C. Heath.

Salamon, Lester M.
1981 "Rethinking Public Management." *Public Policy* 29, no. 3 (Summer): 255–75.

Samuelson, Paul A.
1948 "Consumption Theory in Terms of Revealed Preference." *Economica* N.S. 15, no. 60 (November): 243–53.
1954 "The Pure Theory of Public Expenditure." *Review of Economics and Statistics* 36, no. 4 (November): 387–89.
1970 *Economics*. 8th ed. New York: McGraw-Hill.

Sarason, Seymour B.
1974 *The Psychological Sense of Community*. San Francisco: Jossey-Bass.

Scanzoni, John
1965 "A Reinquiry into Marital Disorganization." *Journal of Marriage and the Family* 27, no. 4 (November): 483–91.

Scarr, Sandra, and Weinberg, Richard A.
1978 "The Influence of 'Family Background' on Intellectual Attainment." *American Sociological Review* 43, no. 5 (October): 674–92.

Schachter, Stanley
1968 "Cohesion, Social." In *International Encyclopedia of the Social Sciences*, vol. 2, edited by David L. Sills, pp. 542–46. New York: Macmillan.

Schattschneider, Elmer E.
1960 *The Semisovereign People*. Hinsdale, Ill.: Dryden.

Schlozman, Kay Lehman, and Verba, Sidney
1978 "The New Unemployment: Does it Hurt?" *Public Policy* 26, no. 3 (Summer): 333–58.

Schneider, Mark
1976 "The Quality of Life and Social Indicators Research." *Public Administration Review* 36, no. 3 (May/June): 297–305.

Schneider, Stephen H., and Chen, Robert S.
1980 "Carbon Dioxide Warming and Coastline Flooding: Physical Factors and Climatic Impact." *Annual Review of Energy* 5: 107–40.

Scitovsky, Tibor
1976 *The Joyless Economy*. New York: Oxford University Press.

Scott, Marvin B., and Lyman, Stanford M.
1968 "Accounts." *American Sociological Review* 33, no. 1 (February): 46–62.

Scott, Robert A., and Shore, Arnold R.
1979 *Why Sociology Does Not Apply: A Study of the Use of Sociology in Public Policy*. New York: Elsevier.

Scott, Wolf, with Argalias, Helen and McGranahan, Donald V.
1973 "The Measurement of Real Progress at the Local Level." Geneva: United Nations Research Institute for Social Development, Report No. 73.3.

Scriven, Michael
1967 "The Methodology of Evaluation." In *Perspectives of Curriculum*

Evaluation, edited by Ralph W. Tyler, Robert M. Gagné, and Michael Scriven, pp. 39–83. Chicago: Rand McNally.

Sears, Pauline S., and Barbee, Ann H.
1977 "Career and Life Satisfactions Among Terman's Gifted Women." In *The Gifted and the Creative: A Fifty-Year Perspective*, edited by Julian W. Stanley, William C. George, and Cecilia H. Solano, pp. 28–65. Baltimore, Md.: Johns Hopkins University Press.

Sears, Robert R.
1977 "Sources of Life Satisfactions of the Terman Gifted Men." *American Psychologist* 32, no. 2 (February): 119–28.

Seidman, David
1980 "Numbers that Count: The Law and Policy of Population Statistics Used in Formula Grant Allocation Programs," *George Washington Law Review* 48, no. 2 (January): 229–67.

Sellin, Thorstein, and Wolfgang, Marvin E.
1964 *The Measurement of Delinquency*. New York: Wiley.

Selznick, Philip
1957 *Leadership in Administration*. New York: Row, Peterson.
1960 *The Organizational Weapon*. New York: Free Press. Originally published 1952.

Sen, Amartya
1980 "Description as Choice." *Oxford Economic Papers* N.S. 32, no. 3 (November): 353–69.

Sewell, William H.
1967 Review of Coleman et al., *Equality of Educational Opportunity. American Sociological Review* 32, no. 3 (June): 475–79.

Sharpe, L. J.
1978 "The Social Scientist and Policy-Making in Britain and America: A Comparison." In *Social Policy Research*, edited by Martin Bulmer, pp. 302–12. London: Macmillan.

Sheldon, Eleanor Bernert, and Freeman, Howard E.
1970 "Notes on Social Indicators: Promises and Potential." *Policy Sciences* 1, no. 1 (Spring): 97–111.

Sheldon, Eleanor Bernert, and Land, Kenneth C.
1972 "Social Reporting for the 1970's: A Review and Programmatic Statement." *Policy Sciences* 3, no. 2 (Summer): 137–51.

Sheldon, Eleanor Bernert, and Parke, Robert
1975 "Social Indicators." *Science* 188 (16 May): 693–99.

Sherman, Lawrence W., and Berk, Richard A.
1984 "The Specific Deterrent Effects of Arrest for Domestic Assault." *American Sociological Review* 49, no. 2 (April): 261–72.

Shils, Edward A.
1956 *The Torment of Secrecy*. New York: Free Press.
1977 "Social Science as Public Opinion." *Minerva* 15, no. 3–4 (Autumn/Winter): 273–85.

Shils, Edward A., and Janowitz, Morris
1948 "Cohesion and Disintegration in the Wehrmacht in World War II." *Public Opinion Quarterly* 12, no. 2 (Summer): 280–315.

Shin, Doh C., and Johnson, D. M.
1978 "Avowed Happiness as an Overall Assessment of the Quality of Life." *Social Indicators Research* 5, no. 4 (October): 475–92.

Simon, Herbert A.
1969 *The Sciences of the Artificial*. Cambridge: M.I.T. Press.
1980 "The Behavioral and Social Sciences." *Science* 209 (4 July): 72–78.
1982 "Are Social Problems Problems That Social Science Can Solve?" In *The Social Sciences: Their Nature and Uses*, edited by William H. Kruskal, pp. 1–20. Chicago: University of Chicago Press.

Simutis, Leonard J.
1980 "Development of Social Indicators and Neighborhood Analysis from Computerized Records." In *Computers in Local Government: Urban and Regional Planning*, edited by Auerbach Publishers, Module 2.5.2 S2. Pennsauken, N.J.: Auerbach Publishers.

Sirageldin, Ismail A.-H.
1969 *Non-Market Components of National Income*. Ann Arbor, Mich.: Institute for Social Research, University of Michigan.

Smith, Tom W.
1982 "College Dropouts: An Analysis of the Psychological Well-Being and Attitudes of Various Educational Groups." *Social Psychology Quarterly* 45, no. 1 (March): 50–53.

Sociology of Education
1982 *Sociology of Education* 55, nos. 2/3 (April/July).

Solomon, Richard L., and Corbit, John D.
1974 "An Opponent-Process Theory of Motivation: I. Temporal Dynamics of Affect." *Psychological Review* 81, no. 2 (March): 119–45.

Speare, Alden, Jr., Kobrin, Frances, and Kingkade, Ward
1982 "The Influence of Socioeconomic Bonds and Satisfaction on Interstate Migration." *Social Forces* 61, no. 2 (December): 551–74.

Starbuck, William H.
1975 "Information Systems for Organizations of the Future." In *Information Systems and Organizational Structure*, edited by Erwin Grochla and Norbert Szyperski, pp. 217–28. Berlin: Walter de Gruyter.

Steiner, Gilbert Y.
1981 *The Futility of Family Policy*. Washington, D.C.: Brookings.

Steiner, Jürg
1974 *Amicable Agreement Versus Majority Rule: Conflict Resolution in Switzerland*. Chapel Hill: University of North Carolina Press.

1983 "Decision Process and Policy Outcome: An Attempt to Conceptualize the Problem at the Cross-National Level." *European Journal of Political Research* 11, no. 3 (September): 309–18.

Steiner, Jürg, and Dorff, Robert H.
1980 *A Theory of Political Decision Modes*. Chapel Hill: University of North Carolina Press.

Steinfels, Peter
1977 "The Place of Ethics in Schools of Public Policy." A report from the Hastings Center of Society, Ethics, and Life Sciences to the Ford Foundation, April.

Steinhausen, Jörg
1975 *Soziale Indikatoren als Elemente eines Gesellschaftlichen Planungs- und Steuerungssystems*. Meisenheim am Glan: Anton Hain.

Stigler, George J., and Becker, Gary S.
1977 "De Gustibus Non Est Disputandum." *American Economic Review* 67, no. 2 (March): 76–90.

Stinchcombe, Arthur L., and Wendt, James C.
1975 "Theoretical Domains and Measurement in Social Indicator Analysis." *Social Indicator Models*, edited by Kenneth C. Land and Seymour Spilerman, pp. 37–73. New York: Russell Sage Foundation.

Stipak, Brian
1979 "Are There Sensible Ways to Analyze and Use Subjective Indicators of Urban Service Quality?" *Social Indicators Research* 6, no. 4 (October): 421–38.
1980 "Local Governments' Use of Citizen Surveys." *Public Administration Review* 40, no. 5 (September/October): 521–25.
1983 "Interpreting Subjective Data for Program Evaluation." *Policy Studies Journal* 12, no. 2 (December): 305–14.

Stokes, Donald E.
1982 "Basic Inquiry and Applied Use in the Social Sciences." Paper presented at the Lasswell Symposium, at the meeting of the American Political Science Association, Denver, Colorado, September 1982.

Stone, Julius
1979 "Justice Not Equality." In *Justice*, edited by Eugene Kamenka and Alice Erh-Soon Tay, pp. 97–115. London: Edward Arnold.

Stone, Philip J.
1972 "Child Care in Twelve Countries." In *The Use of Time*, edited by Alexander Szalai, pp. 249–64. The Hague: Mouton.

Stouffer, Samuel A.
1955 *Communism, Conformity, and Civil Liberties*. Garden City, N.Y.: Doubleday.

Stouffer, Samuel A., Lumsdaine, Arthur A., Lumsdaine, Marion Harper, Williams, Robin M., Jr., Smith, M. Brewster, Janis, Irving L., Star, Shirley A., and Cottrell, Leonard S., Jr.
1949 *The American Soldier; Studies in Social Psychology in World War II, Vol. 2: Combat and Its Aftermath*. Princeton: Princeton University Press.

Straus, Murray A.
1979 "Measuring Intrafamily Conflict and Violence: The Conflict Tactics (CT) Scales." *Journal of Marriage and the Family* 41, no. 1 (February): 75–88.

Stucker, James P.
1977 "The Distributional Implications of a Tax on Gasoline." *Policy Analysis* 3, no. 2 (Spring): 171–86.

Sweden, National Central Bureau of Statistics (NCBS)
1981 *Social Report on Inequality in Sweden*. Stockholm: Official Statistics of Sweden, National Central Bureau of Statistics.

Szalai, Alexander, ed.
1972 *The Use of Time*. The Hague: Mouton.

Szanton, Peter
1981 *Not Well Advised*. New York: Russell Sage Foundation.

Taylor, Charles L., Jodice, David A., and Koonce, Wayne L.
1980 "A Systematic Approach to Political Indicators." In *Indicator Systems for Political, Economic, and Social Analysis*, edited by Charles L. Taylor, pp. 117–33. Cambridge: Oelgeschlager, Gunn, & Hain.

Thoits, Peggy A.
1982 "Conceptual, Methodological, and Theoretical Problems in Studying Social Support as a Buffer Against Life Stress." *Journal of Health and Social Behavior* 23, no. 2 (June): 145–59.

Thoits, Peggy, and Hannan, Michael
1979 "Income and Psychological Distress: The Impact of an Income-Maintenance Experiment." *Journal of Health and Social Behavior* 20, no. 2 (June): 120–38.

Tideman, J. Nicolaus, and Tullock, Gordon
1976 "A New and Superior Process for Making Social Choices." *Journal of Political Economy* 84, no. 6 (December): 1145–59.

Tiryakian, Edward A.
1981 "Sexual Anomie, Social Structure, Societal Change." *Social Forces* 59, no. 4 (June): 1025–53.

Tracy, Paul E., and Fox, James Alan
1981 "The Validity of Randomized Response for Sensitive Measurements." *American Sociological Review* 46, no. 2 (April): 187–200.

Turner, R. Jay
1981 "Social Support as a Contingency in Psychological Well-Being." *Journal of Health and Social Behavior* 22, no. 4 (December): 357–67.

U.S. Commission on Civil Rights
1978 *Social Indicators of Equality for Minorities and Women.* Washington, D.C.: U.S. Commission on Civil Rights.
U.S. Department of Commerce, Bureau of the Census
1980 *Social Indicators III.* Washington, D.C.: U.S. Government Printing Office.
U.S. Department of Health, Education, and Welfare
1969 *Toward a Social Report.* Washington, D.C.: U.S.Government Printing Office.
Varga, Karoly
1972 "Marital Cohesion as Reflected in Time-Budgets." In *The Use of Time,* edited by Alexander Szalai, pp. 357–75. The Hague: Mouton.
Vaupel, James W.
1976 "Early Death: An American Tragedy." *Law and Contemporary Problems* 40, no. 4 (Autumn): 73–121.
Veatch, Robert M.
1979 "Justice and Valuing Lives." In *Life Span: Values and Life-Extending Technologies,* edited by Robert M. Veatch, pp. 197–224. New York: Harper & Row.
Veroff, Joseph, Douvan, Elizabeth, and Kulka, Richard A.
1981 *The Inner American: A Self-Portrait from 1957 to 1976.* New York: Basic Books.
Vogel, Joachim
1982 "The Swedish Annual Level-of-Living Surveys: Social Indicators and Social Reporting as an Official Statistics Program." Paper presented at Tenth World Congress of Sociology, Mexico City, August.
von Wright, Georg Henrik
1963 *The Varieties of Goodness.* London: Routledge & Kegan Paul.
Wallis, W. Allen
1971 Letter of transmittal. In President's Commission on Federal Statistics, *Federal Statistics,* vol. 1, Washington, D.C.: U.S. Government Printing Office.
Walster, Elaine H., Walster, G. William, and Berscheid, Ellen
1978 *Equity, Theory and Research.* Boston: Allyn & Bacon.
Walters, Pamela Barnhouse, and Rubinson, Richard
1983 "Educational Expansion and Economic Output in the United States, 1890–1969: A Production Function Analysis." *American Sociological Review* 48, no. 4 (August): 480–93.
Watts, Harold W.
1977 "An Economic Definition of Poverty." In *Improving Measures of Economic Well-Being,* edited by Marilyn Moon and Eugene Smolensky, pp. 19–32. New York: Academic Press.
Watts, Harold W., and Hernandez, Donald J., eds.
1982 *Child and Family Indicators: A Report with Recommendations.* New York: Social Science Research Council.

Webb, Eugene J., Campbell, Donald T., Schwartz, Richard D., and Sechrest, Lee
1966 *Unobtrusive Measures: Nonreactive Research in the Social Sciences*. Chicago: Rand McNally.

Webb, Kenneth, and Hatry, Harry P.
1973 *Obtaining Citizen Feedback: The Application of Citizen Surveys to Local Governments*. Washington, D.C.: Urban Institute.

Weber, Max
1978 *Economy and Society*. Edited by Guenther Roth and Claus Wittich. Berkeley: University of California Press. Translated from 1956 German edition.

Weinstein, Milton C., and Stason, William B.
1976 *Hypertension: A Policy Perspective*. Cambridge: Harvard University Press.
1977 "Foundations of Cost-Effectiveness Analysis for Health and Medical Practices." *New England Journal of Medicine* 296, no. 13 (March 31): 716–21.

Weisbrod, Burton A.
1968 "Income Redistribution Effects and Benefit-Cost Analysis." In *Problems in Public Expenditure Analysis*, edited by Samuel B. Chase, pp. 177–209. Washington, D.C.: Brookings.

Weisbrod, Burton A., Andreano, Ralph L., Baldwin, Robert E., Epstein, Erwin H., and Kelley, Allen C.
1973 *Disease and Economic Development: The Impact of Parasitic Diseases in St. Lucia*. Madison: University of Wisconsin Press.

Weisbrod, Burton A., and Hansen, W. Lee
1977 "An Income-Net Worth Approach to Measuring Economic Welfare." In *Improving Measures of Economic Well-Being*, edited by Marilyn Moon and Eugene Smolensky, pp. 33–50. New York: Academic Press.

Weiss, Carol H.
1972 *Evaluation Research*. Englewood Cliffs, N.J.: Prentice-Hall.
1978 "Improving the Linkage between Social Research and Public Policy." In *Knowledge and Policy: The Uncertain Connection*, National Research Council, Study Project on Social Research and Development, pp. 23–81. Washington, D.C.: National Academy of Sciences.

Wellman, Barry
1979 "The Community Question: The Intimate Networks of East Yorkers." *American Journal of Sociology* 84, no. 5 (March): 1201–31.
1982 "Studying Personal Communities." In *Social Structure and Network Analysis*, edited by Peter V. Marsden and Nan Lin, pp. 61–80. Beverly Hills, Calif.: Sage.

Wessman, Alden E., and Ricks, David F.
1966 *Mood and Personality*. New York: Holt, Rinehart, and Winston.

Whitaker, Gordon P.
1980 "Coproduction: Citizen Participation in Service Delivery." *Public Administration Review* 40, no. 3 (May/June): 240–46.

White, Harrison C.
1982 "Review Essay: Fair Science?" Review of Jonathan R. Cole, *Fair Science. American Journal of Sociology* 87, no. 4 (January): 951–56.

White, Harrison C., Boorman, Scott A., and Breiger, Ronald L.
1976 "Social Structure from Multiple Networks: I. Blockmodels of Roles and Positions." *American Journal of Sociology* 81, no. 4 (January): 730–80.

White, Howard D.
1983 "A Cocitation Map of the Social Indicators Movement." *Journal of the American Society for Information Science* 34, no. 5 (September): 307–12.

White, Lynn K.
1979 "Sex Differentials in the Effect of Remarriage on Global Happiness." *Journal of Marriage and the Family* 41, no. 4 (November): 869–76.

Whittington, Dale, and Grubb, W. Norton
1984 "Economic Analysis in Regulatory Decisions: The Implications of Executive Order 12291." *Science, Technology, & Human Values* 9, no. 1 (Winter): 63–71.

Wildavsky, Aaron
1979 *Speaking Truth to Power*. Boston: Little, Brown.

Wilensky, Harold L.
1961 "The Uneven Distribution of Leisure: The Impact of Economic Growth on 'Free Time.'" *Social Problems* 9, no. 1 (Summer): 32–56.
1964 "The Professionalization of Everyone?" *American Journal of Sociology* 70, no. 2 (September): 137–58.
1975 *The Welfare State and Equality: Structural and Ideological Roots of Public Expenditures*. Berkeley: University of California Press.

Williams, Mark E., and Hadler, Nortin M.
1983 "The Illness as the Focus of Geriatric Medicine." *New England Journal of Medicine* 308, no. 22 (June 2): 1357–59.

Willig, Robert T.
1976 "Consumer's Surplus Without Apology." *American Economic Review* 66, no. 4 (September): 589–97.

Wilson, James Q.
1975 *Thinking About Crime*. New York: Basic Books.

Wilson, Robert L.
1962 "Livability of the City: Attitudes and Urban Development." In *Urban Growth Dynamics*, edited by F. Stuart Chapin, Jr., and Shirley F. Weiss, pp. 359–99. New York: Wiley.

Wilson, Warner
1967 "Correlates of Avowed Happiness." *Psychological Bulletin* 67, no. 4 (April): 294–306.

Winsborough, Hallinan H.
1975 "Age, Period, Cohort, and Education Effects on Earnings by Race." In *Social Indicator Models*, edited by Kenneth C. Land and Seymour Spilerman, pp. 201–17. New York: Russell Sage Foundation.

Wohlstetter, Albert
1968 "Theory and Opposed-Systems Design." In *New Approaches to International Relations*, edited by Morton A. Kaplan, pp. 19–53. New York: St. Martin's.

Wohlstetter, Albert, Hoffman, Fred S., Lutz, R. J., and Rowen, Henry S.
1954 *The Selection and Use of Strategic Air Bases*, R-266. Santa Monica, Calif.: Rand Corporation.

Wolf, Charles, Jr.
1979 "A Theory of Non-Market Failures." *The Public Interest* 55: 114–33.

Zapf, Wolfgang
1972 "Social Indicators: Prospects for Social Accounting Systems." *Social Science Information* 11, nos. 3/4 (June/August): 243–77.
1974 "Social Indicators, 1973: Comparison with Social Reports of Other Nations." In *Social Indicators, 1973: A Review Symposium*, edited by Roxann A. Van Dusen, pp. 20–40. Washington, D.C.: Social Science Research Council.
1977 "Gesellschaftliche Dauerbeobachtung und aktive Politik." In *Sozialpolitik und Sozialberichterstattung*, edited by Hans-Jürgen Krupp and W. Zapf, pp. 210–30. Frankfurt/New York: Campus. English translation presented as "The Polity as Monitor of the Quality of Life," at meeting of the International Political Science Association, Montreal, 19–25 Aug. 1973.
1979 "Lebensbedingungen und wahrgenommene Lebensqualität." In *Sozialer Wandel in Westeuropa*, edited by J. Matthes, pp. 767–90. Frankfurt: Campus.
1982 "Welfare Production: Public versus Private." Paper presented at Tenth World Congress of Sociology, Mexico City, 16–20 August 1982.

Zarnowitz, Victor
1980 "Econometric Models are Shapely, But Rather Dense." *New York Times* (26 October 1980).

Zetterberg, Hans L.
1962 *Social Theory and Social Practice*. New York: Bedminster.

Ziman, John
1968 *Public Knowledge*. Cambridge, England: University Press.

Zuckerman, Harriet
1977 *Scientific Elite: Nobel Laureates in the United States*. New York: Free Press.

Index

www.ingramcontent.com/pod-product-compliance
Lightning Source LLC
LaVergne TN
LVHW090300210926
848480LV00001B/83